Graduate Texts in Physics

Graduate Texts in Physics

Graduate Texts in Physics publishes core learning/teaching material for graduate- and advanced-level undergraduate courses on topics of current and emerging fields within physics, both pure and applied. These textbooks serve students at the MS- or PhD-level and their instructors as comprehensive sources of principles, definitions, derivations, experiments and applications (as relevant) for their mastery and teaching, respectively. International in scope and relevance, the textbooks correspond to course syllabi sufficiently to serve as required reading. Their didactic style, comprehensiveness and coverage of fundamental material also make them suitable as introductions or references for scientists entering, or requiring timely knowledge of, a research field.

More information about this series at http://www.springer.com/series/8431

Peter Hertel

Quantum Theory and Statistical Thermodynamics

Principles and Worked Examples

Springer

Peter Hertel
Universität Osnabrück Fachbereich Physik
Osnabrück
Germany

ISSN 1868-4513 ISSN 1868-4521 (electronic)
Graduate Texts in Physics
ISBN 978-3-319-86437-2 ISBN 978-3-319-58595-6 (eBook)
DOI 10.1007/978-3-319-58595-6

Softcover reprint of the hardcover 1st edition 2017

Cover illustration: courtesy of Thorsten Schneider

Printed on acid-free paper

This Springer imprint is published by Springer Nature
The registered company is Springer International Publishing AG
The registered company address is: Gewerbestrasse 11, 6330 Cham, Switzerland

Preface

This small textbook attempts to present the vast field of quantum physics, including statistical thermodynamics, by focusing on a few basic concepts. Matter is made of particles which, like waves, interfere. There is a limiting case, namely classical mechanics and the various fields of continuum physics, where the interference of particles plays no role. Although the classical disciplines have been studied earlier, their laws and rules are to be derived from quantum physics.

Hence, quantum physics provides the foundation of classical physics and thermodynamics as well as of numerous applications in electronics, engineering, and so on. Moreover, with the advent of nanometer manipulations, useful devices directly based on quantum effects are already a reality or will soon become feasible.

Aside from touching on fundamental questions on the nature of things, the book concentrates on examples which demonstrate the principles as clearly as possible. All examples have been chosen carefully. They illustrate the principles and provide an overview over historically or intrinsically interesting topics and problem-solving strategies. Statistical thermodynamics, as quantum physics at nonvanishing temperature, fits quite naturally.

The mathematical level is intermediate, as low as possible and as high as necessary. The mathematical framework is coherently presented in a chapter on Mathematical Aspects, from topological spaces, measure theory, probability, generalized functions to linear operators.

Any theory should be based on carefully chosen principles which are either stated explicitly or implicitly by embedding the new in an already established theory. Science in general and physics in particular have to stand another test besides logical correctness, usefulness, and simplicity. Their rules and laws must also pass the test of experimental verification.

The presentation is neither purely axiomatic nor historic nor pedagogic, but a mixture of these approaches. Following history would require a deeper understanding of what contemporary scientists would think about nature. A purely pedagogic approach would depend too much on the educational background of the readers which certainly is far from homogeneous. And the problems of the axiomatic method in the sciences are well known. Axioms can only be formulated vaguely at first by using everyday language. They are subsequently employed to derive statements such that the terms in which the axioms were formulated gain precision and technical meaning. Then the wording of the axioms must be sharpened, and so on: an iterative approach.

Our main goal is a readable text with a few principles as guidelines. The many worked examples are discussed at various levels of abstraction and mathematical rigor.

Not only advanced students of general physics, chemistry, molecular biology, or materials science will profit from reading the book, but also philosophicalminded or curious people will like this treatise.

We recommend that the reader study at least the Introduction Sect. 1.1 and Sect. 1.3 on the Quantum Framework and Sect. 1.4 on Time and Space. She or he may then decide where to continue: reading Sect. 1.2 on the Classical Framework as well, Chap. 2 with Simple Examples or Chap. 7 on Mathematical Aspects. Although the book can be studied as a web, it is conceived to be read consecutively from the first to the last page. Enjoy!

Osnabrück, Germany
June 2017

Peter Hertel

Contents

Chapter 1
Basics

In this chapter we present the basic physical and mathematical concepts as well as the vocabulary of quantum theory. The text is necessarily compact and dense. Depending on the reader's background, it may serve as a summary of mathematical tools and an introduction into what one may call quantum parlance. It might also be that thinking in terms of observables, states and expectation values is rather new to someone who has previously studied topics in classical mechanics or electromagnetism only, or quantum mechanics in a hands-on manner. The text should make easy reading to her or him as well. However, a sound knowledge of analysis and algebra is assumed, on an undergraduate level at least.

The mathematical reasoning in this chapter on Basics is rather sloppy—on purpose. Mathematical subtleties might rather distract from understanding the framework.

In the first section Sect. 1.1, an Introduction, some startling observations are recapitulated. Physicists were lead to assume that particles propagate as waves and waves are created or absorbed like particles. It was recognized rather soon that certain properties cannot be measured simultaneously with arbitrarily high precision. Put otherwise: some pairs of observable properties are incompatible. Something which cannot even be formulated in classical theory.

In order to better understand this we present, in Sect. 1.2 on the Classical Framework, a condensed description of classical mechanics which has evolved in hundreds of years with ever greater clarity. However, not all facts fitted into the classical framework: it had to be modified.

The comprising framework has to permit objects which do not commute, so called q-numbers. The fundamental properties of numbers should be retained, such as multiplication with complex numbers, addition, multiplication and a star operation similar to complex conjugation. All the rules for ordinary numbers should apply, except that AB and BA are not necessarily the same.

The linear mappings of a suitable Hilbert space into itself, linear operators, are the searched-for objects. We discuss in Sect. 1.3 on the Quantum Framework their definition, that they form a $C^\star$ algebra, and investigate in particular normal operators.

P. Hertel, *Quantum Theory and Statistical Thermodynamics*,
Graduate Texts in Physics, DOI 10.1007/978-3-319-58595-6_1

Observables are identified with self-adjoint operators and states with probability operators.

Quantum processes are ruled by probability, not by either-or logic. A single measurement cannot be predicted with certainty. Only a long series of repeated identical measurements, an experiment, will provide an average value which converges with the number of repetitions. It converges towards an expectation value. Calculating expectation values is the task of quantum theory, not more, not less.

The final section is devoted to operations in space and time: waiting, spatial translations and rotations. They give rise to the most important observables like energy, linear momentum, location and angular momentum. The latter should be split into orbital angular momentum and spin. We derive a set of invariant characterizations by commutation relations. Whatever the realization for a concrete situation: these commutation relations must be observed, they characterize the properties of Time and Space.

The remainder of this book is a loose collection of topics which attempt an overview over quantum theory including statistical thermodynamics. They serve as illustrations to what was put forward in this more abstract chapter.

1.1 Introduction

We bring forward some effects and considerations which demand a new framework for physics. Max Planck and Albert Einstein successfully explained the black body radiation spectrum and the photoelectric effect. They postulated that light is made up of particles the energy of which is proportional to their frequency. On the other hand, electrons are diffracted by crystals, a phenomenon which is typical for waves. Light, when emitted or absorbed, appears to be a stream of particles; and particles, when propagating in space, behave as if they were waves. Luis de Broglie summarized the relationship by the statement that energy and momentum on the one hand are proportional to angular frequency and wave vector on the other. The proportionally constant $\hbar$ is the same for all particles. This behavior has far reaching consequences. As Werner Heisenberg pointed out, location and momentum cannot be sharply defined simultaneously. The product of localization and momentum uncertainty is always larger than $\hbar/2$. Another consequence of the particle-wave duality is that single events occur only with a certain probability, they are not predictable.

1.1.1 The Quantum of Light

At the age of sixteen Max Planck passed the final examination on leaving high-school with distinction. He now had to decide about his future. His preferences were classical philology, music, or science. He finally chose physics, although discouraged by his teacher Philipp von Holly at the university of Munich who maintained that *physics is such a mature science that only a few gaps would have to be filled*. This then was the opinion of many physicists. A few years later the situation was totally different, and Max Planck was among the revolutionaries who had shaken up physics. He solved the riddle of black body radiation, one of the *few gaps to be filled*, by treating light of angular frequency ω as consisting of particles with energy $\hbar\omega$, with $\hbar$ being a constant of nature.

Already in 1887, Heinrich Hertz discovered that electrons leave a metallic surface more easily under illumination. The kinetic energy of such photoelectrons grows linearly with the inverse wave length, their current is proportional to the intensity of illumination. In 1905 Albert Einstein came forward with a simple explanation of the photoelectric effect. The ingredients were energy conservation and the quantum hypothesis: light of angular frequency $\omega = 2\pi c/\lambda$ is made up of particles with energy $\hbar\omega$. Denote by W the work to be spent for setting free a single electron. It leaves the metal with speed v. The simple equation

$$\hbar\omega = W + \frac{m}{2}v^2 \tag{1.1}$$

explains the electron energy spectrum. It is a balance equation for a single photon absorption, electron creation process. Moreover, the photon current, or light intensity, and the electric current are proportional. The single photon absorption/electron creation processes are independent, at least for weak illumination.

Louis de Broglie in 1924 summarized this and many previous findings in his thesis. Light may come in electromagnetic waves, a superposition of plane waves with wave vector $\boldsymbol{k}$ and angular frequency ω. It also appears to consist of particles with energy E and momentum $\boldsymbol{p}$. Both aspects are related by

$$\boldsymbol{p} = \hbar\boldsymbol{k} \text{ and } E = \hbar\omega. \tag{1.2}$$

The particle-wave duality pertains to all matter.

1.1.2 Electron Diffraction

De Broglie's conjecture was tested immediately with electrons instead of photons. Clinton Davisson and Lester Germer reported in 1927 on an experiment where a beam of electrons was directed on a nickel crystal. Deflected electrons could be detected, the experiment being performed in vacuum. They observed a diffraction pattern just as in an X ray diffraction experiment.

Electron diffraction proved that particles, when propagating, behave as waves which may interfere. There must be something, for instance real or complex valued amplitudes, which add when two waves are superimposed. In fact, there is an amplitude ϕ_r for the passage of the electron from the source via lattice ion r to the detector. The amplitude for the passage of the electron from the source to the detector via the entire crystal then is

$$\phi = \sum_r \phi_r . \tag{1.3}$$

The contributions to ϕ practically always cancel unless Laue's condition is met:

$$\boldsymbol{a}_j \cdot \Delta \boldsymbol{k} = 2\pi \nu_j . \tag{1.4}$$

Here the three vectors $\boldsymbol{a}_1, \boldsymbol{a}_2, \boldsymbol{a}_3$ span the lattice. $\Delta \boldsymbol{k} = \boldsymbol{k}_{\text{out}} - \boldsymbol{k}_{\text{in}}$ is the difference between the wave vector of the outgoing and incoming electron. Laue's condition is met if there are three integer numbers ν_1, ν_2, ν_3 such that (1.4) holds true. The intensity of the diffracted electron beam is proportional to the absolute square $|\phi|^2$ of the amplitude.

1.1.3 Heisenberg Uncertainty Principle

Denote by $f = f(x)$ the amplitude for finding a particle within a small region $\mathrm{d}x$ at position x. The probability for this is $\mathrm{d}x|f|^2$. Since the particle must be somewhere we may write

$$\int \mathrm{d}x \, |f(x)|^2 = 1. \tag{1.5}$$

The average position and position squared are

$$\langle X \rangle = \int \mathrm{d}x \, |f|^2 \, x \text{ and } \langle X^2 \rangle = \int \mathrm{d}x \, |f|^2 \, x^2 \tag{1.6}$$

Hence, the position is defined with accuracy

$$\sigma(X) = \sqrt{\langle X^2 \rangle - \langle X \rangle^2}. \tag{1.7}$$

We regard the wave function f to be a superposition of plain waves with wave number k. This Fourier transform reads

$$f(x) = \int \frac{\mathrm{d}k}{2\pi} \, g(k) \, \mathrm{e}^{\,\mathrm{i}kx} . \tag{1.8}$$

Condition (1.5) implies

$$\int \frac{\mathrm{d}k}{2\pi} \, | \, g(k) \, |^2 = 1. \tag{1.9}$$

The average wave number and wave number squared are

$$\langle K \rangle = \int \frac{\mathrm{d}k}{2\pi} \, | \, g(k) \, |^2 \, k \text{ and } \langle K^2 \rangle = \int \frac{\mathrm{d}k}{2\pi} \, | \, g(k) \, |^2 \, k^2 . \tag{1.10}$$

Again, the wave number is determined with accuracy

$$\sigma(K) = \sqrt{\langle K^2 \rangle - \langle K \rangle^2}. \tag{1.11}$$

Both uncertainties are not independent. In fact,

$$\sigma(X)\,\sigma(K) \geq \frac{1}{2}. \tag{1.12}$$

holds true for an arbitrary function $f = f(x)$. The proof of this finding is deferred to the chapter on Mathematical Aspects.

However, if combined with the de Broglie relation (1.2), we arrive at Heisenberg's uncertainty relation

$$\sigma(X)\,\sigma(P) \geq \frac{\hbar}{2}, \tag{1.13}$$

Location and momentum of a particle cannot be determined simultaneously with arbitrary precision. The more accurate the momentum, the less certain the location, and vice versa. In the next sections we shall discuss the far reaching consequences. Certain pairs of observable properties cannot be measured simultaneously.

1.1.4 Does God Play Dice?

Particle-wave duality, aptly described by de Broglie's relation (1.2), has another disquieting logical implication. Particles can be counted: none, one, two and so on. The particle detector is triggered not at all, once, twice, and so on. Waves, when created or absorbed, behave like particles.

On the other hand, when propagating from the source to the detector, the same particle behaves as if it would be a wave. In particular, if there are many intermediate stages, the individual amplitudes add, they interfere from maximal amplification to complete annihilation with anything in-between. How to reconcile this finding with the fact that particles come as whole entities only?

Imagine a setup as in an electron refraction experiment. The geometry shall be such that there is just one direction of refraction. Electrons not being refracted are transmitted in forward direction. For simplicity, assume detectors which register an electron with probability one. The intensities, here as particles per unit time, are I_{in}, I_{fw} and I_{rf}, and there is the balance equation

$$I_{\text{in}} = I_{\text{fw}} + I_{\text{rf}}. \tag{1.14}$$

The refraction ratio is defined as

$$R = \frac{I_{\text{rf}}}{I_{\text{in}}}. \tag{1.15}$$

If there are N incoming electrons, one expects RN refracted particles and $(1-R)N$ electrons in forward direction.

What if there is just one electron? Obviously, R must be interpreted as a refraction probability. One electron is emitted, and with probability $1-R$ it runs on in forward direction, and with probability R it suffers refraction.

Who decides the fate of this electron? Pure chance, some said. Albert Einstein objected: *Der Alte würfelt nicht.*[1] There are libraries of literature discussing this question, and even some experiments. A large school of physicists postulate so called hidden parameters. However, the overwhelming majority of the physics community supports the now well established opinion that single quantum events happen by chance, respecting certain probabilities. The outcome of a single measurement cannot be predicted. If in an experiment the measurement is repeated again and again, the sequence of relative frequencies converges towards a probability. Such probabilities are the subject of theoretical quantum physics.

1.1.5 Summary

The black body radiation spectrum and the photoelectric effect could be explained by Max Planck and Albert Einstein. They postulated that light, evidently a wave phenomenon, is made up of particles, namely photons. On the other hand, electrons which were known to be particles, propagate through a crystal as if they were waves. They are diffracted with intensity peaks as described by Laue's condition for equivalent X ray beams. The similarity between particles and waves is summarized by the de Broglie relations. We discuss two far reaching consequences. The Heisenberg uncertainty principle states that there are pairs of observable properties which cannot be determined simultaneously with arbitrary precision. And, even more important, single events are not predictable, they are characterized by their probability only. Most physicists agree that chance is a core element of nature, not just a manifestation of complexity or incomplete knowledge of the particular state. There are no hidden variables, God seems to play dice!

1.2 Classical Framework

We present the conceptual and mathematical framework of classical physics, in particular for closed systems with a finite number of degrees of freedom. A physical system, in the classical approximation, is described by its phase space. Observables and states are represented by real valued functions or probability densities, respectively. Phase space points move like an incompressible fluid.

---

[1] German for 'the Old One does not throw dice'.

1.2.1 Phase Space

Imagine a system with a finite number f of degrees of freedom. There is a set of generalized coordinates $q_1, q_2, \dots, q_f$ which uniquely characterize the possible configurations of the system under discussion. With each generalized coordinate q_i one associates a generalized velocity $\dot{q}_i$. The generalized coordinate q_i and its velocity $\dot{q}_i$ are considered to be independent variables.

The kinetic energy T and the potential energy V are to be expressed in terms of generalized coordinates and their generalized velocities. The Lagrangian $L = T - V = L(q, \dot{q})$ describes the dynamics of our system, the evolution with time t.

The generalized momenta are defined by

$$p_i = \frac{\partial L(q, \dot{q})}{\partial \dot{q}_i}, \tag{1.16}$$

they obey the following equations of motion:

$$\dot{p}_i = \frac{\partial L(q, \dot{q})}{\partial q_i}. \tag{1.17}$$

Lagrangian dynamics is based on the observation that constraining forces do not perform work.

The Hamiltonian of our system is

$$H = H(q, p) = \sum_i p_i \dot{q}_i - L. \tag{1.18}$$

The right hand side has to be rewritten into a function of generalized coordinates and momenta. Now the coordinates and the momenta are considered to be independent variables. They must obey the following equations:

$$\dot{q}_i = \frac{\partial H(q, p)}{\partial p_i} \text{ and } \dot{p}_i = -\frac{\partial H(q, p)}{\partial q_i}, \tag{1.19}$$

for $i = 1, 2, \dots, f$.

By the way, the kinetic energy is almost always a quadratic form in the $\dot{q}_i$, the coefficients possibly depending on coordinates. In this case the Hamiltonian is $H = T + V$, the energy.

The points

$$\omega = (q, p) = (q_1, p_1, q_2, p_2, \dots, q_f, p_f) \in \Omega \tag{1.20}$$

form the $2f$ dimensional phase space which we denote by Ω. These points flow on trajectories as prescribed by (1.19).

1.2.2 Observables

Consider the set $\mathfrak{A}$ of sufficiently smooth complex valued functions living on the phase space Ω. Such functions can be multiplied by a complex number, and they may be added. Thus, $\mathfrak{A}$ is a liner space. Moreover, they can be multiplied according to $(AB)(\omega) = A(\omega)B(\omega)$. Multiplication with scalars, addition and multiplication obey the usual laws such that $\mathfrak{A}$ is an algebra.

There is a norm $\| \,.\, \|$ such that for $z \in \mathbb{C}$ and $A, B \in \mathfrak{A}$ the following relations hold true:

$$\begin{aligned} \| zA \| &= | z | \| A \| \\ \| A + B \| &\leq \| A \| + \| B \| \\ \| BA \| &\leq \| B \| \| A \|. \end{aligned} \tag{1.21}$$

There are many such norms. This one may serve as an example:

$$\| A \| = \sup_{\omega \in \Omega} | A(\omega) |. \tag{1.22}$$

In addition, there is a star operation $A \to A^\star$ defined by

$$(A^\star)(\omega) = (A(\omega))^*. \tag{1.23}$$

$A^\star$ is also called the adjoint of A. The star operation obeys

$$A^{\star\star} = A \text{ and } (AB)^\star = B^\star A^\star. \tag{1.24}$$

It is compatible with the norm since

$$\| A^\star A \| = \| A \|^2. \tag{1.25}$$

A normed algebra with a star operation, such as $\mathfrak{A}$, is a C* algebra.

Note that the norm induces a topology. One may define Cauchy sequences and converging sequences. A C* algebra should be closed, i.e. all Cauchy sequences have a limit.

A self-adjoint element M of the C* algebra is an observable. To each phase space point ω belongs a real number $M(\omega) = M^*(\omega)$.

1.2.3 Dynamics

The points in phase space Ω move on trajectories $\omega_t(\omega)$ where $\omega_0(\omega) = \omega$. This induces a mapping in the C* algebra $\mathfrak{A}$ by

$$A \to A_t \text{ where } A_t(\omega) = A(\omega_t(\omega)). \tag{1.26}$$

The rate of change is

$$\frac{\mathrm{d}}{\mathrm{d}t} A_t = \sum_{i=1}^{f} \left(\frac{\partial A}{\partial q_i} \frac{\partial H}{\partial p_i} - \frac{\partial A}{\partial p_i} \frac{\partial H}{\partial q_i} \right). \tag{1.27}$$

We have inserted the equations of motion (1.19). With the Poisson bracket

$$\{A, B\}(\omega) = \sum_{i=1}^{f} \left(\frac{\partial A(\omega)}{\partial q_i} \frac{\partial B(\omega)}{\partial p_i} - \frac{\partial A(\omega)}{\partial p_i} \frac{\partial B(\omega)}{\partial q_i} \right) \tag{1.28}$$

one may write

$$\frac{\mathrm{d}}{\mathrm{d}t} A_t = \{A_t, H\}. \tag{1.29}$$

We wrote H and not H_t because the Hamiltonian obviously does not depend on time. Note that the Poisson bracket is linear in both factors and obeys $\{A, B\} = -\{B, A\}$. Therefore $\{H, H\} = 0$.

For a fixed time t the mapping $\omega \to \bar{\omega} = \omega_t(\omega)$ is a coordinate transformation. For simplicity, we restrict ourselves to an infinitesimal time $\mathrm{d}t$ and one degree of freedom:

$$\bar{q} = q + \mathrm{d}t \frac{\partial H}{\partial p} \text{ and } \bar{p} = p - \mathrm{d}t \frac{\partial H}{\partial q}. \tag{1.30}$$

The functional determinant of this mapping is

$$J(t + \mathrm{d}t) = \det \begin{pmatrix} 1 + \mathrm{d}t \dfrac{\partial^2 H}{\partial q \partial p} & +\mathrm{d}t \dfrac{\partial^2 H}{\partial p \partial p} \\ -\mathrm{d}t \dfrac{\partial^2 H}{\partial q \partial q} & 1 - \mathrm{d}t \dfrac{\partial^2 H}{\partial p \partial q} \end{pmatrix} = 1. \tag{1.31}$$

Since the determinant evaluates to $1 + \mathrm{d}t\{\partial^2 H/\partial p \partial q - \partial^2 H/\partial q \partial p\} + \mathrm{d}t^2 \ldots$ and since the order of performing partial differentiations does not matter we conclude $\dot{J}(t) = 0$ and $J(t) = 1$. This can easily be generalized to an arbitrary number f of degrees of freedom. The natural choice for the phase space volume element is

$$\mathrm{d}\Omega = \frac{\mathrm{d}q_1 \mathrm{d}p_1}{2\pi\hbar} \frac{\mathrm{d}q_2 \mathrm{d}p_2}{2\pi\hbar} \cdots \frac{\mathrm{d}q_f \mathrm{d}p_f}{2\pi\hbar}. \tag{1.32}$$

The $\hbar$ is a relict of quantum theory in the classical limit. Moreover, it serves to have a dimension-less expression. This volume element does not change when the phase space points move along their trajectories. Put otherwise, the flow of phase space points is incompressible. This finding is known as Liouville's theorem.

1.2.4 States

Le $\varrho = \varrho(\omega)$ be a probability distribution on the phase space:

$$\varrho(\omega) \geq 0 \text{ and } \int \mathrm{d}\Omega\, \varrho(\omega) = 1. \tag{1.33}$$

Such a normalized measure describes a state of the system.

We define the expectation value of an observable A in state ϱ by

$$\varrho(A) = \int \mathrm{d}\Omega\, \varrho(\omega) A(\omega). \tag{1.34}$$

The expectation value has the following properties:

- it is a linear functional,

$$\varrho(zA) = z\varrho(A) \text{ and } \varrho(A+B) = \varrho(A) + \varrho(B) \text{ for } z \in \mathbb{C}, \tag{1.35}$$

- it obeys

$$\varrho(A^\star) = \varrho(A)^*, \tag{1.36}$$

- it is positive,

$$\varrho(A^\star A) \geq 0 \tag{1.37}$$

- it is normalized,

$$\varrho(1) = 1. \tag{1.38}$$

1 with $1(\omega) = 1$ is the one-observable. Expectation values of observables are real. Note that a positive observable can always be represented as $A^\star A$.

1.2.5 Properties of Poisson Brackets

The Poisson bracket $\{A, B\}$ as defined by (1.28) is bi-linear in both factors and antisymmetric. This implies $\{A, A\} = 0$. Moreover it is compatible with the star operation, $\{A, B\}^\star = \{A^\star, B^\star\}$. From this we conclude that the Poisson bracket of two observables is again an observable:

$$\{A, B\}^\star = \{A^\star, B^\star\} = \{A, B\}. \tag{1.39}$$

The product rule of differentiation amounts to

$$\{A, BC\} = B\{A, C\} + \{A, B\}C, \tag{1.40}$$

the Jacobi identity says

$$\{A, \{B, C\}\} + \{B, \{C, A\}\} + \{C, \{A, B\}\} = 0. \tag{1.41}$$

With this and (1.29) we calculate

$$\{A_{t+\mathrm{d}t}, B_{t+\mathrm{d}t}\} = \{A_t, B_t\} \tag{1.42}$$

and conclude that the Poisson bracket of two elements of $\mathfrak{A}$ does not change in the course of time. This applies in particular to observables.

1.2.6 Canonical Relations

Recall that ω stands for the collection of arguments $(q_1, p_1, q_2, p_2, \dots, q_f, p_f)$. We define the generalized coordinates and momenta as observables:

$$Q_i(\omega) = q_i \text{ and } P_i(\omega) = p_i. \tag{1.43}$$

Inserting this into (1.28) we obtain

$$\{Q_j, P_k\}(\omega) = \sum_{i=1}^{f} \delta_{ji}\delta_{ki} = \delta_{jk}, \tag{1.44}$$

or

$$\{Q_j, P_k\} = \delta_{jk} 1. \tag{1.45}$$

As we know already, this so-called canonical relation does not depend on time, it is an invariant. It holds for any parametrization q_i of the system configuration provided the p_i are the associated generalized momenta as defined by (1.16).

1.2.7 Pure and Mixed States

Two states ϱ_1 and ϱ_2 may be mixed with weights α_1 and α_2 according to

$$\varrho(\omega) = \alpha_1 \varrho_1(\omega) + \alpha_2 \varrho_2(\omega). \tag{1.46}$$

Since $\alpha_1, \alpha_2 \geq 0$ and $\alpha_1 + \alpha_2 = 1$, the mixture is in fact a state since it is again a probability distribution. The sum in (1.46) may easily be generalized to a suitable integral.

A state is pure if it cannot be written as a mixture. The states

$$\varrho_{\bar{\omega}}(\omega) = 2\pi\hbar\,\delta(q_1 - \bar{q}_1)\delta(p_1 - \bar{p}_1)\dots 2\pi\hbar\,\delta(q_f - \bar{q}_f)\delta(p_f - \bar{p}_f) \tag{1.47}$$

are pure. Every state may be represented as a mixture thereof:

$$\varrho(\omega) = \int \mathrm{d}\bar{\Omega}\; \varrho(\bar{\omega})\; \varrho_{\bar{\omega}}(\omega). \tag{1.48}$$

1.2.8 Summary

We have sketched very briefly the conceptual and mathematical framework of classical mechanics. Expressions for the kinetic and potential energy allow to associate, with each generalized coordinate, a generalized momentum. The collection of generalized coordinates and momenta constitute the phase space of our system.

Observables are represented by real valued functions living on the phase space. They are embedded into a $C^\star$ algebra. The equations of motion for observables are expressed in terms of Poisson brackets.

Probability densities living on the phase space describe states and give rise to expectation values.

The properties of Poisson brackets are discussed, they do not change with time. Generalized coordinates and momenta are observables, their Poisson brackets are of particular importance.

Finally, a classification of states into pure or mixed is mentioned.

All these findings will show up again when quantum physics is discussed—with one small modification. One seemingly small detail cannot be maintained if particles and their interference shall be described.

1.3 Quantum Framework

In the preceding section we have summarized the mathematical framework of classical mechanics. Arbitrarily chosen generalized coordinates together with their associated generalizes momenta parametrize the phase space. Real valued functions living on this phase space are identified with the observable properties of the system under discussion. These functions, or observables, are embedded into the set of complex valued functions which can be added and multiplied. Such functions form a $C^\star$ algebra.

The observables describe properties of a system. Whether property A is observed first and B immediately afterwards or B first and then A—is irrelevant. BA and AB are the same. The $C^\star$ algebra is commutative, in the context of classical mechanics.

Heisenberg's uncertainty principle says something else, though: certain pairs of observables cannot be measured simultaneously. They will not commute with each other. The C* algebra of observables cannot consist of functions on a phase space. A theory for quantum phenomena must be formulated with non-commuting objects. Such objects, or q-numbers, describe the observable properties and the states of a physical system. Since the framework of classical mechanics must be a limiting case, not much has to be modified.

1.3.1 q-Numbers

There are natural, integer, rational and real numbers. Natural numbers are the result of counting. Integer numbers are introduced in order to solve equations like $a + x = b$. Fractions, or rational numbers are required for solving $ax = b$. By adding the limits of Cauchy converging series one arrives at the set $\mathbb{R}$ of real numbers. Algebraically, $\mathbb{R}$ is a field. There is a commutative addition and a commutative multiplication and a number of compatibility rules. Every non-vanishing real number has an inverse.

However, $\mathbb{R}$ is not the biggest such field. The set of real numbers can be enlarged to the field $\mathbb{C}$ of complex numbers such that $\mathbb{R} \subset \mathbb{C}$. The disadvantage is that $\mathbb{C}$ cannot be ordered in a natural way. The advantage is that any polynomial of degree n has at least n roots. For example, the roots of $z^2 + 1 = 0$ are $z = \pm\mathrm{i}$.

There are no larger fields which extend the complex numbers. Richer structures cannot be proper fields, they have to be realized with a non-commutative multiplication. $a + b = b + a$ will be maintained, but $ab = ba$ is no more required. Classical numbers, or c-numbers, do not suffice. We must deal with quantum, or q-numbers. As it turns out, these q-numbers are linear mappings of a Hilbert space into itself.

1.3.2 Hilbert Space

A Hilbert space $\mathcal{H}$ is a linear space with scalar product; it is complete with respect to its natural norm. Its elements are called vectors.

Since it is a linear space, one may add vectors and multiply them by scalars which we assume to be complex numbers. If $f, g \in \mathcal{H}$ are vectors and $\alpha, \beta \in \mathbb{C}$ scalars, then the vector $h = \alpha f + \beta g$ is well defined. The usual rules for addition apply.

For each pair of vectors there is a scalar product, a complex number. The scalar product (g, f) obeys the following rules:

$$\begin{aligned}
&(f, f) = 0 \text{ if and only if } f = 0\\
&(f, f) \geq 0 \text{ for } f \in \mathcal{H}\\
&(h, f + g) = (h, f) + (h, g) \text{ for } f, g, h \in \mathcal{H}\\
&(g, \alpha f) = \alpha(g, f) \text{ for } f, g \in \mathcal{H} \text{ and } \alpha \in \mathbb{C}\\
&(g, f) = (f, g)^* \text{ for } f, g \in \mathcal{H}.
\end{aligned} \tag{1.49}$$

By the way, $(f+g, h) = (f, h)+(g, h)$ and $(\alpha f, g) = \alpha^*(f, g)$ are consequences.

The norm of a vector is defined as

$$\| f \| = \sqrt{(f, f)}. \tag{1.50}$$

It obeys the Cauchy–Schwarz inequality

$$|(g, f)| \leq \| g \| \| f \|. \tag{1.51}$$

From this follows the triangle inequality

$$\| f + g \| \leq \| f \| + \| g \|. \tag{1.52}$$

$\| f - g \|$ therefore measures the distance between f and g and allows statements on convergence. We require that the Hilbert space $\mathcal{H}$ be complete: any Cauchy convergent sequence of vectors has a limit in $\mathcal{H}$.

Two vectors f, g are orthogonal if their scalar product vanishes, $(g, f) = 0$. More, the Cauchy–Schwarz inequality allows to define an angle γ between two non-vanishing vectors:

$$\cos \gamma = \frac{\mathrm{Re}(g, f)}{\| g \| \| f \|}. \tag{1.53}$$

Rez is the real part of the complex number z.

Let $e_1, e_2, \dots \in \mathcal{H}$ be a system of normalized and mutually orthogonal vectors:

$$(e_j, e_k) = \delta_{jk}. \tag{1.54}$$

The system is complete if an arbitrary vector $f \in \mathcal{H}$ may be decomposed as

$$f = \sum \alpha_i e_i \text{ with } \alpha_i = (e_i, f). \tag{1.55}$$

We speak of a complete orthonormal system, or an orthonormal base. Linear combinations (1.55) of the base vectors span the entire Hilbert space.

1.3.3 Linear Operators

We now discuss linear mappings $A : \mathcal{H} \to \mathcal{H}$ of the Hilbert space into itself. Linear means: linear combinations are mapped into linear combinations:

$$A(\alpha f + \beta g) = \alpha A(f) + \beta A(g) \tag{1.56}$$

for $f, g \in \mathcal{H}$ and $\alpha, \beta \in \mathbb{C}$. It is an established practice not to include the argument in round brackets so that (1.56) reads

$$A(\alpha f + \beta g) = \alpha A f + \beta A g. \tag{1.57}$$

Here the bracket says: perform the linear combination first, then map. A operates on the expression to the right of it. Linear mappings are also called linear operators.

Linear operators can be multiplied by scalars,

$$(\alpha A) f = \alpha (A f) \tag{1.58}$$

for $\alpha \in \mathbb{C}$ and for all $f \in \mathcal{H}$.

Linear operators may be added,

$$(A + B) f = A f + B f \tag{1.59}$$

for all $f \in \mathcal{H}$.

It is also possible to multiply linear operators by applying them one after another:

$$(BA) f = B(A f). \tag{1.60}$$

However, BA and AB are, in general, not the same. Multiplication of linear operators is not commutative.

The linear operators of a particular Hilbert space form an algebra $\mathfrak{A}$ with unity. The neutral element of addition is the zero operator 0 as defined by $0f = 0$. The neutral element of multiplication is the unity operator I as defined by $If = f$.

Associated with a linear operator A is an adjoint linear operator $A^\star$. For fixed $f \in \mathcal{H}$, the functional $g \to \Phi(g) = (g, Af)$ is linear. The Riesz representation theorem says that there is a vector $h \in \mathcal{H}$ such that $\Phi(g) = (h, f)$. This vector depends linearly on g so that we may write $h = A^\star g$, where $A^\star$ is a linear operator. To summarize,

$$(A^\star g, f) = (g, A f) \tag{1.61}$$

holds true for all $f, g \in \mathcal{H}$. It is easy to prove the following properties of adjoining:

$$\begin{aligned} A^{\star\star} &= A \\ (\alpha A)^\star &= \alpha^* A^\star \\ (A + B)^\star &= A^\star + B^\star \\ (AB)^\star &= B^\star A^\star. \end{aligned} \tag{1.62}$$

Adjoining is in fact a star operation.

Let us define an operator norm as follows. The linear operator A is bounded by κ if $\| Af \| \le \kappa \| f \|$ holds true for all $f \in \mathcal{H}$. The smallest bound κ is the norm $\| A \|$. This may also be expressed as

$$\| A \| = \sup_{f \neq 0} \frac{\| Af \|}{\| f \|}. \tag{1.63}$$

It has the following properties:

$$\begin{aligned}
&\| A \| = 0 \text{ if and only if } A = 0 \\
&\| \alpha A \| = |\, \alpha \,| \| A \| \\
&\| A + B \| \leq \| A \| + \| B \| \\
&\| AB \| \leq \| A \| \| B \| \\
&\| A \| = \| A^\star \|.
\end{aligned} \tag{1.64}$$

The last line is a little tricky. For a normalized vector f we find $\| Af \|^2 = (Af, Af) = (f, A^\star Af) \leq \| A^\star Af \| \leq \| A^\star A \|$. We first invoke the Cauchy–Schwarz inequality and then the definition of the operator norm. Hence, $\| Af \| \leq \sqrt{\| A^\star A \|}$. Now, take the supremum over normalized vectors f and square the expressions. This results in $\| A \|^2 \leq \| A^\star A \|$. By applying the rule for the norm of a product one obtains $\| A \|^2 \leq \| A^\star \| \| A \|$. Interchanging A with $A^\star$ finally results in $\| A \| = \| A^\star \|$.

The norm is compatible with the star operation, hence $\mathfrak{A}$ is a $\mathrm{C}^\star$ algebra. Note that we speak of bounded linear operators for which (1.63) is finite.

1.3.4 Projectors

A projector is a linear operator Π with the following properties:

$$\Pi = \Pi^\star \text{ and } \Pi^2 = \Pi. \tag{1.65}$$

It is self-adjoint and idempotent.

Define the set

$$\mathcal{L} = \{g \in \mathcal{H} \mid g = \Pi f,\, f \in \mathcal{H}\}. \tag{1.66}$$

Put otherwise, $\mathcal{L}$ is the range of Π the domain of which is $\mathcal{H}$. Since Π is a linear operator, the set $\mathcal{L}$ is a linear space. $\mathcal{L}$ is equipped with a scalar product which is inherited from $\mathcal{H}$. We abbreviate (1.65) by $\mathcal{L} = \Pi\mathcal{H}$.

In general, a Cauchy sequence in the linear subspace $\mathcal{L}$ does not converge in $\mathcal{L}$. It has a limit in $\mathcal{H}$, but not necessarily in $\mathcal{L}$. $\mathcal{L}$ is not necessarily a Hilbert space.

If $g = \Pi f$ then $\Pi g = \Pi^2 f = \Pi f = g$. Vectors in the linear subspace $\mathcal{L} = \Pi\mathcal{H}$ are unaffected by a repeated projections. Restricted to $\mathcal{L}$, the projector Π acts as the unit operator.

Choose an arbitrary $f \in \mathcal{H}$. We define $g = \Pi f$ and $h = f - g$. These two vectors are orthogonal;

$$(g, h) = (\Pi f, f - \Pi f) = (f, \Pi^\star f - \Pi^\star \Pi f) = 0, \tag{1.67}$$

because of $\Pi^\star = \Pi$ and $\Pi^2 = \Pi$. This finding may also be formulated as $\mathcal{H} = \mathcal{L} + \mathcal{L}_\perp$ where $\mathcal{L} = \Pi\mathcal{H}$, $\mathcal{L}_\perp = (I - \Pi)\mathcal{H}$ and $\mathcal{L} \perp \mathcal{L}_\perp$. All the above arguments apply because, with Π, the linear operator $\Pi_\perp = I - \Pi$ is also a projector. $\Pi\Pi_\perp = \Pi_\perp\Pi = 0$ means that the two projectors are orthogonal. This may be generalized to a decomposition of the Hilbert space into more than two linear subspaces.

Let $\Pi_1, \Pi_2 \dots$ be a set of mutually orthogonal projectors. This means

$$\Pi_i = \Pi_i{}^\star \text{ and } \Pi_i \Pi_j = \delta_{ij} \Pi_i. \tag{1.68}$$

We speak of a decomposition of unity if there is a set of mutually orthogonal projectors such that

$$I = \Pi_1 + \Pi_2 + \dots \tag{1.69}$$

holds true. Clearly, this makes sense for bounded operators only which are defined on the entire Hilbert space.

Recall the definition of dimension. A linear subspace $\mathcal{L}$ may be spanned by normalized and mutually orthogonal vectors $e_1, e_2, \dots$. If it can be spanned by at most n base vectors, we say that $\mathcal{L}$ is n-dimensional. If infinitely many base vectors are required, we speak of an infinite-dimensional linear subspace.

Each linear subspace $\mathcal{L}$ is generated by an associated projector Π such that $\mathcal{L} = \Pi\mathcal{H}$. The projector is n-dimensional if the linear subspace $\mathcal{L} = \Pi\mathcal{H}$ has dimension n.

Consider a single normalized vector e spanning a one-dimensional subspace, a line. The corresponding one-dimensional projector is described by

$$\Pi f = (e, f)\, e \text{ for all } f \in \mathcal{H}. \tag{1.70}$$

A complete set of mutually orthogonal normalized vectors $e_1, e_2, \dots$ then defines a decomposition of unity, namely

$$I = \sum_i \Pi_i \text{ where } \Pi_i f = (e_i, f)\, e_i. \tag{1.71}$$

The decomposition is into one-dimensional projectors.

1.3.5 Normal Linear Operators

Let $\nu_1, \nu_2, \dots$ a set of complex numbers and $\Pi_1, \Pi_2 \dots$ a decomposition of unity. Define a linear operator N by

$$N = \nu_1 \Pi_1 + \nu_2 \Pi_2 + \dots \tag{1.72}$$

with a set of complex numbers $\nu_1, \nu_2, \ldots$ One easily shows

$$N N^\star = N^\star N. \tag{1.73}$$

A normal operator commutes with its adjoint. In Sect. A.8 of the chapter on Mathematical Aspects we will sketch a proof why the converse is true as well. Any linear operator commuting with its adjoint is normal according to (1.72).

Normal operators are easy to handle. The entire Hilbert space is decomposed into mutually orthogonal subspaces, and in each subspace the linear operator simply shrinks or stretches the vectors of this subspace. $g \in \Pi_i \mathcal{H}$ is mapped into $\nu_i g$. These factors ν_i are called eigenvalues.

Let $f : \mathbb{C} \to \mathbb{C}$ be a bounded complex function of one complex argument. If the normal operator N is decomposed according to (1.72) one may define its function by

$$f(N) = f(\nu_1)\Pi_1 + f(\nu_2)\Pi_2 + \ldots. \tag{1.74}$$

Obviously, $f(N)$ is normal as well.

Unitary Operators

A unitary operator U is a mapping of $\mathcal{H}$ onto itself which preserves the scalar product, i.e. $(f, g) = (Uf, Ug)$ for all $f, g \in \mathcal{H}$. We deduce

$$U^\star U = I. \tag{1.75}$$

A unitary operator has an inverse, since $Uf = 0$ has no other solution than $f = 0$. This follows from $0 = \| Uf \|^2 = (Uf, Uf) = (f, U^\star Uf) = \| f \|^2$ which implies $f = 0$. Multiplying (1.75) from the right with U^{-1} and from the left with U results in $UU^\star = I$.

One sees that a unitary operator is normal. It can be written as (1.72). Note that $U^\star$ has the same decomposition with ν_i^* instead of ν_i. Therefore,

$$I = U^\star U = |\, \nu_1 \,|^2\, \Pi_1 + |\, \nu_2 \,|^2\, \Pi_2 + \ldots, \tag{1.76}$$

hence $|\, \nu_i \,| = 1$. The eigenvalues of a unitary operator are phase factors, complex numbers on the unit circle.

Unitary operators are similar to rotation matrices in $\mathbb{R}^3$. They map one orthogonal base into another. Assume such a base $e_1, e_2, \ldots$ of mutually orthogonal unit vectors. If U is unitary, the system $Ue_1, Ue_2, \ldots$ is an orthogonal base as well since $(Ue_j, U_k) = (e_j, U^\star U e_k) = (e_j, e_k) = \delta_{jk}$.

The converse is also true. If $(\bar{e}_j, \bar{e}_k) = \delta_{jk}$ and $(e_j, e_k) = \delta_{jk}$ then there is a unitary operator such that $\bar{e}_i = Ue_i$.

Self-adjoint Operators

A self-adjoint operator M is described by

$$M^\star = M. \tag{1.77}$$

It is evidently normal and may be represented as

$$M = m_1 \Pi_1 + m_2 \Pi_2 + \dots, \tag{1.78}$$

its eigenvalues m_i are real numbers. Recall that the Π_i are mutually orthogonal projectors summing up to unity.

The self-adjoint operators are identified with the observables of the physical system under discussion. The system is described by a Hilbert space which carries the linear operators. They can be added, multiplied and starred, forming a C* algebra. Some operators are normal, and some normal operators are self-adjoint. They describe the observable properties of the physical system. The eigenvalues m_i are the possible values the observable M may assume.

Positive Operators

A non-negative operator P is self-adjoint and has non-negative eigenvalues. We write $P \geq 0$. There are various ways to characterize a non-negative operator:

- $(f, Pf) \geq 0$ for all $f \in \mathcal{H}$
- there is a decomposition $I = \Pi_1 + \Pi_2 + \dots$ of unity into mutually orthogonal projectors such that $P = p_1 \Pi_1 + p_2 \Pi_2 + \dots$ with $p_i \geq 0$.
- There is a linear operator A such that $P = A^\star A$.

Projection Operators

Because of $\Pi = \Pi^\star$ a projector is self-adjoint. Its real eigenvalues are zero or one. This follows from $\Pi^2 = \Pi$.

Probability Operators

A probability operator ϱ is self-adjoint, and its eigenvalues w_j are probabilities. This means

$$0 \leq w_j \text{ and } \sum_j w_j \dim(\Pi_j) = 1. \tag{1.79}$$

Each eigenvalue must be counted with its multiplicity $\dim(\Pi_j)$.

1.3.6 Trace of an Operator

Let A be a linear mapping of an n-dimensional Hilbert space into itself. Select an orthonormal base $e_1, e_2, \dots, e_n$, i.e. with the property $(e_j, e_k) = \delta_{jk}$. We describe the linear mapping A by its matrix $A_{jk} = (e_j, Ae_k)$. The trace of this matrix is defined as

$$\mathrm{Tr}\,A = \sum_{i=1}^{n} A_{ii}. \tag{1.80}$$

Chose another orthonormal base $\bar{e}_1, \bar{e}_2, \dots, \bar{e}_n$. The relation between old and new base vectors is $e_i = \sum_j U_{ij}\bar{e}_j$ where U_{ij} is some unitary matrix. The mapping A is now described by the matrix $\bar{A}_{jk} = (\bar{e}_j, A\bar{e}_k)$. We find

$$\sum_i A_{ii} = \sum_{irs} U^*_{ir} U_{is} (\bar{e}_r, A\bar{e}_s) = \sum_{rs} \delta_{rs} \bar{A}_{rs} = \sum_r \bar{A}_{rr}. \tag{1.81}$$

Although the trace is defined and calculated with respect to some orthonormal base, the result is the same for all such bases. It follows that the trace is a property of the linear mapping A, as indicated in (1.80).

It is an easy task to show the following properties of the trace functional:

$$\begin{aligned} \mathrm{Tr}(\alpha A + \beta B) &= \alpha \mathrm{Tr}\,A + \beta \mathrm{Tr}\,B \\ \mathrm{Tr}\,A^\star &= (\mathrm{Tr}\,A)^* \\ \mathrm{Tr}\,AB &= \mathrm{Tr}\,BA \end{aligned} \tag{1.82}$$

For an infinite-dimensional Hilbert space the definition of the trace is more delicate. The sum over diagonal matrix elements may not converge. We therefore restrict the definition to self-adjoint operators. The eigenvalues α_i of a self-adjoint operator A shall be absolutely summable, i.e. $\sum_i | \alpha_i | < \infty$. If this is true, the trace

$$\mathrm{Tr}\,A = \sum_i \alpha_i \tag{1.83}$$

is well defined and finite. It can be shown that

$$\mathrm{Tr}\,A = \sum_i (e_i, Ae_i) \tag{1.84}$$

holds true for an arbitrary complete orthonormal system of base vectors. Properties (1.82) hold true as well. In particular, if A has a trace and if B is bounded, then both AB and BA have traces which are equal.

1.3.7 Expectation Values

We have discussed at great length the concept of an observable. An observable M is represented by a linear operator which maps the system's Hilbert space into itself. It is moreover self-adjoint: there is a decomposition of the Hilbert space into mutually orthogonal linear subspaces $\mathcal{L}_i = \Pi_i \mathcal{H}$ such that

$$M = m_1 \Pi_1 + m_2 \Pi_2 + \dots \tag{1.85}$$

holds true with real eigenvalues m_i. If the observable M is measured, the possible result is one of these eigenvalues.

The state of the system is described by a probability operator ϱ. Its eigenvalues are weights: $w_i \geq 0$ and $\sum_i w_i = 1$. This may be expressed in terms of the operator itself,

$$\varrho \geq 0 \text{ and } \operatorname{Tr} \varrho = 1. \tag{1.86}$$

Guided by plausibility considerations and the classical analog we write

$$\langle M \rangle = \varrho(M) = \operatorname{Tr} \varrho M \tag{1.87}$$

for the expectation value of the observable M if the system is in state ϱ. As it should be, the functional $A \to \varrho(A)$ is linear and real for self-adjoint operators. The observable "perform a measurement" is adequately described by I. Its expectation value is $\varrho(I) = 1$. It is not difficult to show that non-negative operators P have a non-negative expectation value. Just take the eigenvectors e_i of ϱ as a complete orthonormal set.

$$\operatorname{Tr} \varrho A = \sum_i (e_i, \varrho A\, e_i) = \sum w_i (e_i, A\, e_i) \geq 0. \tag{1.88}$$

The inequality

$$\varrho(A) \leq \varrho(B) \text{ if } A \leq B \tag{1.89}$$

is a consequence. A is less or equal to B if $B - A$ is non-negative.

1.3.8 Summary

We began with a discussion why the searched for observables must be represented by objects which do not necessarily commute. Physics has fared well with linear operators. They live on a Hilbert space, a linear space with scalar product. Therewith norms for vectors and linear operators enter the game. This allows to discuss convergence of infinite series and of continuity. The scalar product also allows to define the adjoint of an operator which is the analog of a complex conjugate number.

We then turned our attention to normal operators. They commute with their own adjoint. Normal operators are simple to handle. There is a decomposition of the Hilbert space into mutually orthogonal linear subspaces such that a normal operator maps these subspaces into themselves by simply scaling all vectors by the same factor. The scaling factor ν_i for the linear subspace $\mathcal{L}_i = \Pi_i\mathcal{H}$ is the associated eigenvalue.

We briefly discuss unitary operators the eigenvalues of which are phase factors, complex numbers of modulus one. Self-adjoint operators have real eigenvalues. They are identified with the system's observables. The eigenvalues are the possible values the observable may attain.

The state of a system is described by a probability operator. Its eigenvalues, counted with their multiplicities, are weights, or probabilities: non-negative numbers which sum up to one.

Defining the trace is straightforward for finite-dimensional Hilbert space. Although calculated with respect to an orthonormal base, the result does not depend on which base was chosen. The trace therefore characterizes the mapping. The same applies to self-adjoint operators of an infinite-dimensional Hilbert space provided the eigenvalues (with multiplicities) are summable.

We finally put all this together and present an expression for the expectation value $\varrho(M)$ of the observable M in state ϱ. Note the analogs:

- self-adjoint function $M = M(\omega)$ on a phase space Ω
- probability function $\varrho = \varrho(\omega)$
- integral $\int \mathrm{d}\omega$ over phase space

and

- self-adjoint operator $M : \mathcal{H} \to \mathcal{H}$ of a Hilbert space $\mathcal{H}$
- probability operator ϱ
- trace Tr over Hilbert space.

States are characterized by $\varrho \geq 0$ and $\int \mathrm{d}\omega\, \varrho(\omega) = 1$ or $\mathrm{Tr}\, \varrho = 1$, respectively.

With this section we have summarized the mathematical framework of quantum physics. It has been written with a first orientation in mind, in casual language, avoiding subtleties. Finer points are deferred to the chapter on Mathematical Aspects.

1.4 Time and Space

We introduce time as the time span between preparing the state of the system for measurement and measuring a property. The waiting operators form an Abelian group the members of which are indexed by the time span. This group of unitary operators is generated by a self-adjoint operator which, by definition, is the energy observable H.

The three components of linear momentum are brought into play by a similar procedure. After preparing the state, the measuring device, the observable, is spatially shifted. This situation is again described by an Abelian group of unitary operators which is generated by three commuting observables, the components P_i of total linear momentum.

Instead of translated, the entire measuring apparatus may be rotated. Again, the corresponding group of unitary operators is generated by three self-adjoint operators, the three components J_i of total angular momentum. However, since rotations in general do not commute, the components of angular momentum will not commute either. We shall derive the angular momentum commutation rules.

We also introduce location observables and derive their commutation relations with other observables. The canonical commutation relations are a by-product. Already at this abstract level the notion of spin enters the stage.

What we have described so far is in the spirit of Heisenberg: states remain unchanged, the observables change if translated in time or space or being rotated. Schrödinger's view was different. Preparing a system in a certain state and then waiting—this is another preparation prescription and defines another state. We show how to transform between these two aspects, the Heisenberg and the Schrödinger picture.

1.4.1 Measurement and Experiment

Idealized, a measurement proceeds in two steps. First, the system is prepared with a suitable apparatus and by a detailed prescription. Second, a certain property of the system is measured. The first step leads to a definite state ϱ, the second involves an observable M. If the system has been prepared in state ϱ, and if property M is measured, the expected result is predicted to be $\varrho(M)$.

In order to check this prediction, the measurement must be repeated very often. We call such a long series of identical measurements an experiment. The first measurement yields the result r_1, the second r_2, and so on. Note that these measured values are possible results as described by the set $\{m_1, m_2, \dots\}$ of M-eigenvalues.

The average measurement result is

$$\bar{M} = \frac{1}{N}\sum_{n=1}^{N} r_n, \tag{1.90}$$

and with $N \to \infty$ the expressions (1.90) should converge towards the expectation value $\varrho(M)$. Usually, a single measurement makes no sense, only the average values of experiments should provide reliable and reproducible information.

The measured values $r_1, r_2, \dots, r_N$ of an experiment will fluctuate. In general, they are not all the same. One might think at first that this behavior is due to an imprecisely prepared state or to an imperfect measurement. But endless discussions and even experiments have revealed: it is impossible to predict the outcome of a single measurement; it is dictated by chance. Only the average value of an entire experiment is a reproducible result. The limit of such an average for a longer and longer series of measurements converges towards the expectation value which can be calculated as $\varrho(M)$. Quantum theory must be based on the concept of states, observables, and expectation values.

1.4.2 Time Translation

We just repeated that a measurement consist of preparing the system in a certain state ϱ and performing a measurement of the observable M. Assume now that, before measuring M, we are to wait for a time span τ. One may also say that, after having prepared the state ϱ, we wait for the time span τ. The first formulation amounts to modify the observable M to M_τ. The second formulation insinuates that we modify the state ϱ into ϱ_τ. The former picture is due to Heisenberg, the latter to Schrödinger. For the moment we adopt Heisenberg's view.

The observables $M = M_0$ and M_τ have the same possible values. Their eigenvalues coincide. It follows that they must be related by a unitary transformation,

$$M_\tau = U^\star M U \text{ where } U^\star U = U U^\star = I. \tag{1.91}$$

The waiting operator depends on the time span τ, $U = U_\tau$. Now, waiting a time τ' and then waiting τ'' is the same as waiting a time $\tau' + \tau''$. Therefore

$$U_{\tau''} U_{\tau'} = U_{\tau''+\tau'} \tag{1.92}$$

has to hold. The solution, with $U_0 = I$, is an exponential function:

$$U_\tau = \mathrm{e}^{-\mathrm{i}\tau\Omega}. \tag{1.93}$$

The minus sign is a convention. The imaginary unit has been split off so that the linear operator Ω is self-adjoint. Its eigenvalues are angular frequencies. However, because the classical limit has been investigated before quantum theory, the energy $H = \hbar\Omega$ was used instead. Therefore, the waiting operator can also be written as

$$U_\tau = \mathrm{e}^{-\frac{\mathrm{i}}{\hbar}\tau H}. \tag{1.94}$$

H is a self-adjoint operator, an observable. Its eigenvalues are the possible energy levels of the system.

The equation of motion of an observable reads

$$\frac{\mathrm{d}M_\tau}{\mathrm{d}\tau} = \{M_\tau, H\}. \tag{1.95}$$

We have introduced the abbreviation

$$\{A, B\} = -\frac{\mathrm{i}}{\hbar}(AB - BA) = -\frac{\mathrm{i}}{\hbar}[A, B]. \tag{1.96}$$

Note that the expression $\{A, B\}$ is self-adjoint if both A and B are.

If an observable M commutes with the Hamilton operator, $\{M, H\} = 0$, it is a constant of motion. An example is the energy itself. Energy is conserved.

1.4.3 Space Translation

Now another situation. The apparatus for preparing the state ϱ is kept fixed, as before, but the measuring equipment for the observable M shall be translated in space by $\boldsymbol{a}$. In the spirit of Heisenberg, this defines a new observable $M_{\boldsymbol{a}}$. The new observable has the same eigenvalue as M, they are therefore related by a transformation with unitary operators,

$$M_{\boldsymbol{a}} = U_{\boldsymbol{a}}{}^{\star} M U_{\boldsymbol{a}}. \tag{1.97}$$

Note that $U_{\boldsymbol{a}} = U_{a_1} U_{a_2} U_{a_3}$ is a product of three commuting unitary operators since translations commute. With similar arguments as above we may write

$$U_{\boldsymbol{a}} = \mathrm{e}^{\displaystyle \mathrm{i}\boldsymbol{a} \cdot \boldsymbol{K}}. \tag{1.98}$$

The sign is dictated by relativity: $k_0 = \omega/c$ and $\boldsymbol{k}$ shall be a four-vector. The imaginary unit guarantees that the three operators K_1, K_2, K_3 are self-adjoint. They are observables, namely the three wave vector components of the system. They commute with each other.

$$\boldsymbol{P} = \hbar \boldsymbol{K} \tag{1.99}$$

is, by definition, the self-adjoint total linear momentum. The linear momentum operators commute among themselves and with the energy,

$$\{P_j, P_k\} = \{P_j, H\} = 0. \tag{1.100}$$

We conclude that the total linear momentum is conserved.

The analog of Heisenberg's equation of motion (1.95) is easily worked out:

$$\frac{\partial M_{\boldsymbol{a}}}{\partial a_j} = \{P_j, M_{\boldsymbol{a}}\}. \tag{1.101}$$

1.4.4 Location

Imagine a single particle. Its location is an observable $\boldsymbol{X}$ with components X_1, X_2, X_3. Upon translation by $\boldsymbol{a}$ these observables transform as

$$U_{\boldsymbol{a}}{}^{\star} X_i U_{\boldsymbol{a}} = X_i - a_i I. \tag{1.102}$$

We differentiate this according to (1.101) and find

$$\left.\frac{\partial (X_i)_{\boldsymbol{a}}}{\partial a_j}\right|_{\boldsymbol{a}=0} = -\{X_i, P_j\} = -\delta_{ij} I. \tag{1.103}$$

This result, namely

$$\{X_i, P_j\} = \delta_{ij} I \tag{1.104}$$

is the famous canonical commutation rule.

1.4.5 Rotation

Rotations in a three-dimensional space do not commute in general and are more difficult to handle.

Let us begin with the xy-plane. A rotation about the z-axis by an angle α is described by

$$\begin{pmatrix} x' \\ y' \end{pmatrix} = \begin{pmatrix} \cos\alpha & -\sin\alpha \\ \sin\alpha & \cos\alpha \end{pmatrix} \begin{pmatrix} x \\ y \end{pmatrix}. \tag{1.105}$$

The rotation matrix may be written as an exponential,

$$R_\alpha = \begin{pmatrix} \cos\alpha & -\sin\alpha \\ \sin\alpha & \cos\alpha \end{pmatrix} = \mathrm{e}^{-\mathrm{i}\alpha M} \text{ with } M = \begin{pmatrix} 0 & -\mathrm{i} \\ \mathrm{i} & 0 \end{pmatrix}. \tag{1.106}$$

This follows from $M^2 = I$. The matrix elements of the generator are given by the Levi-Civita symbol in two dimensions, $M_{jk} = -\mathrm{i}\epsilon_{jk}$.

In three dimensions we have three such planes and consequently three generators M_i, now 3×3 matrices, with elements $(M_i)_{jk} = -\mathrm{i}\epsilon_{ijk}$. Likewise the axis of rotation is now a unit vector $\hat{\boldsymbol{n}}$. With $\boldsymbol{\alpha} = \alpha\hat{\boldsymbol{n}}$ one generalizes to three spatial dimensions:

$$R_{\boldsymbol{\alpha}} = \mathrm{e}^{-\mathrm{i}\boldsymbol{\alpha}\cdot\boldsymbol{M}}. \tag{1.107}$$

Because of $M_i = M_i^{\star}$ the rotation matrix is unitary, $R_{\boldsymbol{\alpha}} R_{\boldsymbol{\alpha}}^{\star} = I$. Since it is also real, we speak of an orthogonal matrix. Orthogonal matrices cause rotations in real space.

The determinant of an orthogonal matrix R is either plus or minus one. The former describes proper rotations. Since $\det R_{\boldsymbol{\alpha}}$ depends continuously on $\boldsymbol{\alpha}$, the determinant of (1.107) is the same as that of $R_0 = I$, namely $+1$. Equation (1.107) describes proper rotations.

These commutation rules characterize rotations:

$$[M_1, M_2] = \mathrm{i}M_3\ ,\ [M_2, M_3] = \mathrm{i}M_1\ ,\ [M_3, M_1] = \mathrm{i}M_2, \tag{1.108}$$

or[2]

$$[M_i, M_j] = \mathrm{i}\epsilon_{ijk}M_k. \tag{1.109}$$

By rotating the measuring device, the possible outcomes are not changed. Therefore, rotated observables are given by a transformation

$$M_\alpha = U_\alpha M U_\alpha{}^\star \tag{1.110}$$

with a unitary operator

$$U_\alpha = \mathrm{e}^{-\frac{\mathrm{i}}{\hbar}\,\boldsymbol{\alpha}\cdot\boldsymbol{J}}. \tag{1.111}$$

Splitting off the imaginary unit guarantees that the three linear operators J_i are self-adjoint. The minus sign is a convention. The inverse $\hbar$ is analogous to energy and linear momentum. The three observables J_i are the system's total angular momentum. As we have shown above, the commutation rules are

$$[J_1, J_2] = \mathrm{i}\hbar J_3 \text{ or } \{J_1, J_2\} = J_3 \tag{1.112}$$

and cyclic permutations of indexes. To summarize:

$$\{J_i, J_j\} = \epsilon_{ijk}J_k. \tag{1.113}$$

1.4.6 Orbital Angular Momentum and Spin

Define $\boldsymbol{L} = \boldsymbol{X} \times \boldsymbol{P}$, the vector product of location and linear momentum. It is the system's orbital angular momentum. By inserting the canonical commutation relations (1.104) one calculates the following commutation rules:

$$\{L_i, L_j\} = \epsilon_{ijk}L_k, \tag{1.114}$$

formally the same as (1.113). However, total angular momentum and orbital angular momentum are not the same, $\boldsymbol{J} = \boldsymbol{L} + \boldsymbol{S}$. The internal angular momentum $\boldsymbol{S}$, or spin, should be understood as the angular momentum of the particle at rest in its own center of mass frame. It is an internal property of particles and does not depend on location

[2]With Einstein's summation convention.

or linear momentum. Therefore we have $\{X_i, S_j\} = \{P_i, S_j\} = 0$. $\{L_i, S_j\} = 0$ is a consequence as well as

$$\{S_i, S_j\} = \epsilon_{ijk} S_k. \tag{1.115}$$

Three observable V_i form a vector if they commute with the total angular momentum according to

$$\{J_i, V_j\} = \epsilon_{ijk} V_k. \tag{1.116}$$

The total angular momentum J_i, the orbital angular momentum L_i, the spin S_i form a vector, but also the location X_i. The square of a vector commutes with angular momentum,

$$\{J_i, V_j V_j\} = \epsilon_{ijk} V_j V_k + \epsilon_{ijk} V_k V_j = 0, \tag{1.117}$$

it is a scalar. The energy H is a scalar, $\{J_i, H\} = 0$. The same equation says that the total angular momentum is conserved.

1.4.7 Schrödinger Picture

Up to now we have argued like Heisenberg. The state ϱ is prepared and the transformed observable $M' = UMU^\star$ is measured. We have discussed the unitary operators U for waiting, spatial translation and rotation. The expectation value is given by

$$\varrho(M') = \mathrm{Tr}\,\varrho\, UMU^\star. \tag{1.118}$$

Now, within a trace the factors may be cyclically permuted,

$$\mathrm{Tr}\,\varrho\, UMU^\star = \mathrm{Tr}\,U^\star \varrho\, U\, M = \mathrm{Tr}\,\varrho' M. \tag{1.119}$$

This is the expectation value of M in the transformed state $\varrho' = U^\star \varrho U$. Today we speak of Schrödinger's picture. The transformed state ϱ' is prepared and the observable M will be measured. Clearly, both views lead to the same prediction for the result of an experiment.

Time translation, or waiting, is now described by a time dependent state $\varrho = \varrho_\tau$, namely

$$\varrho_\tau = \mathrm{e}^{-\frac{\mathrm{i}}{\hbar}\tau H}\, \varrho\, \mathrm{e}^{\frac{\mathrm{i}}{\hbar}\tau H}. \tag{1.120}$$

Differentiation results in the following equation of motion:

$$\frac{\mathrm{d}\varrho_\tau}{\mathrm{d}\tau} = \{H, \varrho_\tau\}. \tag{1.121}$$

A probability operator commuting with energy is stationary. It does not change in the course of time. Equation (1.121) is known as van Neumann's equation.

Let us specialize to a pure state. Its probability operator is a one-dimensional projector, $\varrho = \Pi(f)$. $f \in \mathcal{H}$ should be normalized, the projector is defined by $\Pi(f)g = (f, g)f$. It projects on the linear subspace spanned by f. For any unitary operator U, $U\Pi(f)U^{\star}$ is a one-dimensional projector as well, namely on Uf. With $U = U_\tau = \mathrm{e}^{-\frac{\mathrm{i}}{\hbar}\tau H}$ we write $f_\tau = U_\tau f$ and calculate

$$-\frac{\hbar}{\mathrm{i}}\frac{\mathrm{d}f_\tau}{\mathrm{d}\tau} = Hf_\tau. \tag{1.122}$$

This is the famous Schrödinger equation for the time-evolution of pure states, formulated in the Schrödinger picture. Note that $\| f_\tau \|^2$ does not change with time, and the vector f_τ is normalized as well.

If we calculate a solution of the time-independent Schrödinger equation

$$Hf = Ef, \tag{1.123}$$

we have found a stationary state. f_τ then changes with time according to

$$f_\tau = \mathrm{e}^{-\frac{\mathrm{i}}{\hbar}\tau E} f \tag{1.124}$$

and defines the same one-dimensional projector as f. The time dependency in (1.124) is by a mere phase factor, f_τ and f span the same one-dimensional subspace.

It should be clear by now why the eigenvalues and eigenvectors (1.123) of the energy play such a dominant role in textbooks on quantum mechanics. These eigenvectors correspond to stationary pure states which may be combined to non-stationary and mixed states.

1.4.8 Summary

In this section we have exploited the properties of space and time. The most important observables and their commutation rules have been derived: energy H, linear momentum $\boldsymbol{P}$, location $\boldsymbol{X}$ as well as three types of angular momentum: total $\boldsymbol{J}$, orbital $\boldsymbol{L}$, and internal or spin $\boldsymbol{S}$. We collect these rules in Table 1.1, they reflect the structure of the space-time continuum. We did not touch the problem of an observable to be boosted, or put in motion. Problems with special relativity will therefore not arise.

Table 1.1 Quantum Poisson brackets $\{A, B\}$. There is a row for each A and a column for each B. $\{X_i, P_j\} = \delta_{ij} I$ is an example. An asterisk says that the entry depends on the Hamiltonian under study

$\{A, B\}$	H	P_j	X_j	J_j	L_j	S_j
H	0	0	$*$	0	$*$	$*$
P_i	0	0	$-\delta_{ij} I$	$\epsilon_{ijk} P_k$	$\epsilon_{ijk} P_k$	0
X_i	$*$	$\delta_{ij} I$	0	$\epsilon_{ijk} X_k$	$\epsilon_{ijk} X_k$	0
J_i	0	$\epsilon_{ijk} P_k$	$\epsilon_{ijk} X_k$	$\epsilon_{ijk} J_k$	$\epsilon_{ijk} L_k$	$\epsilon_{ijk} S_k$
L_i	$*$	$\epsilon_{ijk} P_k$	$\epsilon_{ijk} X_k$	$\epsilon_{ijk} L_k$	$\epsilon_{ijk} L_k$	0
S_i	$*$	0	0	$\epsilon_{ijk} S_k$	0	$\epsilon_{ijk} S_k$

We also derived the time-dependent Schrödinger equation and have commented on the role of the time-independent equation for the energy eigenvectors. The latter describe stationary pure states, and they can be combined to non-stationary and mixed states. This will become clear in the following chapters which are dedicated to various applications.

Chapter 2
Simple Examples

In this chapter we present a few simple examples. Examples how to make use of the conceptual and mathematical framework sketched in Chap. 1 on Basics.

One of the most simple systems concerns the ground state of the ammonia molecule NH_3. There is a unilateral triangle of hydrogen ions and a nitrogen ion on the central axis either below or above. Thus, there are two equivalent configurations, up or down, with rather interesting implications. We can explain the ground state level splitting giving rise to the microwave frequency standard. An external electric field may further shift the energy levels, a phenomenon which goes under the name of Stark effect. It is possible to separate a beam into ground state and excited molecules such that extremely weak microwave signals may be amplified (maser).

Another simple example for quantum theoretic reasoning is the Hopping model. It explains the origins of electronic band structure, the nature of quasi-particles and the scattering and trapping of quasi-electrons on an impurity.

We then turn to neutron scattering. Even if the target consist of randomly oriented molecules, its arrangement of nuclei can be reconstructed from the differential cross section.

The example of a free particle which we will study next is helpful in two respects. It turns out that it moves just like a free particle in classical mechanics, with the expectation value of the location observable replacing the position coordinate. In addition, the particle spreads out more and more in the course of time.

Small oscillations about the equilibrium configuration are quantized, so another simple example. The excited states have discrete, equidistant energies.

2.1 Ammonia Molecule

Ammonia, or azan, is a colorless gas with a pungent odor. It is synthesized in large quantities mainly as raw material of fertilizers. Ammonia is the common English name for NH_3. There is also a completely different $NH_4{}^-$ ion which in English is

P. Hertel, *Quantum Theory and Statistical Thermodynamics*,
Graduate Texts in Physics, DOI 10.1007/978-3-319-58595-6_2

Fig. 2.1 Sketch of an ammonia molecule. There is a equilateral *triangle* of protons (*smaller spheres*) *above* or *below* the nitrogen ion (*larger sphere*). The ground state is a symmetric superposition of this and the mirror-reflected one

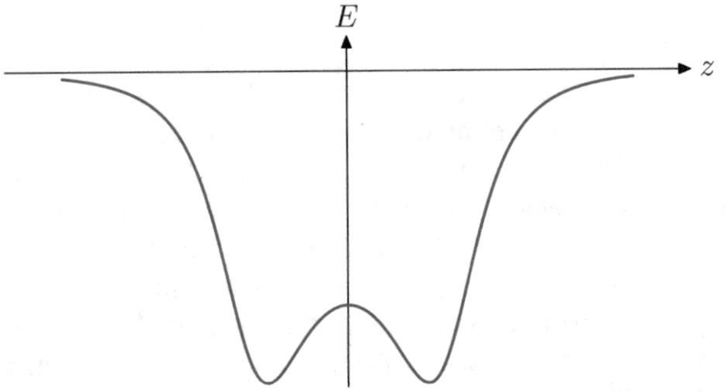

Fig. 2.2 Ammonia molecule potential energy. The potential energy E of an ammonia molecule is plotted versus dislocation z. The protons form a unilateral triangle, and an axis through its center and perpendicular to the proton plane is a natural site for the N^{3-} ion. z denotes the deviation from an in-plane position

called ammonium. The latter ion (see Fig. 2.1) consists of a tetrahedron of protons with a nitrogen ion at its center, a highly symmetric and firmly bound ion. Here we discuss the ammonia molecule because it is not so symmetric, a fact which gives rise to interesting effects.

There is an N^{3-} ion, three protons, or H^+, and three shared electrons which bind this to a molecule. One might think at first that the three protons form a equilateral triangle with the nitrogen ion at its center. Calculations and neutron diffraction experiments however result in another configuration. The three protons which repel each other form a unilateral triangle in fact. The nitrogen ion is located at an axis perpendicular to the hydrogen triangle passing through its center, as expected. But the nitrogen ion is displaced. It is either above or below the hydrogen ions plane. See Fig. 2.2 for a sketch of the potential energy. We disregard here all other possible configurations and concentrate only on this up-down degree of freedom.

2.1.1 Hilbert Space and Energy Observable

Let us construct a mathematical model. The two base vectors

$$f_{\uparrow} = \begin{pmatrix} 1 \\ 0 \end{pmatrix} \text{ and } f_{\downarrow} = \begin{pmatrix} 0 \\ 1 \end{pmatrix} \tag{2.1}$$

shall describe the up- and down configuration of the molecule. They span the Hilbert space $\mathbb{C}^2$ of two-component vectors of complex numbers. For $f, g \in \mathbb{C}^2$ the scalar product is

$$(g, f) = g_1^* f_1 + g_2^* f_2. \tag{2.2}$$

Evidently, the up- and down vectors are normalized and orthogonal.

Observables are represented by self-adjoint complex 2×2 matrices. For the Hamiltonian in particular we may write

$$H = \begin{pmatrix} E & -V \\ -V & E \end{pmatrix}. \tag{2.3}$$

The two diagonal elements must be equal because the up and down configurations are equivalent. One of the off-diagonal elements should be the complex conjugate of the other. By tuning the phase between the base states, they can be chosen equal and real, even positive. The minus sign is there for convenience.

The time-independent Schrödinger equation $H f = \epsilon f$ is solved by

$$f_{\text{lo}} = \frac{1}{\sqrt{2}} \begin{pmatrix} 1 \\ 1 \end{pmatrix} \text{ with } \epsilon_{\text{lo}} = E - V \tag{2.4}$$

and by

$$f_{\text{hi}} = \frac{1}{\sqrt{2}} \begin{pmatrix} 1 \\ -1 \end{pmatrix} \text{ with } \epsilon_{\text{hi}} = E + V. \tag{2.5}$$

The eigenvectors f_{lo} and f_{hi} correspond to the orthogonal projectors

$$\Pi_{\text{lo}} = \frac{1}{2} \begin{pmatrix} 1 & 1 \\ 1 & 1 \end{pmatrix} \text{ and } \Pi_{\text{hi}} = \frac{1}{2} \begin{pmatrix} 1 & -1 \\ -1 & 1 \end{pmatrix}. \tag{2.6}$$

By construction, the energy observable is

$$H = \epsilon_{\text{lo}} \Pi_{\text{lo}} + \epsilon_{\text{hi}} \Pi_{\text{hi}}. \tag{2.7}$$

There are two stationary energy states, one an even, the other an odd superposition of the up and down configurations $f_{\uparrow}$ and $f_{\downarrow}$.

If ammonia molecules are exposed to microwave radiation, photons of energy

$$\hbar\omega_0 = \epsilon_{\text{hi}} - \epsilon_{\text{lo}} = 2V \tag{2.8}$$

will be absorbed. They cause transitions from the ground state Π_{lo} to the excited state Π_{hi}. We will study the mechanism for this effect later. Equation (2.8) defines the microwave standard of f = 23.87012 GHz. This frequency corresponds to a vacuum wave length of λ = 1.255932 cm.

2.1.2 Ammonia Molecule in an External Electric Field

The electric charge of an ammonia molecule vanishes, but not its dipole moment. If the molecule has the $\uparrow$ configuration, its dipole moment is $d = ae$, where e denotes the unit charge, that of a proton. The length a = 0.019 nm is the average displacement of charges. Hence the electric dipole moment operator is

$$D = \begin{pmatrix} d & 0 \\ 0 & -d \end{pmatrix}. \tag{2.9}$$

In an external constant electric field $\boldsymbol{\mathcal{E}}$, the energy contribution of an electric dipole $\boldsymbol{D}$ is $-\boldsymbol{D} \cdot \boldsymbol{\mathcal{E}}$. In our case this reads

$$H = \begin{pmatrix} E - d\mathcal{E} & -V \\ -V & E + d\mathcal{E} \end{pmatrix}. \tag{2.10}$$

$\mathcal{E} = \mathcal{E}_z$ is the electric field strength along the axis of orientation of the ammonia molecule. Its eigenvalues are

$$\epsilon_{\text{lo}}(\mathcal{E}) = E - \sqrt{V^2 + d^2\mathcal{E}^2} \text{ and } \epsilon_{\text{hi}}(\mathcal{E}) = E + \sqrt{V^2 + d^2\mathcal{E}^2}. \tag{2.11}$$

A plausible result. For $\mathcal{E} = 0$ it reduces to $\epsilon_{\text{lo}} = E - V$ and $\epsilon_{\text{hi}} = E + V$, as discussed above. For $V = 0$ we obtain $\epsilon_{\text{lo}}(\mathcal{E}) = E - |d\mathcal{E}|$ and $\epsilon_{\text{hi}}(\mathcal{E}) = E + |d\mathcal{E}|$ which becomes obvious by inspecting (2.10). See Fig. 2.2 for a visualization (Fig. 2.3).

We have just explained the Stark effect. All kinds of spectral lines are affected if a probe of atoms or molecules is exposed to an external electric field. Atomic, molecular or solid state physics is governed by the motion of electrons due to Coulomb forces. Therefore, Planck's constant $\hbar$, the elementary charge e, the electron mass m and the factor $4\pi\epsilon_0$ come into play. We will later discuss the atomic system of units based on these four constants of nature. While the dipole moment d of an ammonia molecule is 0.40 in atomic units, the atomic unit for electric field strength is E_* = 5.151×10^{11} V m^{-1}. Laboratory field strengths are typically a few kilovolts per centimeter, i.e. very much smaller than $\mathcal{E}_*$. Therefore, $d\,\mathcal{E}$ is small and we may

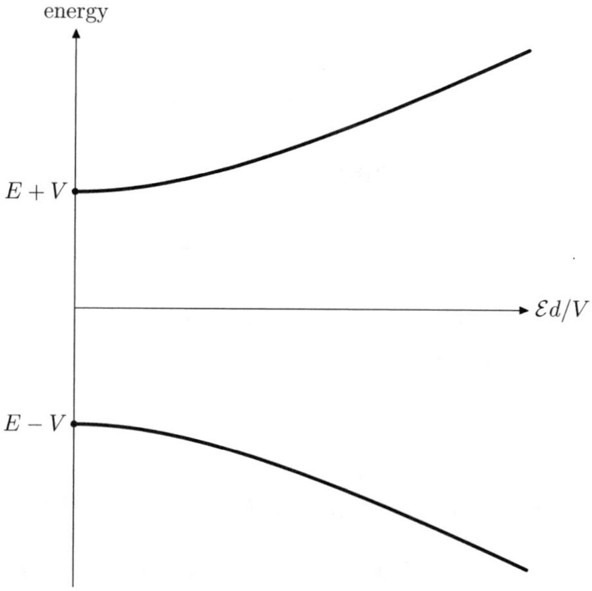

Fig. 2.3 Stark effect of the ammonia molecule. The molecule with dipole moment d is exposed to an external electric field strength $\mathcal{E}$. The energy difference between excited and ground state varies quadratically for small field strengths and linearly for large values

safely assume that the Stark effect is quadratic in the external electric field strength. Only if V vanishes or is exceptionally small, the linear Stark effect can be observed.

2.1.3 Dipole Moment Expectation Value

Let us work out the eigenstates of the ammonia molecule in the presence of an electric field. We know the solution for $\mathcal{E} = 0$, namely $\varrho_{\text{lo}} = \Pi_{\text{lo}}$ and ϱ_{hi} of (2.9). we rotate these states by an angle α,

$$\Pi_{\text{lo,hi}}(\mathcal{E}) = \begin{pmatrix} \cos\alpha & -\sin\alpha \\ \sin\alpha & \cos\alpha \end{pmatrix} \Pi_{\text{lo,hi}} \begin{pmatrix} \cos\alpha & \sin\alpha \\ -\sin\alpha & \cos\alpha \end{pmatrix}, \tag{2.12}$$

which results in

$$\Pi_{\text{lo,hi}}(\mathcal{E}) = \frac{1}{2} \begin{pmatrix} 1 \mp \sin 2\alpha & \pm\cos 2\alpha \\ \pm\cos 2\alpha & 1 \pm \sin 2\alpha \end{pmatrix}. \tag{2.13}$$

The negative sign refers to the ground state (lo), the plus to the excited state (hi). The trace (sum of eigenvalues) of both matrices is one, their determinants (product of eigenvalues) vanish. Hence, both Π_{lo} and Π_{hi} are projectors, and by construction, they are orthogonal. The angle α should be such that

$$H = \begin{pmatrix} E - d\mathcal{E} & -V \\ -V & E + d\mathcal{E} \end{pmatrix} = \epsilon_{\text{lo}}\Pi_{\text{lo}} + \epsilon_{\text{hi}}\Pi_{\text{hi}} \tag{2.14}$$

holds true, where the energy eigenvalues $\epsilon_{\text{lo,hi}}(\mathcal{E})$ are (2.10). The solution is

$$\sin 2\alpha = \frac{d\,\mathcal{E}}{\sqrt{V^2 + d^2\,\mathcal{E}^2}}. \tag{2.15}$$

In fact a plausible result. If there is no external electric field, the previous situation is recovered. If the field is sufficiently weak, the approximation

$$\sin\alpha \approx \frac{d\mathcal{E}}{2V} \tag{2.16}$$

applies which says that the energy eigenstates $\varrho_{\text{lo,hi}}(\mathcal{E})$ deviate but little from $\varrho_{\text{lo,hi}}$ without a field.

We are now ready to calculate the dipole moment expectation value. The observable D is defined by (2.12), the energy eigenstates are given by (2.12). Here is the result:

$$d_{\text{lo}} = \text{Tr}\,\varrho_{\text{lo}}(\mathcal{E})D = -\frac{d^2\,\mathcal{E}}{\sqrt{V^2 + d^2\,\mathcal{E}^2}}, \tag{2.17}$$

and a similar expression for d_{hi} (opposite sign).

For $V \ll d\,\mathcal{E}$, i.e. a large electric field, the dipole moments coincide with $\mp d$. In the realistic case $d\mathcal{E} \ll V$ the dipole moment expectation value is proportional to the electric field strength. $\alpha = d^2/V$ is the polarizability of a single molecule.

2.1.4 Ammonium Maser

MASER stands for Microwave Amplification by Stimulated Emission of Radiation. Its counterpart is LASER, Light Amplification by Stimulated Emission of Radiation. Both methods to generate radiation have in common that there is

- a system with at least two energy levels,
- a mechanism for preferring the excited energy level (pumping),
- a triggering causing, with a small amount of energy, a phase-synchronized transition from the high to the low energy level.

We shall discuss the latter effect in section Sect. 4.1 on Forced transitions. Normally, in thermal equilibrium, the low energy level is populated with a higher probability than the high energy level, and stimulation will have no effect. Let us here concentrate on the pumping mechanism: a way to create a situation where the high energy level is populated with a higher probability than the low energy level. For lasers, pumping is usually achieved by forcing transitions to excited states which decay to long lived intermediate states. The ammonia maser is technically rather complicated, but more straightforward to understand.

A beam of ammonia molecules passes through a spatially inhomogeneous electric field. Initially, their polarization is random, because for reasonable temperatures the difference between ϵ_{hi} and ϵ_{lo} is small as compared with $k_B T$. However, the energies $\epsilon_{\text{lo}}(\mathcal{E}(\boldsymbol{x}))$ and $\epsilon_{\text{hi}}(\mathcal{E}(\boldsymbol{x}))$ differ and depend on location. In an inhomogeneous external electric field there are forces

$$\boldsymbol{F}_{\text{lo,hi}} = -\nabla_x \epsilon_{\text{lo,hi}}(\mathcal{E}(\boldsymbol{x})) \tag{2.18}$$

which will deflect molecules with different dipole moment in different directions. With this mechanism it is possible to select a beam of excited molecules. When entering a resonator cavity without electric field, practically all ammonia molecules are in the excited ϱ_{hi} state. A feeble microwave field with just the right frequency will stimulate phase-synchronized transitions to the ϱ_{lo} state and thereby generate a large and coherent output signal.

The 1964 Nobel prize in physics was awarded to Charles Hard Townes, Nicolay Gennadiyevich Basov and Aleksandr Mikhailovich Prokhorov 'for fundamental work in the field of quantum electronics, which has led to the construction of oscillators and amplifiers based on the maser-laser principle'.

2.1.5 *Summary*

The ammonia molecule comes in two configurations, one being the mirror of the other. The Hilbert space of this system is $\mathbb{C}^2$ and the associated $C^\star$ algebra consists of complex 2×2 matrices.

The Hamiltonian is set up and diagonalized. The eigenstates correspond to superpositions of the two basic configurations, and there is an energy difference between the ground state and the excited state. This energy difference corresponds to a transition frequency of about 24 GHz and a wave length of 1.26 cm.

An external electric field affects the energy eigenvalues of a molecule with intrinsic electric dipole moment, such as the ammonia molecule. This Stark effect, the shift of spectral lines, is normally small and depends quadratically on the electric field strength unless there is a degeneracy which produces a linear effect. We have calculated the net electric dipole moment of polarized molecules.

A beam of randomly polarized ammonia molecules splits, in an inhomogeneous electric field, into two beams. One is made up of molecules in the ground state, the other consists of excited molecules. The excited molecules are injected into a resonance cavity where a tiny signal may cause the coordinated, in-phase transition to the ground state. This is, in a nutshell, the principle of microwave amplification by the stimulated emission of radiation, or MASER.

2.2 Quasi-Particles

We will discuss a very much simplified model for the behavior of electrons or holes in a crystal. This model crystal is a regular one-dimensional chain of identical ions plus a charge which can hop from one site to its next neighbors. The charge may be a loosely bound electron or a missing electron, a hole. As we shall see, such charges will move freely in the model crystal. However, their energy-momentum relation depends on the interaction with the surrounding, and the same is true for velocity and mass. The latter, the effective mass, is not a constant property of the particle!

There are two approaches. One may model the crystal by a finite, periodic ring of n sites and, in the end, send n to infinity. But it is also possible to describe the infinite crystal straightaway. In the latter case we have to cope with unbounded operators and continuously distributed quasi-eigenvalues. In this context we will discuss wave packets.

Electrons or holes moving in the crystal are quasi-particles because their energy-momentum relation differs from $\epsilon = p^2/2m$ of non-relativistic free particles. The dispersion relation $\epsilon = \epsilon(p)$ or $\omega = \omega(k)$ is determined by the interaction of the particle with its surrounding.

2.2.1 *Hilbert Spaces* $\mathbb{C}^n$ *and* ℓ^2

Denote by u_r the configuration that the electron is located at site r. An arbitrary configuration will be described by $c = \sum c_r u_r$ where r is an integer index.

We describe a periodic structure by $r = -m, -m+1, \dots, m-1, m$. There are $n = 2m + 1$ sites. The structure becomes periodic if u_{m+1} and u_{-m} are considered to be identical. In this case the crystal lattice has no boundary which is essential for an infinite lattice. The n-tuples $\{c_r\}_{r=-m,\dots,m}$ of complex numbers form a linear space $\mathbb{C}^n$. Addition and multiplication by scalars is defined componentwise. With the scalar product

$$(d, c) = \sum_{r=-m}^{m} d_r^* \, c_r \tag{2.19}$$

$\mathbb{C}^n$ becomes a Hilbert space . The squared vector norm is $\| c \|^2 = (c, c)$.

Sending $n \to \infty$ gives the linear space $\mathbb{C}^\infty$ of complex sequences $\{c_r\}_{r \in \mathbb{Z}}$. Again, addition and multiplication by scalars is defined componentwise. However, the usual scalar product can only be defined if

$$\| c \|^2 = \sum_{r \in \mathbb{Z}} |c_r|^2 < \infty. \tag{2.20}$$

Then,

$$(d, c) = \sum_{r\in\mathbb{Z}} d_r^* \, c_r \tag{2.21}$$

converges for all admissible sequences c and d because of the Cauchy–Schwarz inequality

$$|(d, c)| \le \| c \| \, \| d \| \, . \tag{2.22}$$

The corresponding Hilbert space is

$$\ell^2 = \{c \in \mathbb{C}^\infty \bigm| \| c \|^2 < \infty\}, \tag{2.23}$$

the scalar product being (2.21).

Linear mappings $\mathbb{C}^n \to \mathbb{C}^n$ are represented by $n \times n$ matrices of complex numbers. For an arbitrary unit vector $c \in \mathbb{C}^n$ and a matrix A we estimate

$$\| Ac \|^2 = \sum_r | \sum_s A_{rs} c_s |^2 \le \sum_r \sum_s |A_{rs}|^2 |c_s|^2 \le n^2 \, A^2_{\text{max}} . \tag{2.24}$$

We have estimated the matrix elements $|A_{rs}|^2$ by their maximal value and $|c_s|^2$ by 1. Equation (2.24) says that the matrix A is bounded. Addition and multiplication with scalars is componentwise. Multiplication of matrices is matrix multiplication: first A then B defines BA. There is an operator norm $A \to \| A \|$ because all matrices are bounded. Adjoining $A_{rs} \to {A^\star}_{rs} = A^*_{sr}$ has all properties of a star operation. The set of complex $n \times n$ matrices is a $C^\star$ algebra.

Let us now turn to the linear mappings $\ell^2 \to \ell^2$.

The bounded operators form a $C^\star$ algebra, with standard addition, multiplication, operator norm and the adjoint as star operation. Unfortunately, however, we also have to do with observables which are unbounded, such as the location operator X. It cannot be defined on the entire Hilbert space, but only on a linear subspace, its domain of definition.

With $\mathcal{L} = \mathbb{C}^\infty$ as the initial linear space, $\mathcal{H} = \ell^2$ as Hilbert space and $\mathcal{D}$ as domain of definition we have

$$\mathcal{D} \subset \mathcal{H} \subset \mathcal{L}. \tag{2.25}$$

The linear operator $X : \mathcal{L} \to \mathcal{L}$ is initially defined by $(Xc)_r = arc_r$ where a denotes the distance between adjacent sites. Restricting it to the Hilbert space results in a mapping $X : \mathcal{H} \to \mathcal{L}$. Images Xc of Hilbert space vectors will not have a finite norm, in general. We therefore restrict X further to a linear subspace $\mathcal{D}$ which is defined by

$$\mathcal{D} = \{c \in \mathcal{H} \bigm| \| Xc \| < \infty\}. \tag{2.26}$$

X now is a linear operator $X : \mathcal{D} \to \mathcal{H}$. What about its adjoint?

Chose $c \in \mathcal{D}$ and $d \in \mathcal{H}$. Then

$$(d, Xc) = a \sum_r d_r^* (r c_r) = a \sum_r (r d_r)^* c_r = (b, c) \qquad (2.27)$$

converges. Per definition, $b = X^\star d$ is the adjoint of X. $X^\star$ is defined on $\mathcal{D}$ just as X and does there the same as X. We conclude $X = X^\star$. The location observable, although unbounded, is self-adjoint.

2.2.2 *Hopping Model*

Back to physics. We want to describe a loosely localized electron which may hop to its next neighbor to the left or to the right. We describe this by the energy observable

$$H = E\, I - V(R + L). \qquad (2.28)$$

E is the energy level at which the electron is loosely bound. R and L are right and left shifts, respectively:

$$R u_r = u_{r+1} \text{ and } L u_r = u_{r-1}\,. \qquad (2.29)$$

The hopping amplitude is V. As we shall see, $V > 0$ is a sensible choice. Note $RL = LR = I$. Right and left shift commute, $R^{-1} = L$ and $L^{-1} = R$. From

$$(d, Rc) = \sum_r d_r^* c_{r-1} = \sum_s d_{s+1}^* c_s = (Ld, c) \qquad (2.30)$$

we learn $R^\star = L$ and $L^\star = R$. Consequently the operators R and L are unitary, hence bounded.

What are the energy eigenvectors and eigenvalues? Because of translational symmetry we try a kind of plane waves:

$$w_r \propto \mathrm{e}^{\,\mathrm{i} kar}\,. \qquad (2.31)$$

This vector is in fact an eigenvector of H with eigenvalue

$$\epsilon(k) = \hbar\omega(k) = E - 2V \cos ka. \qquad (2.32)$$

For the finite periodic ring, sites u_{m+1} and u_{-m} should be considered the same. Therefore, the phase kan must be an integer multiple of 2π. We conclude

$$k_s = \frac{2\pi s}{na} \text{ for } s = -m, -m+1, \ldots, m-1, m. \qquad (2.33)$$

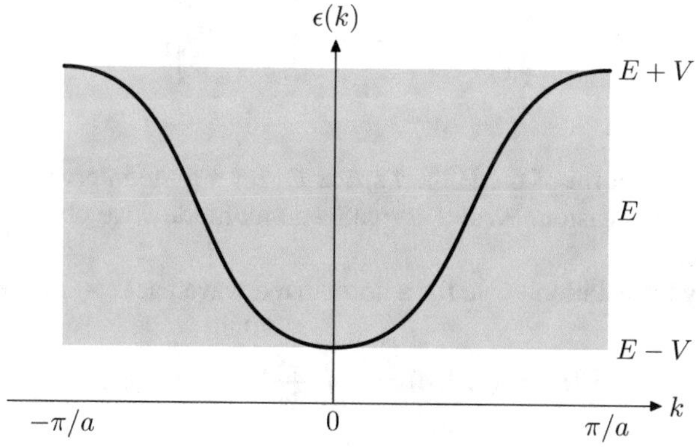

Fig. 2.4 Dispersion function of the nearest neighbor hopping model. The function $E(k) = \hbar\omega(k)$ is plotted versus the wave number k in the Brillouin zone. The n-fold degenerate level E for a localized electron splits into an energy band (*shaded*). In a crystalline solid there is one band for each energy level

There are $n = 2m + 1$ equally distributed wave numbers k_s in the interval $k \in [-\frac{\pi}{a}, +\frac{\pi}{a}]$. This is the Brillouin zone (Fig. 2.4).

2.2.3 *Wave Packets*

Our arguments for the infinite model crystal remain the same. The difference is that wave numbers are not quantized, but continuously distributed within the Brillouin zone. The dispersion function is a real valued function of a real argument.

We require another modification. The eigenvectors (2.31) belong to $\mathcal{L} = C^\infty$, but not to $\mathcal{H} = \ell^2$. They cannot be normalized, they are quasi-eigenvectors.

In order to construct normalizable vectors we have to superimpose quasi-eigenvectors by

$$c_r = a \int_{\mathrm{BZ}} \frac{\mathrm{d}k}{2\pi} \hat{c}(k)\, \mathrm{e}^{\,\mathrm{i}\, akr}\,, \tag{2.34}$$

where BZ stands for the Brillouin zone. We calculate

$$\sum_r |c_r|^2 = \int_{\mathrm{BZ}} \frac{a\mathrm{d}q}{2\pi} \int_{\mathrm{BZ}} \frac{a\mathrm{d}k}{2\pi} \sum_r \hat{c}^*(q)\hat{c}(k)\, \mathrm{e}^{\,\mathrm{i}\,(k-q)ar}\,. \tag{2.35}$$

Because of

$$\sum_r \mathrm{e}^{\,\mathrm{i}\,(q-k)ar} = \frac{2\pi}{a}\delta(q-k) \tag{2.36}$$

one obtains

$$\| c \|^2 = a \int_{\text{BZ}} \frac{\mathrm{d}k}{2\pi} |\hat{c}(k)|^2 = \| \hat{c} \|^2 . \tag{2.37}$$

$\| \hat{c} \|$ is the vector norm of the Hilbert space $\mathcal{L}_2$ for a finite interval which we shall discuss at a later occasion. With a normalized amplitude $\hat{c}$ we obtain a normalized wave package.

The energy expectation value for a normalized wave packet—a pure state—is

$$\langle H \rangle = (c, Hc) = a \int_{\text{BZ}} \frac{\mathrm{d}k}{2\pi} |\hat{c}(k)|^2 \, \hbar\omega(k), \tag{2.38}$$

with the dispersion function (2.32). If $\hat{c} = \hat{c}(k)$ is sharply peaked at $k = \bar{k}$, we have realized a wave packet with average energy $E \approx E(\bar{k})$. However, this is not an eigenvector.

2.2.4 Group Velocity and Effective Mass

Recall that R and L commute and note $[R, X] = -aR$ and $[L, X] = aL$.

The quasi-particle's velocity is the observable

$$\dot{X} = \{H, X\} = -\frac{\mathrm{i}}{\hbar} \frac{V}{2} [R + L, X] = -\frac{\mathrm{i}}{\hbar} \frac{Va}{2} (R - L), \tag{2.39}$$

which, if applied to the wave package (2.34), gives

$$\langle \dot{X} \rangle = a \int_{\text{BZ}} \frac{\mathrm{d}k}{2\pi} |\hat{c}(k)|^2 v_{\text{gr}}(k), \tag{2.40}$$

where

$$v_{\text{gr}}(k) = \frac{Va}{\hbar} \sin ka = \omega'(k). \tag{2.41}$$

You should convince yourself that $Va/\hbar$ is a velocity. We speak of normal dispersion because velocity and wave vector point in the same direction.

And what about the acceleration? Now, for $H = EI - V(R + L)$ we calculate

$$\ddot{X} = \{H, \dot{X}\} = 0, \tag{2.42}$$

because R and L commute. This explains why the mobile electrons in a solid are called quasi-free: they are quasi-particles because their dispersion function is determined by the surrounding, and they move as free particles, i.e. unaccelerated. How do they respond to an external force?

If an electric field $\mathcal{E}$ is applied the potential energy $-q\mathcal{E}X$ has to be added. q is the particle's charge: $-e$ for electrons, $+e$ for holes. The following considerations refer to electrons.

We now have to do deal with the energy

$$H(\mathcal{E}) = EI - V(R+L) + e\mathcal{E}X. \tag{2.43}$$

The expression for the velocity remains the same because the location observable X commutes with the potential energy. The acceleration is given by

$$\ddot{X} = -\frac{Va^2}{\hbar^2} e\mathcal{E}(L+R). \tag{2.44}$$

For a wave packet (2.34) we find

$$m_{\text{eff}}\,\langle\,\ddot{X}\,\rangle = -e\,\mathcal{E} \text{ where } \frac{1}{m_{\text{eff}}} = a \int\limits_{\text{BZ}} \frac{\mathrm{d}k}{2\pi} |\,\hat{c}(k)\,|^2\, \frac{1}{\hbar}\omega''(k). \tag{2.45}$$

The effective mass m_{eff} is a constant only if energy depends on momentum quadratically which is the case for free particles. Quasi-electrons sufficiently close to the bottom of the energy band have a positive effective mass. The true electron mass m_{e} does show up only indirectly if the hopping amplitude V is calculated within a more basic theory of solids.

By the way, our model can be extended by the possibility to hop directly not only to the next neighbor, but also to next but one, and so on. The dispersion curve then becomes a Fourier series $V_1 \cos ka + V_2 \cos 2ka + \dots$. The expressions for the group velocity and effective mass, however, remain unchanged if expressed as derivatives of the dispersion curve.

2.2.5 *Scattering at Defects*

Crystal properties, like conductivity, are determined to a large extend by the interaction of mobile quasi-free electrons with imperfections such as lattice vibrations, dislocations, interstitials, vacancies, or impurities. This is not the right place to discuss imperfections because our crude one-dimensional model cannot describe most of them. We here concentrate on a defect caused by an impurity. At site $r = 0$ the ordinary ion is replaced by a different one. The energy level of it will differ from E as well as the hopping amplitude V. We concentrate on the essential and leave the hopping amplitude untouched.

The following energy observable models the situation:

$$(Hc)_r = E_r c_r - V(c_{r+1} + c_{r-1}) \text{ with } E_r = E - \Delta\,\delta_{r,0}. \tag{2.46}$$

For the moment we do not bother whether $\{c_r\}_{r\in\mathbb{Z}}$ is normalizable or not. The energy shift Δ may be positive or negative.

A single plane wave will not do because the problem lacks translational symmetry. Guided by intuition we try an incoming plane wave plus an outgoing spherical wavelindexWave!spherical. For one dimension this reads

$$w_r = \mathrm{e}^{\mathrm{i}kar} + f\,\mathrm{e}^{\mathrm{i}ka|r|} \text{ with } k > 0. \tag{2.47}$$

$f = f(k)$ is the scattering amplitude. We have to solve

$$(Hw)_0 = (E - \Delta)w_0 - V(w_1 + w_{-1}) = E(k)w_0 \tag{2.48}$$

and

$$(Hw)_r = E - V(w_{r+1} + w_{r-1}) = E(k)w_r \text{ for } r \neq 0. \tag{2.49}$$

Equation (2.49) is satisfied by (2.47) if $E(k) = E - V\cos ka$ is chosen, the eigenvalue from the preceding discussion. From (2.48) one obtains

$$f(k) = -\frac{\Delta}{\Delta + 2\mathrm{i}V\sin ka}. \tag{2.50}$$

What does this mean?

Our *ansatz* (2.47) describes an incoming wave (amplitude 1), a reflected wave (amplitude f) and a transmitted wave (amplitude $1 + f$). Hence, the quasi-electron is reflected with probability R and transmitted with probability T; these are given by

$$R = |f(k)|^2 = \frac{\Delta^2}{\Delta^2 + 4V^2\sin^2 ka} \tag{2.51}$$

and

$$T = |1 + f(k)|^2 = \frac{4V^2\sin^2 ka}{\Delta^2 + 4V^2\sin^2 ka}. \tag{2.52}$$

We list some properties of the scattering amplitude:

- The restriction $k > 0$ may be dropped. Our results are valid for plane waves running from left to right or from right to left.
- $R + T = 1$; no particle gets lost and none is created.
- With $\Delta \to 0$ we find $R \to 0$ and $T \to 1$. No defect, no scattering.

- Very slow quasi-particles, i.e. with $k \approx 0$, will not be transmitted.
- Surprisingly, reflectivity R and transmittivity T do not depend on the sign of the energy level difference Δ. As we shall see, a defect with positive Δ may trap a quasi-electron. $\Delta < 0$ means repulsion. Attractive and repulsive impurities scatter alike.

2.2.6 *Trapping by Defects*

We rewrite (2.47) as

$$w_r \propto \frac{1}{f(k)} \mathrm{e}^{\mathrm{i}kar} + \mathrm{e}^{\mathrm{i}ka|r|} \quad \text{with } k > 0. \tag{2.53}$$

We look for a pole of $k \to f(k)$, which is analytic in the complex k-plane. If such a pole exists with $\mathrm{Im}\, k > 0$, (2.53) describes a bound state. Now, with $k = \mathrm{i}\kappa$, there is a pole $1/f(k) = 0$ if

$$\sinh \kappa a = \frac{\Delta}{2V} \tag{2.54}$$

holds true. $\mathrm{Im}\, k > 0$ requires $\kappa > 0$, and there is always a solution provided the energy difference Δ is positive. The energy level of the defective ion must be lower than the one of the bulk ion. In this case the otherwise mobile quasi-electron is trapped, or immobilized.

Although trapped, it still may be found in its vicinity. It is not difficult to normalize the bound state solution (2.53):

$$w_r = \frac{\mathrm{e}^{-\kappa a|r|}}{\sqrt{\tanh \kappa a}}. \tag{2.55}$$

The probability to find the trapped quasi-electron at site $|r|$ is

$$p_r = |w_r|^2 = \frac{\mathrm{e}^{-2\kappa a|r|}}{\tanh \kappa}. \tag{2.56}$$

Small Δ means small κa. Consequently, the extension of the bound state is large. There is a smooth transition from not bound at all to weakly bound to strongly localized.

2.2.7 Summary

We have discussed the motion of an electron (or whole) in a one-dimensional model crystal. An electron is loosely localized at an ion and may hop to its next neighbors. We contrast two approaches.

The crystal has finitely many lattice sites and is periodic. This avoids ends, the surface, which would spoil translational symmetry. All linear operators of the corresponding finite-dimensional Hilbert space $\mathbb{C}^n$ are bounded. More realistic results are obtained if at the end of a calculation n is sent to infinity

It is also possible to discuss the infinite chain right from the beginning. However, not all vectors of $\mathbb{C}^\infty$ are normalizable. The subset ℓ_2 of normalizable sequences of complex numbers may be equipped with a scalar product to become a Hilbert space. Then, however, we have to deal with unbounded observables, like location X.

In the case of infinitely many lattice sites, the eigenvector of the energy observable are not normalizable, and one must form normalized wave packets.

The energy eigenvalues, as a function $\hbar\omega(k)$ of wave number k, is known as the dispersion function. The dispersion function differs from that of free particles, one therefore speaks of quasi-particles. The electron moves with group velocity v_{gr} which is the first derivative of the dispersion function. If an external electric field is applied, the charged quasi-electron reacts with an acceleration. The proportionality factor is the effective mass m_{eff} which is not a constant. In fact, it is produced by the electron's interaction with the environment.

Our model even allows to take a crystal defect into account. A normal ion may be replaced by an impurity the energy level of which is $E - \Delta$ instead of E. We encounter scattering: an incoming plane wave excites an outgoing spherical wave. The probability T to be transmitted and the probability R to be reflected are calculated; they add up to 1.

If the energy level of the impurity is lower than normal ($\Delta > 0$) the otherwise mobile quasi-electron is trapped. There is a smooth transition from not bound at all to loosely bound to strongly bound.

It is generally true that poles of the scattering amplitude in the complex plane indicate bound states.

2.3 Neutron Scattering on Molecules

Neutrons are responsive to nucleons only, i.e. protons and neutrons. They are blind to electrons. The shared electrons of two or more nuclei will bind these nuclei to a compound, a molecule. Such a molecule or a molecule ion is seen by neutrons as a rigid configuration of nuclei. The shared electrons are required to glue that configuration, but a neutron will not see them. In this subsection we want to study the scattering on simple and more complex molecules and on a crystal which is nothing else but a giant molecule.

We do so in a formulation which goes back to Richard Feynman.

2.3.1 Feynman's Approach

In his famous lectures[1] Feynman postulates three simple rules:

1. The probability of an event in an ideal experiment is given by the square of the absolute value of a complex number ϕ which is called the probability amplitude:

$$\begin{aligned} P &= \text{probability} \\ \phi &= \text{probability amplitude} \\ P &= |\phi|^2 \end{aligned} \tag{2.57}$$

2. When an event can occur in several alternative ways, the probability amplitude for the event is the sum of the probability amplitudes for each way considered separately. There is interference:

$$\begin{aligned} \phi &= \phi_1 + \phi_2 \\ P &= |\phi_1 + \phi_2|^2 \end{aligned} \tag{2.58}$$

3. If an experiment is performed which is capable of determining whether one or another alternative is actually taken, the probability of the event is the sum of the probabilities for each alternative. Interference is lost:

$$P = P_1 + P_2 \tag{2.59}$$

There is no Hilbert space, no observable, and no state, just probabilities, probability amplitudes, and rules for combining them. Feynman got rather far with this concept, but failed to describe something so basic as the hydrogen atom.

The Hilbert space comes into play through the back door. "If an event can occur in several alternative ways" is rather vague. It means that several intermediate mutually orthogonal states are involved. Two states (in the sense of projection on a Hilbert space vector) are either parallel, orthogonal or something in-between. Parallel means they are the same, alternative is associated with orthogonal, and the general case "finding g in f" has a probability amplitude $\phi = (g, f)/\parallel g \parallel \parallel f \parallel$.

Let us see how far we will get with the amplitude approach.

[1] http://www.feynmanlectures.caltech.edu/.

2.3.2 Spherical and Plain Waves

The event "a neutron with wave number k is emitted at the source $\boldsymbol{x}_S$ and detected at location $\boldsymbol{x}$" has a probability amplitude $\phi(\boldsymbol{x}_S, \boldsymbol{x}, k)$. No direction is mentioned, so we assume spherical symmetry. Translational invariance—the same amplitude for $\boldsymbol{x}_S, \boldsymbol{x}$ and $\boldsymbol{x}_S + \boldsymbol{a}, \boldsymbol{x} + \boldsymbol{a}$—requires a dependency on both arguments via r=$|\boldsymbol{x}_S - \boldsymbol{x}|$. The probability to be at any place on a sphere around the source does not depend on its radius. All this amounts to

$$\phi(\boldsymbol{x}_S, \boldsymbol{x}, k) = \frac{1}{\sqrt{4\pi}} \frac{\mathrm{e}^{\mathrm{i}kr}}{r}. \tag{2.60}$$

Indeed, the integral

$$\int_{|\boldsymbol{x}-\boldsymbol{x}_S|=R} \mathrm{d}\Omega \, |\phi(\boldsymbol{x}_S, \boldsymbol{x}, k)|^2 = 1 \tag{2.61}$$

does not depend on R.

Let us write $\boldsymbol{x} = \boldsymbol{x}_S + R\,\boldsymbol{n} + \boldsymbol{\xi}$ where $\boldsymbol{n}$ is a unit vector pointing from the source in direction $\boldsymbol{x}$. The local coordinates with origin $\boldsymbol{x}$ are called $\boldsymbol{\xi}$. For $|\boldsymbol{\xi}| \ll R$ the spherical wave is described as

$$\phi(\boldsymbol{x}_S, \boldsymbol{x} + \boldsymbol{\xi}, k) \approx \frac{1}{\sqrt{4\pi}} \frac{\mathrm{e}^{\mathrm{i}kR}}{R} \mathrm{e}^{\mathrm{i}k\,\boldsymbol{n}\cdot\boldsymbol{\xi}}. \tag{2.62}$$

We have approximated

$$r = |R\boldsymbol{n} + \boldsymbol{\xi}| = \sqrt{R^2 + 2R\boldsymbol{n}\cdot\boldsymbol{\xi} + \dots} = R + \boldsymbol{n}\cdot\boldsymbol{\xi} + \dots \tag{2.63}$$

Equation (2.62) says that far away from the source in a restricted area the spherical wavelooks like a plane wave with wave vector $\boldsymbol{k} = k\boldsymbol{n}$.

2.3.3 Neutron Scattering on a Diatomic Molecule

To keep things as simple as possible, we talk about a diatomic molecule with equal nuclei, say N_2. If the distance between the molecules is a we write $\boldsymbol{\xi}_{1,2} = \pm\boldsymbol{a}/2$ for their locations.

We put a neutron detector at $\boldsymbol{x}_D$. The events are

- a neutron is emitted at $\boldsymbol{x}_S$,
- it is absorbed and re-emitted by a nucleus a $\boldsymbol{\xi}_j$,
- and is detected $\boldsymbol{x}_D$.

The probability amplitude for this is

$$\phi(\boldsymbol{x}_\mathrm{S} \to \boldsymbol{\xi}_j \to \boldsymbol{x}_\mathrm{D}) = \frac{\mathrm{e}^{\mathrm{i}kR_\mathrm{S}}}{\sqrt{4\pi}R_\mathrm{S}}\,\mathrm{e}^{\mathrm{i}\boldsymbol{k}_\mathrm{in}\cdot\boldsymbol{a}/2}\, f_j\,\frac{\mathrm{e}^{\mathrm{i}kR_\mathrm{D}}}{\sqrt{4\pi}R_\mathrm{D}}\,\mathrm{e}^{-\mathrm{i}\boldsymbol{k}_\mathrm{out}\cdot\boldsymbol{a}/2}\,. \tag{2.64}$$

f_j is the scattering amplitude for nucleus j; here we set $f_j = f$ because the nuclei are identical. It does not depend on direction because the neutrons we discuss here are slow. More about this in a subsequent chapter.

According to (2.58) we have to sum over $\boldsymbol{x}_1$ and $\boldsymbol{x}_2$. The result is

$$\phi(\boldsymbol{x}_\mathrm{S} \to \boldsymbol{x}_\mathrm{D}) = \frac{2f}{4\pi R_\mathrm{s} R_\mathrm{D}} \cos\frac{\boldsymbol{\Delta}\cdot\boldsymbol{a}}{2}. \tag{2.65}$$

$\boldsymbol{\Delta} = \boldsymbol{k}_\mathrm{out} - \boldsymbol{k}_\mathrm{in}$ is the wave vector transfer from the inbound neutron to the outbound neutron. The probability for a neutron to be scattered by the molecule—the absolute square of (2.65)—

- falls off as $1/4\pi R_\mathrm{S}^2$ with the distance between source and molecule;
- falls off as $1/4\pi R_\mathrm{D}^2$ with the distance between molecule and detector;
- is proportional to $|f|^2$, the absolute square of the amplitude for neutron–nitrogen nucleus scattering; and
- depends on the wave vector transfer $\boldsymbol{\Delta}$ and the molecule orientation $\boldsymbol{a}$.

Before we discuss our result, we rephrase it as an expression for the scattering cross section, and we will also average over the molecule's orientation.

2.3.4 Cross Section

A particle which hits an obstacle may ignore it or will be deflected. One speaks of diffraction if the deflected particles run in certain well defined directions only, as is the case if the target is a large enough crystal. If the target is not a crystal, but a gas, liquid or amorphous solid, the deflected particles show a smooth angular distribution, and we speak of scattering.

Denote by $I = I(x)$ the current density of incident particles.[2] Because of scattering, it depends on the distance x which the particles have traveled through the target which we assume to be homogeneous. In a thin layer $\mathrm{d}x$ there are $n\mathrm{d}x$ obstacles encountered, per unit area. The number of particles which are removed from the incident beam is proportional to the number of the uncorrelated obstacles, and we write

$$\mathrm{d}I = -\sigma\, n\, \mathrm{d}x, \tag{2.66}$$

[2] Particles per unit time per unit area.

with σ as a proportionality constant. It is an area, a cross section. In classical thinking, if a particle hits it, it will be deflected and removed from the incident beam. Equation (2.66) may be integrated to

$$I(x) = I(0)\,\mathrm{e}^{-\sigma n x}. \tag{2.67}$$

The simplest case is that particles are removed from the incident beam because they are merely deflected by an angle ϑ. Unless they are polarized, the azimuth angle φ does not enter. We therefore may write

$$\sigma = \int \mathrm{d}\Omega\, \sigma_{\mathrm{d}}(k, \vartheta). \tag{2.68}$$

The integral over the solid angle is described by

$$\int \mathrm{d}\Omega = \int_0^{\pi} \mathrm{d}\vartheta\, \sin\vartheta \int_0^{2\pi} \mathrm{d}\varphi = 4\pi. \tag{2.69}$$

The differential cross section $\sigma_{\mathrm{d}} = \sigma_{\mathrm{d}}(k, \vartheta)$ normally depends on the wave number of the incident particles and on the deflection angle.

2.3.5 *Orientation Averaged Cross Section*

The differential cross section for neutron–atom scattering is

$$\sigma_{\mathrm{d}}^{\mathrm{A}} = |f|^2, \tag{2.70}$$

as we will discuss in detail in a later section. Therefore, the differential cross section for neutron–molecule scattering is

$$\sigma_{\mathrm{d}}^{\mathrm{M}} = \left\{4\cos^2\frac{\boldsymbol{\Delta}\cdot\boldsymbol{a}}{2}\right\}\sigma_{\mathrm{d}}^{\mathrm{A}}. \tag{2.71}$$

Before discussing this result we should average over the molecule's orientation. Diatomic molecule with equal nuclei, like N_2, cannot have an electric dipole moment which is required for at least partially orienting them in an external electric field. So the target—a gas or liquid—is unpolarized, and we obtain

$$\sigma_{\mathrm{d}}^{\mathrm{M}} = 2\left\{1 + \frac{\sin\Delta a}{\Delta a}\right\}\sigma_{\mathrm{d}}^{\mathrm{A}} \tag{2.72}$$

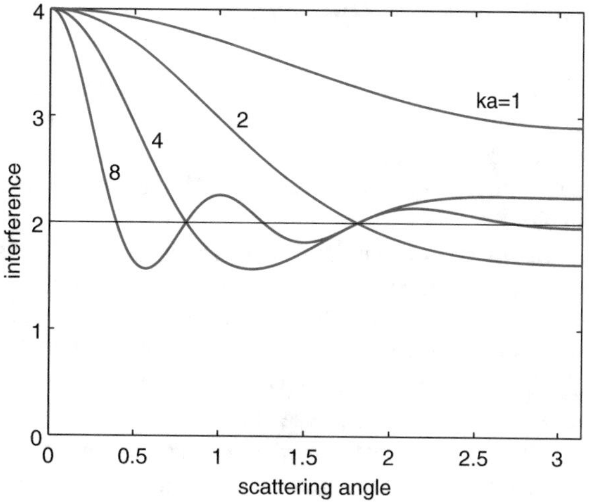

Fig. 2.5 Differential neutron-nitrogen molecule cross section. The interference factor $\sigma_{\mathrm{d}}^{\mathrm{M}}/\sigma_{\mathrm{d}}^{\mathrm{A}}$ is plotted versus the scattering angle ϑ for $ka = 1, 2, 4, 8$ in decreasing order. Values above 2 mean constructive interference while values below 2 signify destructive interference

This result is truly remarkable (Fig. 2.5).

Recall that a is the distance between the two nitrogen nuclei. Δ stands for the wave number transfer from the incoming to the outgoing neutron. It is given by

$$\Delta = 2k \sin \frac{\vartheta}{2}. \tag{2.73}$$

$\Delta\, a$ is dimensionless. A very small value indicates slow neutrons or a small scattering angle. We obtain

$$\sigma^{\mathrm{M}} = 4\sigma^{\mathrm{A}} \text{ for } \Delta\, a \ll 1. \tag{2.74}$$

For large wave vector transfer, however, the result is

$$\sigma^{\mathrm{M}} = 2\sigma^{\mathrm{A}} \text{ for } \Delta\, a \gg 1. \tag{2.75}$$

Equation (2.74) means maximal constructive interference. The amplitude doubles, the cross section becomes four times as large. Equation (2.75) says that there is no interference. The cross section for two nitrogen atoms, although bound to a molecule, is simply twice the cross section for a single nucleus. This is the view of classical mechanics. However, the region in between, $\Delta\, a \approx 1$, is interesting. The measured differential cross section $\sigma_{\mathrm{d}}^{\mathrm{M}} = \sigma_{\mathrm{d}}^{\mathrm{M}}(k, \vartheta)$ may be fitted to (2.72). In this way the structure of the target particle—being diatomic with nucleus distance a—may be determined even though the molecules are oriented at random.

The example shows why one needs bigger and bigger accelerators to unveil the structure of ever tinier elementary particles. Only with head-on collisions of energetic particles a high momentum transfer $\hbar\Delta$ may be achieved. By high-energy electron nucleon scattering, for example, it could be proven that protons and neutron are composite particles of three quarks.

2.3.6 Neutron Diffraction

In the introductory chapter we have mentioned the impact of electron diffraction experiments. It was clear at this time that electrons are particles and X-rays are waves, electromagnetic waves. However, such particles produced the same diffraction pattern as X-rays. Likewise, the electromagnetic wave comes in lumps, so called photons.

In order to substantiate the viewpoint of particle waves we will discuss neutron-crystal scattering which, as it turns out, is diffraction.

We simply replace the nitrogen molecule by a crystal. To keep things simple, we assume a crystal of one species of ions only. We say ions, because part of the electrons is localized around the nuclei, the remainder is shared and glues the crystal together. Neutrons do not see them.

There are three vectors $\boldsymbol{a}^{(1)}$, $\boldsymbol{a}^{(2)}$, and $\boldsymbol{a}^{(3)}$ which translate copies of the unit cell at the origin. The most general unit cell is situated at

$$\boldsymbol{\xi}_j = \boldsymbol{\xi}_{j_1\, j_2\, j_3} = j_1\boldsymbol{a}^{(1)} + j_2\boldsymbol{a}^{(2)} + j_3\boldsymbol{a}^{(3)}, \tag{2.76}$$

where the indexes are integer numbers. j_1 runs in $[-N_1, N_1]$, j_2 and j_3 likewise. There are altogether $(2N_1+1)(2N_2+1)(2N_3+1)$ unit cells which form our crystal.

We generalize (2.64) by replacing $\pm\boldsymbol{a}/2$ with $\boldsymbol{\xi}_j$ and sum over all index triples:

$$I(\boldsymbol{\Delta}) = \left| \sum_j f_j \, \mathrm{e}^{\,\mathrm{i}\,\boldsymbol{\Delta}\cdot\boldsymbol{\xi}_j} \right|^2 . \tag{2.77}$$

$I = I(\boldsymbol{\Delta})$ is the intensity dependency on wave vector transfer. In our example—one species of nuclei only—all scattering amplitudes f_j are equal. The sum extends over $j = (j_1, j_2, j_3)$ as in (2.76). One finds

$$I(\boldsymbol{\Delta}) = |f|^2\, S_1\, S_2\, S_3 \tag{2.78}$$

where the factors S stands for

$$S_1 = \frac{\sin^2 N_1 \boldsymbol{\Delta}\cdot\boldsymbol{a}^{(1)}/2}{\sin^2 \boldsymbol{\Delta}\cdot\boldsymbol{a}^{(1)}/2}, \tag{2.79}$$

with similar expressions for S_2 and S_3. Their order of magnitude is 1 unless the denominator vanishes:

$$\boldsymbol{\Delta}\cdot\boldsymbol{a}^{(r)} = 2\pi\nu_r \text{ with } \nu_1, \nu_2, \nu_3 \in \mathbb{Z}. \tag{2.80}$$

If these three Laue conditions[3] are satisfied simultaneously, a narrow pencil of diffracted neutrons is observed; otherwise there is just a feeble background.

The same conditions apply to X ray diffraction experiments. The incident wave then is a spherical wave which is emitted by the anode of an X ray tube. The crystal ions j get polarized periodically, and the oscillating dipole moment emits secondary spherical waves the power of which is registered at the detector. The emitters are excited by one and the same wave, so they oscillate in phase. By the superposition principle of electromagnetism, the fields of the emitters j sum up and yield an expression (2.77). Instead of the scattering amplitude, the ion's polarizability shows up. No wonder X ray, electron or neutron diffraction are alike.

2.3.7 *Summary*

We have learned that three simple rules for probability amplitudes suffice to describe, for example, the interaction of particles with composite structures. Probabilities P are absolute squares of probability amplitudes ϕ. If there are alternative intermediate stages and if the particular alternative is not registered, amplitudes add up and the sum must then be squared. In contrast, if the alternative leaves a foot-print, then amplitudes must first be squared to probabilities which are to be summed.

We explain how to describe the scattering process by a differential cross section. It depends on the wave number of the incoming neutrons and on the scattering angle.

The scattering of slow neutrons by diatomic molecules with the same nucleus is the simplest possible case. Even if the molecules are randomly oriented, the differential cross section exhibits a characteristic dependency on the wave number of the incoming neutrons and on the scattering angle. There is maximal constructive interference in forward direction or for slow neutrons. The faster the incoming neutrons, the less interference is observed. In the intermediate region there is even destructive interference. By fitting the data to a model, the molecular structure can be elucidated: number of nuclei, whether they are identical, and their distance.

2.4 Free Particles

In this section we consider a single freely moving particle. To simplify even more, we consider one spatial dimension only. The generalization to three dimensions is straightforward although we did not yet discuss angular momentum.

Throughout in this chapter we argue along Schrödinger's line: states change with time, observables remain fixed. We also confine the discussion to pure states to be described by normalized vectors.

---

[3]They are equivalent to the Bragg conditions.

The Hilbert space of square-integrable wave function is presented with a digression on test functions. The most interesting linear operators are not bounded, in particular location, linear momentum and energy. Suitable domains of definition have to be introduced. Although self-adjoint, these operators have continuously distributed quasi-eigenvalues with quasi-eigenfunctions which are not in the Hilbert space.

We derive the well-known explicit expression for the linear momentum observable and the canonical commutation relation.

Only then we focus on freely moving particles. We show that the average location moves with constant speed and that the particle's localization deteriorates in the course of time. We conclude with a remark on the classical limit.

2.4.1 Square Integrable Functions

Think of a particle that has been prepared in some pure state. As we know, such states are described by one-dimensional projections. They project on a linear subspace which is spanned by a normalized vector.

We refer to the probability amplitude $f = f(x)$ for the event that the particle is found at location x. More precisely, the probability $\mathrm{d}w$ to find the particle in the small interval $[x - \mathrm{d}x/2, x + \mathrm{d}x/2]$ is proportional to the probability amplitude squared,

$$\mathrm{d}w = \mathrm{d}x \mid f(x) \mid^2 . \tag{2.81}$$

The probability that the particle is found anywhere is one,

$$\int \mathrm{d}w = \int \mathrm{d}x \mid f(x) \mid^2 = 1. \tag{2.82}$$

This says that f describes a state, since it is normalized.

The Hilbert space under discussion therefore is the linear space of square normalizable functions,

$$\mathcal{H} = \mathcal{L}_2(\mathbb{R}) = \{f : \mathbb{R} \to \mathbb{C} \mid \int \mathrm{d}x \mid f(x) \mid^2 < \infty\}. \tag{2.83}$$

The natural choice for a scalar product is

$$(g, f) = \int \mathrm{d}x \, g^*(x) \, f(x). \tag{2.84}$$

It is finite for all $g, f \in \mathcal{H}$ because of the Cauchy–Schwarz inequality. $\mathcal{L}_2(\mathbb{R})$ denotes the Lebesgue square-integrable functions living on $\mathbb{R}$. Complex valued is silently understood. More on the Lebesgue integral can be found in the chapter on Mathematical Aspects.

2.4.2 Location

Let us investigate the location observable X declared by $(Xf)(x) = xf(x)$, i.e. multiplication of the function with its argument. X is not bounded and therefore not defined for all wave-functions $f \in \mathcal{H}$. Just take the sequence $f_1, f_2, \dots$ where $f_j(x) = 1$ in the interval $j \le x \le j+1$ and vanishes outside. The f_j are normalized, and we estimate $\| Xf_j \| \ge j$. It follows that X is unbounded. However, the location operator can be defined on the following domain of definition:

$$\mathcal{D}_X = \{f \in \mathcal{H} \mid \int \mathrm{d}x \mid xf(x) \mid^2 < \infty\}. \tag{2.85}$$

Note that $X^\star$ does the same as X and is also defined on $\mathcal{D}_X$. Therefore, the location operator is self-adjoint.

There is another way how to cope with physically important, but unbounded operators.

A test function $t : \mathbb{R} \to \mathbb{C}$ is infinitely often continuously differentiable and vanishes at $\mid x \mid \to \infty$, even after it has been multiplied by an arbitrary polynomial. We denote the linear subspace of test functions by $\mathcal{S}$. This test function space $\mathcal{S}$ is dense in $\mathcal{H}$ which we will not prove here. Any Hilbert space vector can be approximated arbitrarily well by test functions. Hence, studying X on the linear space of test functions is nearly the same as studying it on the entire Hilbert space. Clearly, if t is a test function, the image Xt is a test function as well.

Let us summarize these considerations by $X\mathcal{S} \subset \mathcal{S}$ and $X\mathcal{D}_X \subset \mathcal{H}$.

2.4.3 Linear Momentum

As we know, the linear momentum is the generator of translations. Denote by $f_a = f_a(x) = f(x+a)$ the translated wave function, a test function say. We expand this into a Taylor series:

$$f(x+a) = f(x) + \frac{a}{1!}f'(x) + \frac{a^2}{2!}f''(x) + \dots \tag{2.86}$$

This may be rewritten into

$$f_a(x) = \left(f + \frac{\mathrm{i}a}{1!}Kf + \frac{\mathrm{i}a^2}{2!}K^2 f + \dots\right)(x) \tag{2.87}$$

or

$$f_a = \mathrm{e}^{\,\mathrm{i}aK} f \text{ or } f_a = \mathrm{e}^{\frac{\mathrm{i}}{\hbar}aP} f. \tag{2.88}$$

The wave number operator K and the linear momentum P are

$$K = -\mathrm{i}\,\nabla \text{ and } P = -\mathrm{i}\,\hbar\,\nabla, \text{ respectively.} \tag{2.89}$$

∇ stands for ordinary differentiation.

Chapter 1 on Basics is about an arbitrary system described by an abstract Hilbert space. There we discussed spatial translations and its generator, the linear momentum. Now we talk about a specific system with its specific Hilbert space $\mathcal{L}_2(\mathbb{R})$ of wave functions. No wonder why we now obtain a concrete representation of the momentum operator, an expression which operates on wave functions.

In which sense is the linear momentum operator self-adjoint? It is certainly not defined on the entire Hilbert space since not all square-integrable functions are differentiable.

A function $f = f(x)$ is absolutely continuous if it can be written as an integral from $-\infty$ to x over a square-integrable function g. We declare the derivative of it by $f'(x) = g(x)$. The largest domain of definition for the momentum operator therefore is

$$\mathcal{D}_P = \{f \in \mathcal{H} \mid f(x) = c + \int_{-\infty}^{x} \mathrm{d}y\, g(y) \text{ where } g \in \mathcal{H}\}. \tag{2.90}$$

∇ is well defined on $\mathcal{D}_P$, and we show in the chapter on Mathematical Aspects that the momentum operator is self-adjoint with this domain of definition.

The momentum operator evidently maps test functions into test functions. It is a simple exercise to prove, by partial integration, that

$$(Pu, t) = (u, Pt) \text{ for } t, u \in \mathcal{S} \tag{2.91}$$

holds true. This is no proof that P is self-adjoint, however, rather close to it.

We summarize these domain considerations by $P\mathcal{S} \subset \mathcal{S}$ and $P\mathcal{D}_P \subset \mathcal{H}$.

By the way, the commutation relation of location and momentum is calculated by

$$[X, P]f(x) = x\frac{\hbar}{\mathrm{i}}f'(x) - \frac{\hbar}{\mathrm{i}}(xf(x))' = -\frac{\hbar}{\mathrm{i}}f(x), \tag{2.92}$$

i.e.

$$[X, P] = \mathrm{i}\hbar I \text{ or } \{X, P\} = I. \tag{2.93}$$

The latter formula can be found in Sect. 1.2 on the Classical Framework as well as in Sect. 1.3 on the Quantum Framework in Chap. 1 on Basics.

2.4.4 Wave Packets

Formally, the eigenfunctions of the location operator are

$$\xi_a(x) = \delta(x - a), \tag{2.94}$$

because of $X\xi_a(x) = x\delta(x - a) = a\delta(x - a) = a\xi_a(x)$. The eigenvalue is a, a real number. The δ-distribution, however, cannot be squared and integrated. It is not in the Hilbert space; and it is not a proper eigenfunction. In the following sense, however, it serves as such:

$$f(x) = \int \mathrm{d}a \; f_a \, \xi_a(x). \tag{2.95}$$

The integral replaces the sum over eigenvalues, and each quasi-eigenfunction ξ_a is multiplied by a coefficient $f_a = f(a)$. The decomposition (2.95) is of no use.

The situation is different for the momentum observable. The eigenvalue equation for $\pi_q(x)$ reads

$$P\pi_q = q\pi_q \text{ or } \frac{\hbar}{\mathrm{i}}\pi_q'(x) = q\pi_q(x). \tag{2.96}$$

The solution is

$$\pi_q(x) = \mathrm{e}^{\frac{\mathrm{i}}{\hbar}q\,x}, \tag{2.97}$$

it exists for all real eigenvalues q. These solutions, however, are not in the Hilbert space since $|\,\pi_q(x)\,|^2 = 1$ cannot be integrated over $x \in \mathbb{R}$.

The decomposition of an arbitrary wave function into momentum quasi-eigenfunctions (2.97) is

$$f(x) = \int \frac{\mathrm{d}q}{2\pi\hbar} \hat{f}(q)\, \pi_q(x) = \int \frac{\mathrm{d}q}{2\pi\hbar} \hat{f}(q)\; \mathrm{e}^{\frac{\mathrm{i}}{\hbar}qx}. \tag{2.98}$$

This is nothing else but the Fourier transformation $f \to \hat{f}$ of the wave function! The integral replaces the usual sum over eigenvalues. The exponentials are the quasi-eigenfunctions which get multiplied by the Fourier transform $\hat{f}$.

Because of

$$\int \mathrm{d}x \; \mathrm{e}^{\frac{\mathrm{i}}{\hbar}(q'' - q')} = 2\pi\hbar\,\delta(q'' - q') \tag{2.99}$$

one calculates

$$\int \mathrm{d}x \,|\, f(x)\,|^2 = \int \frac{\mathrm{d}q}{2\pi\hbar} |\, \hat{f}(q)\,|^2. \tag{2.100}$$

If f is square-integrable, then its Fourier transform $\hat{f}$ is square-integrable as well.

Momentum quasi-eigenfunctions, or plain waves, are not admissible to represent a pure state, but packets (2.98) of plane waves are.

2.4.5 Motion of a Free Particle

For a free particle, the Hamiltonian depends on the momentum only, and not on location. This may be

$$H = \sqrt{m^2c^4 + P^2c^2} \tag{2.101}$$

for a relativistic particle of mass m or

$$H = mc^2 + \frac{1}{2m}P^2 \tag{2.102}$$

in non-relativistic approximation.

Recall that states $f = f_t$ change with time according to the Schrödinger equations

$$\frac{\mathrm{d}}{\mathrm{d}t} f_t = -\frac{\mathrm{i}}{\hbar} H f_t. \tag{2.103}$$

The average location of the particle at time t is

$$\langle X \rangle_t = \int \mathrm{d}x \; f_t^*(x)\, X\, f_t(x). \tag{2.104}$$

We calculate

$$\frac{\mathrm{d}}{\mathrm{d}t}\langle X \rangle_t = \int \mathrm{d}x \; f_t^*(x)\, \{X, H\}\, f_t(x). \tag{2.105}$$

In non-relativistic approximation (2.102) we have

$$\{X, P^2\} = P\{X, P\} + \{X, P\}P = 2P \tag{2.106}$$

and therefore

$$\frac{\mathrm{d}}{\mathrm{d}t}\langle X \rangle_t = \frac{1}{m}\langle P \rangle_t. \tag{2.107}$$

If we differentiate once more with respect to time, we will encounter the commutator of the momentum with the energy. It vanishes for a free particle since the energy is a function of momentum only. Thus,

$$\frac{\mathrm{d}}{\mathrm{d}t}\langle P \rangle_t = 0 \text{ and } \frac{\mathrm{d}^2}{\mathrm{d}t^2}\langle X \rangle_t = 0. \tag{2.108}$$

A free particle moves with constant velocity, unaccelerated.

2.4.6 *Spreading of a Free Particle*

How well can a particle be localized? We shall answer this question soon for a free particle.

The variance $\sigma^2(X) = \langle X^2 \rangle - \langle X \rangle^2$ is a measure of localization. In order to simplify the calculation we choose an inertial system such that the particle—which moves with constant velocity—is at rest, $\langle X \rangle_t = 0$. Then the momentum expectation value $\langle P \rangle_t$ vanishes as well, see (2.107). It follows that $\langle X^2 \rangle$ and $\langle P^2 \rangle$ are the respective variances. How do they depend on time?

The momentum variance does not change with time. For the location we calculate

$$\frac{\mathrm{d}}{\mathrm{d}t}\langle X^2 \rangle_t = \int \mathrm{d}x \; f_t^*(x)\{X^2, H\} f_t(x), \tag{2.109}$$

in analogy with (2.105). The Poisson bracket gives $(XP + PX)/m$. The second derivative results in (2.109), and the bracket now reads $\{XP + PX, P^2\}/2m^2 = 2P^2/m^2$. Hence:

$$\frac{\mathrm{d}^2}{\mathrm{d}t^2}\langle X^2 \rangle_t = \frac{2}{m^2}\langle P^2 \rangle, \tag{2.110}$$

or

$$\sigma^2(X)_t = \sigma^2(X)_0 + \frac{\sigma^2(P)}{m^2} t^2. \tag{2.111}$$

This result requires interpretation.

Equation (2.110) says that the second derivative with respect to time is a positive constant, the solution therefore a convex parabola. Such a parabola has a minimum at a certain time which we chose to be $t = 0$. The precision with which the particle is localized grows quadratically with time. The minimal spread cannot be made to vanish since

$$\sigma(X)\,\sigma(P) \geq \frac{\hbar}{2} \tag{2.112}$$

holds true for any state f. This is Heisenberg's uncertainty relation which we discussed in the Introduction Sect. 1.1 and in the chapter on Mathematical Aspects.

Often the spread of an observable M is assessed by the standard deviation $\sigma(M) = \sqrt{\langle M^2 \rangle - \langle M \rangle^2}$. Heisenberg's uncertainty principle reads $\sigma(X)\,\sigma(P) \geq \hbar/2$. Equation (2.111) says that the spread of the wave packet ultimately grows linearly with time.

Also note that the spread growth rate is inversely proportional to the mass. The more massive a particle, the longer it behaves classically, i.e. it does not spread at all.

2.4.7 Summary

The probability amplitude $f = f(x)$ for finding the particle at a certain location x is a wave function. Linear combinations of such wave functions span the Hilbert space of square-integrable functions of one real argument with its natural scalar product.

The location and the linear momentum operators have been discussed. The former multiplies the wave function by its argument, the latter amounts to differentiation. Both operators are unbounded and cannot be defined on the entire Hilbert space, and we have said a few words on their respective domains of definitions. We also introduce test functions which form a dense linear subspace. Location and momentum obey the canonical commutation relations.

Instead of discrete eigenvalues and their eigenvectors we encounter a continuum of quasi-eigenvalues and quasi-eigenfunctions which are not in the Hilbert space. Nevertheless, any wave function may be decomposed into such quasi-eigenfunctions, the sum being replaced by an integral. In the case of the momentum observable one has re-discovered the Fourier decomposition.

The energy of a free particle does not depend on location; it depends on momentum only. Therefore, the particle's average location moves with constant velocity. However, it tends to spread. The spread, as assessed by the root mean square deviation increases linearly with time. The more massive the particle, the longer one can ignore the deterioration of location.

As a consequence, although self-adjoint, both observables have eigenvectors which are not in the Hilbert space. Nevertheless, replacing the sum by an integral over the quasi-eigenvalues, any wave function may be expanded into quasi-eigenfunctions. We also re-derive the canonical commutation rules, now for a concrete representation and briefly discuss wave packets.

We then have focused on free particles: the Hamiltonian depends on momentum only, it is translationally invariant. We demonstrated that the particle's average location moves with constant speed, the classical result. In the course of time, however, the particle spreads out more and more. The more massive the particle, the longer this effect can be ignored.

2.5 Small Oscillations

The following situation is encountered rather often. There is a stable equilibrium configuration with X denoting the deviation from it. The potential energy, which has its minimum at the equilibrium configuration, may be expanded around $X = 0$. The linear term vanishes because we assume equilibrium. The quadratic term is positive because the equilibrium is stable. Close-by configurations have a larger potential energy. Higher powers of the deviation, so-called anharmonicities, are neglected. One speaks of an harmonic oscillator. To a good approximation, the vibrational degrees of freedom of molecules and of solids are small oscillations, the latter giving rise to the concept of phonons. We shall present in this section an algebraic approach to diagonalizing the Hamiltonian of the harmonic oscillator. There is a ladder of eigenstates with operators for climbing up and down this ladder. The wave function representations of eigenstates, the eigenfunctions, are calculated as well.

In this book shall come across ladder operators several times, for instance when the irreducible representations of angular momentum are studied or when discussing the particle number as an observable.

2.5.1 The Hamitonian

Let us write

$$H = \frac{1}{2m} P^2 + \frac{m\omega^2}{2} X^2 \tag{2.113}$$

for the Hamiltonian. m is a mass and $m\omega^2$ the positive spring constant. The self-adjoint operators X and P shall satisfy the canonical commutation relations

$$[X, P] = \mathrm{i}\hbar I. \tag{2.114}$$

In the Heisenberg picture we calculate

$$\dot{X} = \{X, H\} = \frac{1}{m} P \text{ and } \ddot{X} = \{\dot{X}, H\} = -\omega^2 X, \tag{2.115}$$

as expected.

We define the linear operator

$$A = \sqrt{\frac{m\omega}{2\hbar}}\, X + \mathrm{i}\sqrt{\frac{1}{2m\omega\hbar}}\, P. \tag{2.116}$$

The following commutation rule is basic for what follows

$$[A, A^\star] = I. \tag{2.117}$$

Here we just state that the energy can be written as

$$H = \hbar\omega \left(N + \frac{1}{2} I \right) \text{ where } N = A^\star A. \tag{2.118}$$

N is self-adjoint while A is not.

2.5.2 Ladder Operators

The self-adjoint operator N has real eigenvalues. Let z be one of them and e the corresponding normalized eigenfunction:

$$Ne = ze. \tag{2.119}$$

Because of

$$[A^\star, N] = [A^\star, A^\star A] = A^\star [A^\star, A] = A^\star \tag{2.120}$$

we conclude

$$NA^\star e = A^\star Ne + A^\star e = (z+1)A^\star e. \tag{2.121}$$

This says that $A^\star e$ is another eigenvector of N with eigenvalue $z+1$. Likewise, Ae is an eigenvector of N with eigenvector $z-1$, because of

$$[N, A] = -A \tag{2.122}$$

and

$$NAe = ANe - Ae = (z-1)Ae. \tag{2.123}$$

By repeating this procedure one might conclude that the eigenvalues of N are z,$z \pm 1$,$z \pm 2$, and so on.

However, this cannot be the case since N is non-negative, and consequently its eigenvalues must also be non-negative. How to resolve this dilemma?

Well, eigenvectors must be normalized or normalizable. When stepping down, one will encounter a vanishing vector which erroneously went for an eigenvector. There must be a normalized vector Ω such that

$$A\Omega = 0, \tag{2.124}$$

the null vector. $N\Omega = 0$ says that this vector $e_0 = \Omega$ is an eigenvector of N with eigenvalue 0. Therefore, all eigenvectors are obtained by stepping up repeatedly,

$$e_n = \frac{(A^\star)^n}{\sqrt{n!}}\,\Omega \text{ for } n = 0, 1, 2, \ldots. \tag{2.125}$$

One can show by complete induction that the e_n are normalized and obey

$$Ne_n = ne_n. \tag{2.126}$$

N is a number operator: its eigenvalues are the natural numbers.

Now we have solved the initial task, namely to diagonalize the energy:

$$H\, e_n = \left(n + \frac{1}{2}\right) \hbar\omega\, e_n. \tag{2.127}$$

The energy levels are equidistant.

2.5.3 *Eigenstate Wave Functions*

Up to now we did not have to specify the Hilbert space. The above spectrum of energy eigenvalues and the corresponding eigenstates have been obtained by purely algebraic means. We have employed the canonical commutation rules and nothing else. We now want to construct the wave functions.

We therefore specialize to the Hilbert space $\mathcal{H} = \mathcal{L}_2(\mathbb{R})$. Now the location operator X and the linear momentum are described by

$$(Xf)(x) = xf(x) \text{ and } (Pf)(x) = \frac{\hbar}{\mathrm{i}} f'(x). \tag{2.128}$$

Stepping down with A is terminated by the condition $A\Omega = 0$ which translates to

$$\Omega'(x) + \kappa^2 x \Omega(x) = 0 \text{ where } \kappa = \sqrt{\frac{m\omega}{\hbar}}. \tag{2.129}$$

Note that κ has the dimension of an inverse length. The solution is

$$e_0(x) = \Omega(x) \propto \mathrm{e}^{-\kappa^2 x^2/2}. \tag{2.130}$$

This Gaussian is the ground state of the harmonic oscillator. It is an even function of x.

The first excited state is generated by applying $A^\star$ to the ground state. One obtains

$$e_1(x) = A^\star \Omega(x) \propto x\Omega(x). \tag{2.131}$$

This is an odd function.

It should have become clear by how to proceed. There is a closed formula

$$e_n(x) \propto H_n(\kappa x)\, \mathrm{e}^{-\kappa^2 x^2/2} \tag{2.132}$$

where the H_n are Hermite polynomials. Unfortunately, we must stop here.

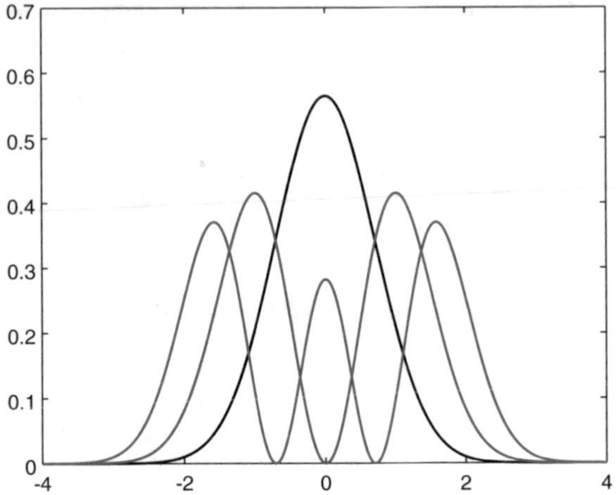

Fig. 2.6 Harmonic oscillator eigenfunctions. The probability densities $|\, e_n(x) \,|^2$ are plotted versus κx for the lowest three eigenstates. The ground state has one peak, the next two, and so on

Note that all harmonic oscillator eigenfunctions are test functions. The may serve as an orthogonal base for the linear subspace $\mathcal{S}$.

We have plotted the probability density $|\, e_n(x) \,|^2$ for the lowest eigenstates $n = 0, 1, 2$ in Fig. 2.6.

2.5.4 *Summary*

We have justified the harmonic oscillator model for small oscillations about a stable equilibrium. The Hamiltonian which is quadratic in both momentum P and deviation X may be re-written in $\hbar\omega\, N$ where $N = A^{\star}A$ is a number operator.

This was established by observing that $A^{\star}$ climbs up and A steps down a ladder of eigenstates by one. Since the harmonic oscillator Hamiltonian is bounded from below, there must be a ground state Ω to be annihilated by A. The energy eigenvalues are $n\hbar\omega$ where $n \in \mathbb{N}$ plus an irrelevant constant energy shift $\hbar\omega/2$.

By choosing a representation in terms of wave-functions, the ground state $e_0 = \Omega(x)$ and all energetically higher excited states can be calculated. The harmonic oscillator eigenfunctions are test functions, the may serve as an orthogonal base.

Chapter 3
Atoms and Molecules

Normal matter consists of neutrons, protons and electrons only. Protons and neutrons form nuclei, they are tightly bound, tiny and massive. The field of Atomic and Molecular Physics focuses on the properties of one or more nuclei interacting with a cloud of electrons. Orders of magnitude are governed by the electron mass m, its charge $-e$, the Coulomb force constant $1/4\pi\epsilon_0$ and, naturally, by Planck's action constant $\hbar$.

The field of study extends to ideally infinitely large molecules with short range or long range order, i.e. to the liquid or solid state of ordinary matter. Molecules, when coming sufficiently close, may also regroup their nuclei. This effect defines the field of Chemistry.

Here we cannot cover even one of the mentioned fields in sufficient depth. We focus instead on some prototypical situations:

- The Hydrogen Atom. There is one nucleus of charge e, the proton, and one electron. The simplest atom.
- The Helium Atom. There is one nucleus of charge $2e$, and two electrons. How to cope with identical particles?
- The Hydrogen Molecule. There are two protons which repel each other and two electrons which also repel each other, but the protons attract the electrons. Two protons and two electrons may be bound to form the simplest molecule H_2.
- How does an external static electric or magnetic field change the energy levels and stationary states of an atom or molecule?

Except for the hydrogen atom, non-relativistic or relativistic, analytical solutions are not feasible. We therefore also talk about well-known approximation schemes for arriving at plausible results.

P. Hertel, *Quantum Theory and Statistical Thermodynamics*,
Graduate Texts in Physics, DOI 10.1007/978-3-319-58595-6_3

3.1 Radial Schrödinger Equation

A simple situation like the motion of a single particle in a field of force already poses a formidable problem. Since the kinetic energy is represented by the Laplacian operator, we have to deal with partial differential equations instead of ordinary. Simple solutions exist only in cases where the problem at hand may be reduced to an ordinary differential equation. We shall discuss how to calculate the bound states of a particle moving in a central field of force.

3.1.1 Spherical Coordinates

The ordinary three-dimensional space is usually parameterized by Cartesian coordinates x_1, x_2, x_3. Another parameterization by spherical coordinates is better suited for problems with spherical symmetry. We define the radius r of a sphere around the origin and two angles for pinning down the point on the surface of such a sphere:

$$\begin{aligned} x_1 &= r\,\sin\theta\,\cos\phi \\ x_2 &= r\,\sin\theta\,\sin\phi \\ x_3 &= r\,\cos\theta. \end{aligned} \tag{3.1}$$

Thus, $\mathbb{R}^3 = [0, \infty) \times [0, \pi] \times [0, 2\pi)$ where ϕ and $\phi + 2\pi$ mean the same. $\theta = 0$ describes the north pole, $\theta = \pi/2$ the equator, and $\theta = \pi$ the south pole. This is not the convention of geography. $\phi = 0$ corresponds to the Greenwich meridian.

The functional determinant is

$$\frac{\partial(x_1, x_2, x_3)}{\partial(r, \theta, \phi)} = r^2\,\sin\theta. \tag{3.2}$$

It is positive except at the origin and at the poles. Clearly, at the origin $r = 0$ no angles are required, and at the poles $\theta = \pm\pi$ the longitude angle ϕ makes no sense. With (3.2) the volume element is

$$\mathrm{d}V = \mathrm{d}x_1\mathrm{d}x_2\mathrm{d}x_3 = \mathrm{d}r\,\mathrm{d}\theta\,\mathrm{d}\phi\; r^2\sin\theta. \tag{3.3}$$

A square-integrable function $f = f(r, \theta, \phi)$ is characterized by

$$\int_0^\infty \mathrm{d}r\, r^2 \int_0^\pi \mathrm{d}\theta\,\sin\theta \int_0^{2\pi} \mathrm{d}\phi\; |f(r, \theta, \phi)|^2 < \infty. \tag{3.4}$$

3.1.2 The Laplacian

The orbital angular momentum is $\boldsymbol{L} = \boldsymbol{X} \times \boldsymbol{P}$, the vector product of location $\boldsymbol{X}$ and linear momentum $\boldsymbol{P}$. The latter commute as $[X_j, P_k] = \mathrm{i}\hbar\delta_{jk}$. We know that the orbital angular momentum obeys the proper commutation relations $[L_i, L_j] = \mathrm{i}\hbar\epsilon_{ijk}L_k$ and that $\boldsymbol{L}^2$ commutes with all components L_i.

The following identity will be used:

$$\epsilon_{ijk}\epsilon_{iab} = \delta_{ja}\delta_{kb} - \delta_{jb}\delta_{ka}, \tag{3.5}$$

which is a plausible result. The left hand side is a sum over i such that four indices remain, i.e. products of Kronecker symbols. δ_{jk} and the like are not possible, they would be symmetric. δ_{ja} must be accompanied by δ_{kb}, and δ_{jb} by δ_{ka}. The value $\epsilon_{i23}\epsilon_{i23} = 1$ fixes the signs.

With (3.5) we derive the following expression for the orbital angular momentum squared:

$$\boldsymbol{L}^2 = L_i L_i = \epsilon_{ijk}\epsilon_{iab} X_j P_k X_a P_b = X_j P_k X_j P_k - X_j P_k X_k P_j. \tag{3.6}$$

The first term on the right hand side of (3.6) can be reshaped into $\boldsymbol{X}^2\boldsymbol{P}^2 + \mathrm{i}\hbar\,\boldsymbol{X}\cdot\boldsymbol{P}$. The second gives $-(\boldsymbol{X}\cdot\boldsymbol{P})^2 - 3\mathrm{i}\hbar\,\boldsymbol{X}\cdot\boldsymbol{P}$. With

$$\boldsymbol{P}^2 = -\hbar^2\Delta \text{ and } \boldsymbol{X}\cdot\boldsymbol{P} = -\mathrm{i}\hbar r\frac{\partial}{\partial r} \tag{3.7}$$

we finally arrive at the following expression for the Laplacian in spherical coordinates:

$$\Delta = \frac{\partial^2}{\partial r^2} + \frac{2}{r}\frac{\partial}{\partial r} - \frac{\hat{\boldsymbol{L}}^2}{r^2}. \tag{3.8}$$

Here $\hat{\boldsymbol{L}} = \boldsymbol{L}/\hbar$ is the dimensionless orbital angular momentum. The expression (3.8) for the Laplacian in spherical coordinates, although often used in quantum theory, is a purely mathematical statement. No $\hbar$ is involved.

3.1.3 Spherical Harmonics

If expressed in spherical coordinates, the angular momentum operators read

$$
\begin{aligned}
L_1 &= \mathrm{i}\hbar \left\{ \sin\phi \frac{\partial}{\partial\theta} + \cot\theta \cos\phi \frac{\partial}{\partial\phi} \right\} \\
L_2 &= \mathrm{i}\hbar \left\{ -\cos\phi \frac{\partial}{\partial\theta} + \cot\theta \sin\phi \frac{\partial}{\partial\phi} \right\} \\
L_3 &= -\mathrm{i}\hbar \frac{\partial}{\partial\phi}
\end{aligned}
\tag{3.9}
$$

and

$$
\boldsymbol{L}^2 = -\hbar^2 \left\{ \frac{1}{\sin\theta} \frac{\partial}{\partial\theta} \left(\sin\theta \frac{\partial}{\partial\theta} \right) + \frac{1}{\sin^2\phi} \frac{\partial^2}{\partial\phi^2} \right\}. \tag{3.10}
$$

These operators act on sufficiently smooth functions $Y = Y(\theta, \phi)$ which are square-integrable according to

$$
\| Y \|^2 = \int_0^{\pi} \mathrm{d}\theta \, \sin\theta \int_0^{2\pi} \mathrm{d}\phi \, | Y(\theta, \phi) |^2 < \infty. \tag{3.11}
$$

Let us briefly recollect what has been discussed in more detail in the chapter on Mathematical Aspects.

Only $\boldsymbol{L}^2$ and one component, say L_3, can be diagonalized simultaneously. The eigenfunctions, or spherical harmonics, fulfill

$$
\boldsymbol{L}^2 \, Y_{lm} = l(l+1)\, \hbar^2 \, Y_{lm} \tag{3.12}
$$

and

$$
L_3 \, Y_{lm} = m \, \hbar \, Y_{lm}. \tag{3.13}
$$

The values $l = 0, 1, 2, \ldots$ are integer numbers, they characterize the quantized orbital angular momentum. For each l, the second label $m = -l, -l+1, \ldots, l$ runs from $-l$ to $+l$ in steps of one. m characterizes the projection of the orbital angular momentum on the 3-axis.

Since all angular momentum operators are self-adjoint, the set of spherical harmonics is a base. They are orthonormal with respect to the scalar product induced by (3.11),

$$
(Y_{l''m''}, Y_{l'm'}) = \delta_{l''l'} \, \delta_{m''m'}, \tag{3.14}
$$

and any square-integrable function Y living on the unit sphere may be decomposed as

$$
Y(\theta, \phi) = \sum_{l=0}^{\infty} \sum_{m=-l}^{+l} c_{lm} \, Y_{lm}(\theta, \phi). \tag{3.15}
$$

Conversely, a set of complex coefficients c_{lm} with

$$\sum_{l=0}^{\infty} \sum_{m=-l}^{+l} | c_{lm} |^2 < \infty \tag{3.16}$$

generates a square-integrable function according to (3.15).

3.1.4 Spherical Symmetric Potential

We will now study the following Schrödinger equation:

$$-\frac{\hbar^2}{2\mu} \Delta f + V(R)\, f = E\, f. \tag{3.17}$$

μ is the mass,[1] $f = f(x_1, x_2, x_3)$ a wave function, and $R = \sqrt{X_1^2 + X_2^2 + X_3^2}$ denotes the observable "distance from origin". E is the searched for energy eigenvalue. The eigensolution f describes a stationary state.

The same equation reads

$$\frac{\hbar^2}{2\mu} \left\{ -\frac{\partial^2}{\partial r^2} - \frac{2}{r}\frac{\partial}{\partial r} + \frac{\hat{\boldsymbol{L}}^2}{r^2} \right\} f + V(r)\, f = E\, f, \tag{3.18}$$

where $f = f(r, \theta, \phi)$.

Let us now expand the wave function into spherical harmonics:

$$f(r, \theta, \phi) = \sum_{l=0}^{\infty} \sum_{m=-l}^{+l} u_{lm}(r)\, Y_{lm}(\theta, \phi). \tag{3.19}$$

The coefficients $u_{lm} = u_{lm}(r)$ now will depend on the distance r from origin. The Schrödinger equation (3.18) becomes an ordinary eigenvalue equation for each pair l, m:

$$H_l\, u = \frac{\hbar^2}{2\mu} \left\{ -u'' - \frac{2}{r} u' + \frac{l(l+1)}{r^2} u \right\} + V(r)\, u = E\, u, \tag{3.20}$$

with $u = u(r) = u_{lm}(r)$. Note that the quantum number m does not appear.

Equation (3.20) is the radial Schrödinger equation for a spherical symmetric potential V. The Hamiltonian H_l consist of four terms. The first three describe kinetic energy. The first two contributions stand for the kinetic energy of radial motion, the third one for transversal motion. It describes the so called centrifugal barrier which is always repulsive and depends on the angular momentum quantum number l. The last term is the potential energy.

[1] The reduced mass of the relative motion of two particles.

3.1.5 Behavior at Origin and Infinity

The radial Schrödinger equation appears to be rather singular close to $r = 0$. In the following discussion we assume a regular, i.e. not-singular potential. The kinetic energy then completely dominates the behavior close to the origin.

Close to $r = 0$ we have to solve

$$-u'' - \frac{2}{r}u' + \frac{l(l+1)}{r^2}u = 0. \tag{3.21}$$

We try a power of r and compute two solutions:

$$u \propto r^l \text{ and } u \propto r^{-(l+1)}. \tag{3.22}$$

The former is fine, the latter is not admissible.

Close to $r = 0$ the second solution squared behaves as $| u |^2 = r^{-2l-2}$. For $l = 1, 2, \ldots$ the normalization integral

$$\| u \|^2 = \int_0^\infty \mathrm{d}r\, r^2 \,| u(r) |^2 \tag{3.23}$$

will not converge at the lower end.

This argument, however, does not hold for vanishing orbital angular momentum $l = 0$. In this case the radial wave function in the vicinity of $r = 0$ is $u(r) \propto 1/r$. Equation (3.21) is fulfilled for $r > 0$. At $r = 0$, however, we encounter the delta-distribution:

$$-\Delta \frac{1}{r} = 4\pi\delta^3(\boldsymbol{x}), \tag{3.24}$$

which means that (3.21) is not satisfied.

To sum it up, at $r \approx 0$ a square-integrable solution of the radial Schrödinger equation runs as r^l where $l = 0, 1, \ldots$ characterizes the orbital angular momentum. The larger l, the less probable to find the particle close to the center.

At infinity only the second derivative matters. We assume that the potential vanishes there. With this convention the following equation has to be solved at $r \to \infty$:

$$-\frac{\hbar^2}{2\mu}u'' = Eu. \tag{3.25}$$

For $E > 0$ the solution is

$$u \propto \mathrm{e}^{\pm ikr} \quad \text{where } k = \frac{\sqrt{2\mu E}}{\hbar}. \tag{3.26}$$

Such sine and cosine solutions are not square-integrable.

For $E < 0$ we find

$$u \propto e^{\pm\kappa r} \quad \text{where} \quad \kappa = \frac{\sqrt{-2\mu E}}{\hbar}. \tag{3.27}$$

One of the two solutions explodes exponentially and will not be square-integrable. The solution with the negative exponent, on the other hand, is normalizable. It is a candidate for a bound state.

3.1.6 Alternative Form

Let us introduce the function

$$w(r) = r\,u(r). \tag{3.28}$$

$\mid w(r) \mid^2 \mathrm{d}r$ is the probability to find the particle outside a sphere of radius r and inside a sphere of radius $r + \mathrm{d}r$. The radial Schrödinger equation now reads

$$H_l\,u = \frac{\hbar^2}{2\mu}\left\{-w'' + \frac{l(l+1)}{r^2}\,w\right\} + V(r)\,w = E\,w. \tag{3.29}$$

Bound state solutions are characterized by

$$\int_0^\infty \mathrm{d}r \mid w(r) \mid^2 < \infty. \tag{3.30}$$

At $r = 0$ they run as

$$w(r) \propto r^{l+1}, \tag{3.31}$$

at infinity they fall off exponentially according to

$$w(r) \propto e^{-\kappa r} \tag{3.32}$$

with

$$\kappa = \frac{\sqrt{-2\mu E}}{\hbar} \tag{3.33}$$

an inverse length.

3.1.7 Summary

A spherically symmetric potential is best dealt with in spherical coordinates. We have demonstrate how the Laplacian Δ may be split into a radial and an angular part.

The latter turns out to be proportional to $\boldsymbol{L}^2$, the orbital angular momentum squared. Its eigenfunctions are the spherical harmonics $Y_{lm} = Y_{lm}(\theta, \phi)$.

Any wave function may be factorized into a radial part $u = u(r)$ and a spherical harmonic. The radial part obeys an ordinary differential equation which depends on angular momentum l but not on the eigenvalue m of L_3.

We also discussed the behavior of potential eigenfunction close to the origin and at infinity. Such eigenfunctions run as $u(r) \propto r^l$ for small r and drop off exponentially at infinity.

We have fixed the energy scale by setting the potential at infinity to zero. With this, the energy levels of bound states are always negative.

3.2 Hydrogen Atom

The hydrogen atom is a bound state of proton and electron kept together by the attractive Coulomb force. It is the most abundant and the simplest of all atoms and molecules and allows for a relatively simple exact solution. It is prototypical for atomic physics if augmented by the Pauli principle.

In order to get rid of inessential factors we employ the system of atomic units. We will see that the equations of atomic and molecular physics simplify considerably.

Since the Coulomb potential is spherically symmetric, we have to solve the radial Schrödinger equation, an ordinary differential eigenvalue equation.

We also mention a fully relativistic description by the Dirac equation and the resulting fine-splitting. The description is far from complete, and you might want to skip this on first reading.

The last subsection is a crude model for a classical hydrogen atom. The orbiting electron emits electromagnetic radiation thus spiraling into the proton. It would live a finite time only. The text could also be part of the introduction because it demonstrates that classical mechanics and electrodynamics cannot describe the structure of atoms and molecules.

3.2.1 Atomic Units

The proton is so massive that the reduced mass μ and the electron mass $m = m_e$ are practically the same; we neglect the difference.[2] The electron then moves in the Coulomb force field:

$$Hf = \frac{\hbar^2}{2m}\Delta f - \frac{e^2}{4\pi\epsilon_0}\frac{1}{|\boldsymbol{x}|} f = Ef \tag{3.34}$$

[2] The error is less than the correction by relativistic effects. All dimensionless energies should be multiplied by $E_*/(1 + m_e/m_p)$, see (3.36).

where $f : \mathbb{R}^3 \to \mathbb{C}$ is a square-integrable and sufficiently smooth complex valued function of three Cartesian coordinates. $-e^2$ is the product of electron and proton charge.

Let us introduce dimension-less quantities which are marked by a tilde. The units they are associated with are products of powers of

- electron mass $m_e = 9.109383 \times 10^{-31}\,\mathrm{kg}$
- unit charge $e = 1.602177 \times 10^{-19}\,\mathrm{s\,A}$
- Planck's quantum of action $\hbar = 1.054572 \times 10^{-34}\,\mathrm{kg\,m^2\,s^{-1}}$
- Coulomb force constant $4\pi\epsilon_0 = 1.112650 \times 10^{-10}\,\mathrm{kg^{-1}\,m^{-3}\,s^4\,A^2}$

The atomic unit of length, or Bohr's radius, is

$$a_* = \frac{m_e\, e^2}{(4\pi\epsilon_0)^2\, \hbar^2} = 0.052918 \text{ nm}. \tag{3.35}$$

The atomic unit of energy, the Hartree, is defined by

$$E_* = \frac{m_e e^4}{(4\pi\epsilon_0)^2 \hbar^2} = 27.211 \text{ eV}. \tag{3.36}$$

With this convention we may write $f(\boldsymbol{x}) = a_*^{3/2} \tilde{f}(\tilde{\boldsymbol{x}})$ such that

$$\int \mathrm{d}^3 x \mid f(\boldsymbol{x}) \mid^2 = \int \mathrm{d}^3 \tilde{x} \mid \tilde{f}(\tilde{\boldsymbol{x}}) \mid^2 = 1 \tag{3.37}$$

holds true. The right hand side is the dimension-less version of the left hand side, and both agree.

Similarly, (3.34) may be brought into its dimension-less form:

$$\tilde{H} \tilde{f} = \frac{1}{2} \tilde{\Delta} \tilde{f} - \frac{1}{\mid \tilde{\boldsymbol{x}} \mid} \tilde{f} = \tilde{E} \tilde{f}. \tag{3.38}$$

Rewriting equations into its dimension-less form boils down to setting $m_e = e = \hbar = 4\pi\epsilon_0 = 1$ and putting a tilde over all variables. From now on we omit the latter and shall write

$$H f = \frac{1}{2} \Delta f - \frac{1}{\mid \boldsymbol{x} \mid} f = E f \tag{3.39}$$

instead of (3.34). This is an equation with only dimension-less quantities. It is simpler to handle. Moreover, it can be programmed since computers know nothing else than dimensionless numbers.

3.2.2 Non-relativistic Hydrogen Atom

Recall the spherical harmonics $Y_{lm} = Y_{lm}(\theta, \phi)$. They obey[3]

$$\boldsymbol{L}^2 Y_{lm} = l(l+1) Y_{lm} \text{ and } L_3 Y_{lm} = m Y_{lm} \tag{3.40}$$

where $l = 0, 1, \dots$ and $m = -l, -l+1, \dots, l$.

Any square-integrable function f may be represented as a series of spherical harmonics:

$$f(r, \theta, \phi) = \sum_{l=0}^{\infty} \sum_{m=-l}^{l} u_{lr}(r)\, Y_{lm}(\theta, \phi). \tag{3.41}$$

The eigenvalue equation (3.39) becomes the radial Schrödinger equation

$$-\frac{1}{2}u'' - \frac{1}{r}u' + \frac{l(l+1)}{2r^2}u - \frac{1}{r}u = Eu. \tag{3.42}$$

$u = u_{lm}(r)$ depends on the angular momentum quantum number l but not on the eigenvalue m of L_3. Bound state solutions exist for negative energy E only, and we write $\kappa = \sqrt{-2E}$.

In the preceding Sect. 3.1 on the Radial Schrödinger Equation we have studied the behavior of the solution at the origin and at infinity. We therefore try the following *ansatz*:

$$u(r) \propto r^l \, \mathrm{e}^{-\kappa r} \sum_{\nu=0}^{\infty} c_\nu r^\nu. \tag{3.43}$$

The third factor is an interpolating power series.

We insert (3.43) into (3.42) and compare the coefficients in front of $r^{\nu+1}$:

$$c_{\nu+1} = 2\frac{\kappa(l+\nu+1) - 1}{(l+\nu+1)(l+\nu+2) - l(l+1)}\, c_\nu. \tag{3.44}$$

For large ν this recurrence relation is

$$c_{\nu+1} \approx \frac{2\kappa}{\nu}\, c_\nu \tag{3.45}$$

the solution of which is proportional to $c_\nu \approx \kappa^\nu / \nu!$. The interpolating power series behaves asymptotically as $\exp 2\kappa r$. Equation (3.43) explodes exponentially, it is not square-integrable.

For almost all values of κ. If, however, a nominator of (3.44) vanishes, the chain of coefficients is broken, and all subsequent c_n vanish. Instead of a power series we have a polynomial, and (3.43) is square-integrable.

[3]Recall $\hbar = 1$.

Now, $\kappa(l + \nu + 1) - 1$ vanishes for $\kappa = 1/n$ where $n = \nu_r + l + 1$. Here

- $l = 0, 1, \ldots$ is the angular momentum quantum number.
- $\nu_r = 0, 1, \ldots$ is the degree of the interpolating polynomials. ν_r is called the radial quantum number because the radial wave function has ν_r zeros, or nodes.
- $n = l + \nu_r + 1$ is a whole number. It is called the principal quantum number. The energy of the corresponding state is

$$E = -\frac{1}{2n^2}. \tag{3.46}$$

Note that the principal quantum number assumes values $n = 1, 2, \ldots$.

The energy levels are degenerate. First, for a given principal quantum number n, there are states for $l = n - 1, n - 2, \ldots, 0$ with the same energy $-1/2n^2$. Moreover, for each admissible n, l pair there are $2l + 1$ degenerate states for $m = -l, -l + 1, \ldots, l$. Since the electron has two internal spin states, there is an additional twofold degeneracy.

Traditionally, the orbital angular momentum is written as

- s for sharp, $l = 0$
- p for principal, $l = 1$
- d for diffuse, $l = 2$
- f for fundamental, $l = 3$
- and so on, alphabetically

These are terms used in early spectroscopy.

Thus, the ground state of the hydrogen atom is 1 s with a binding energy of 1/2 atomic units, or 13.61 eV.

The next two excited states are 2 s and 1p, both with a binding energy of 1/8 a.u., or 3.401 eV.

3.2.3 *Orbitals*

The orbitals, or wave functions, of the hydrogen atom are labeled by three integer quantum numbers: the principal number $n = 1, 2, \ldots$, the angular momentum $l = 0, 1, \ldots, n - 1$, and the magnetic quantum number $m = -l, -l + 1, \ldots, l$. The stationary bound states of an electron-proton system are

$$f_{nlm}(r, \theta, \phi) = u_{nl}(r)\, Y_{lm}(\theta, \phi). \tag{3.47}$$

We have studied the spherical harmonics Y_{lm} before and focus here on the radial part u_{rl} which is a product of r^l (centrifugal barrier), a polynomial, and an exponential. The exponential decreases with an exponent $-\kappa r$ where κ is the inverse of the main quantum number n. The polynomial is of degree $n - l - 1$, its coefficients may be

Table 3.1 Some hydrogen atom radial wave functions

$n\,l$	$u_{nl}(r)$	E_{nl}
1s	$2\,\mathrm{e}^{-r}$	$-1/2$
2s	$\sqrt{1/2}\,(1 - r/2)\,\mathrm{e}^{-r/2}$	$-1/8$
2p	$\sqrt{1/24}\,r\,\mathrm{e}^{-r/2}$	$-1/8$
3s	$\sqrt{4/27}\,(1 - 2r/3 + 2r^2/27)\,\mathrm{e}^{-r/3}$	$-1/18$
3p	$\sqrt{32/2187}\,r\,(1 - r/6)\,\mathrm{e}^{-r/3}$	$-1/18$
3d	$\sqrt{8/98415}\,r^2\,\mathrm{e}^{-r/3}$	$-1/18$

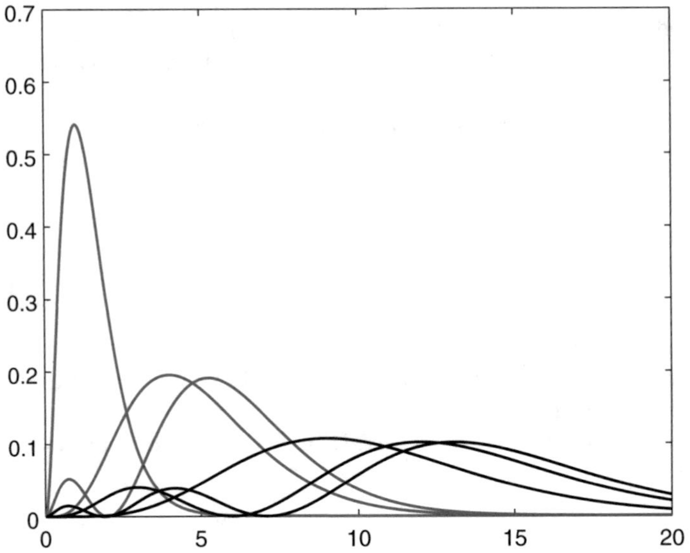

Fig. 3.1 Hydrogen atom orbitals. The probability density $r^2 |\, u_{nl} \,|^2$ to find the electron is plotted versus distance r. The graph with its maximum at $r = 2$ a.u. (*red*) is for the 1 s state. The two curves with maxima near 5 a.u. (*blue*) represent the 2 s and 2p states, respectively. They are degenerate. The same applies to the three orbitals 3s, 3p, and 3d (*black*) the maxima of which are close to 12 a.u. Note that there are $n - l - 1$ nodes, or zeros

read off from (3.44). We list, for 1 s through 3p, the normalized radial wave functions in Table 3.1.

The probability density for finding the electron at a distance r is $r^2 \,|\, u_{nl} \,|^2$. The corresponding curves are plotted in Fig. 3.1.

3.2.4 Relativistic Hydrogen Atom

The results above contain various approximations. First, the kinetic energy $\sqrt{\boldsymbol{P}^2c^2 + m^2c^4} - mc^2$ was replaced by $\boldsymbol{p}^2/2m$. Second, the electron spin and its magnetic moment have been neglected. Third, the electromagnetic interaction is described semi-classically: a quantum particle moves in an external, classical field.

Paul Dirac, a British physicist and mathematician, remedied the first two shortcomings. He invented an equation for the spinning electron which respects the theory of relativity. As it turned out, the bound states of an hydrogen atom can be calculated analytically as well.

Dirac Equation

The correct relation between energy $E = p_0 c$ and momentum $\boldsymbol{p}$ is

$$p_0^2 - \boldsymbol{p}^2 = g^{\mu\nu} p_\mu p_\nu = m^2 c^2. \tag{3.48}$$

$g^{\mu\nu}$ is the diagonal Minkowski tensor with $+1, -1, -1, -1$ on the diagonal. μ, ν assume values 0 (time part) and $1, 2, 3$ (space part). A sum over an upper and a lower index in the same term is assumed (Einstein's convention).

Equation (3.48) is quadratic in the energy p_0 or time derivative. In contrast, the Schrödinger equation is of first order. How to linearize (3.48)? Relativity requires that if the searched for expression is linear in p_0, it must also be liner in the spatial components $\boldsymbol{p}$. We write

$$\gamma^\mu p_\mu = mc. \tag{3.49}$$

Squaring this yields in fact (3.48), provided the four coefficients obey

$$\gamma^\mu \gamma^\nu + \gamma^\nu \gamma^\mu = g^{\mu\nu}. \tag{3.50}$$

Clearly, the γ^μ cannot be numbers. The simplest choice are 4×4 matrices. Among the many possibilities the following standard representation is particularly useful:

$$\gamma^0 = \begin{pmatrix} \sigma^0 & 0 \\ 0 & -\sigma^0 \end{pmatrix} \text{ and } \gamma^j = \begin{pmatrix} 0 & \sigma^j \\ -\sigma^j & 0 \end{pmatrix} \text{ for } j = 1, 2, 3. \tag{3.51}$$

σ^0 is the 2×2 unit matrix, the σ^j are the 2×2 Pauli matrices

$$\sigma^1 = \begin{pmatrix} 0 & 1 \\ 1 & 0 \end{pmatrix} \quad \sigma^2 = \begin{pmatrix} 0 & -\mathrm{i} \\ \mathrm{i} & 0 \end{pmatrix} \quad \sigma^3 = \begin{pmatrix} 1 & 0 \\ 0 & -1 \end{pmatrix}. \tag{3.52}$$

The γ matrices refer to a four-dimensional space the elements of which are spinors ψ. The spinor components are square-integrable wave functions. The Dirac equation for a free particle now reads

$$\mathrm{i}\hbar\gamma^{\mu}(\partial_{\mu}\psi) = mc\psi. \tag{3.53}$$

The operator on the left-hand side is a scalar with respect to Poincaré transformations because the γ matrices transform as a four-vector, by definition.

Some Observables

Consider a spinor wave function ψ, a 4×1 column vector with components $\psi_a = \psi_a(t, \boldsymbol{x})$. By definition, the adjoint spinor is $\bar{\psi} = \psi^{\dagger}\gamma^0$. This is a row vector.[4] Note the relation $(\gamma^{\mu})^{\dagger}\gamma^0 = \gamma^0\gamma^{\mu}$.

The Dirac equation for the adjoint spinor is

$$-\mathrm{i}\hbar(\partial_{\mu}\bar{\psi})\gamma^{\mu} = mc\bar{\psi}. \tag{3.54}$$

We multiply the Dirac equation (3.53) from the left by $\bar{\psi}$ and the adjoint equation from the right by ψ and subtract them. The result is

$$j^{\mu} = \bar{\psi}\gamma^{\mu}\psi \text{ with } \partial_{\mu}j^{\mu} = 0. \tag{3.55}$$

There is a conserved four-current density. Moreover, its temporal component $j^0 = \psi^{\dagger}\psi$ is never negative. We therefore may interpret $j^0/c = \rho$ as the particle number density and $\boldsymbol{j}$ as the particle current density.

The orbital angular momentum $\boldsymbol{L} = \boldsymbol{X} \times \boldsymbol{P}$ acts on each component of the spinor wavefunction in the same way. The spin angular momentum is a 4×4 matrix in spinor space, namely

$$\boldsymbol{S} = \frac{\hbar}{2}\begin{pmatrix} \boldsymbol{\sigma} & 0 \\ 0 & \boldsymbol{\sigma} \end{pmatrix}. \tag{3.56}$$

It acts at each location in the same way. Clearly, both commute:

$$[L_i, S_j] = 0. \tag{3.57}$$

The total angular momentum is $\boldsymbol{J} = \boldsymbol{L} + \boldsymbol{S}$.

The Hamiltonian, or energy, can be read off from (3.49):

$$\begin{aligned} H &= \gamma^0 c\boldsymbol{\gamma}\cdot\boldsymbol{P} + mc^2\gamma^0 \\ &= c\begin{pmatrix} mc & 0 & P_3 & P_1 - \mathrm{i}P_2 \\ 0 & mc & P_1 + \mathrm{i}P_2 & -P_3 \\ P_3 & P_1 - \mathrm{i}P_2 & -mc & 0 \\ P_1 + \mathrm{i}P_2 & -P_3 & 0 & -mc \end{pmatrix} \end{aligned} \tag{3.58}$$

A resting particle has two different energy eigenvalues, namely $+mc^2$ and $-mc^2$. Such negative energy solutions must be interpreted in terms of anti-particles. The

[4] The dagger here indicates transposition and complex conjugation.

Dirac equation does not only describe the electron, but also its twin, the positron. We shall not follow this thread any further.

Minimal Coupling

As is well known, the electromagnetic field strengths have a potential:

$$\boldsymbol{E} = -\nabla\phi - \dot{\boldsymbol{A}} \text{ and } \boldsymbol{B} = \nabla \times \boldsymbol{A}. \tag{3.59}$$

$A^0 = \phi/c$ and $\boldsymbol{A}$ form a four-vector A^μ. There are many such four-potentials so that auxiliary conditions may be imposed. We choose the Lorentz gauge

$$\partial_\mu A^\mu = 0, \tag{3.60}$$

the potential shall be four-divergence free.[5] Maxwell's equations are satisfied if the potential obeys the wave equation

$$\Box A^\mu = \mu_0 \, j^\mu, \tag{3.61}$$

where μ_0 is the vacuum permeability, j^μ the electric current density, with $j^0 = \rho c$ in standard notation. $\Box$, the d'Alembertian, is defined as $\partial_\nu \partial^\nu$.

The principle of minimal coupling describes the interaction of a charged particle with the electromagnetic field. The particle four-momentum P^μ has to be replaced by $P^\mu - qA^\mu$. This prescription is compatible with relativity.

Bound States

With $A^0 \propto 1/r$ and $\boldsymbol{A} = 0$ the Dirac Hamiltonian for the hydrogen atom is well defined. The solution path is similar to what we have described above: expand the wave functions into spherical harmonics, split off the behavior at $r \approx 0$ and $r \approx \infty$, establish a recursion relation for the interpolating power series, which must terminate.

Since now a fourth constant of nature enters the stage, namely the velocity c of light, we introduce the so-called fine structure constant:

$$\alpha = \frac{1}{4\pi\epsilon_0} \frac{e^2}{\hbar c} = \frac{1}{137.036}. \tag{3.62}$$

This is $1/c$ in atomic inverse velocity units.

Here we just write down the result of the calculation. The bound states are characterized by a main quantum number $n = 1, 2, \dots$ and by an angular momentum quantum number $j = 1/2, 3/2, \dots$. The energy levels of the hydrogen atom are

$$E_{nj} = \frac{mc^2}{\sqrt{1 + \left(\dfrac{\alpha}{n - j - 1/2 + \sqrt{(j + 1/2)^2 - \alpha^2}} \right)^2}}, \tag{3.63}$$

[5] Recall $x_0 = ct$.

in atomic units.

An expansion into powers of the fine structure constant begins as

$$E_{nj} - mc^2 = -\frac{1}{2n^2}\left(1 + \frac{\alpha^2}{4n^2}\frac{4n-3}{j+1/2} + \dots\right). \tag{3.64}$$

For $\alpha = 0$ this coincides with our previous result. The α^2 correction to the level scheme goes by the name of fine structure. Levels with the same main quantum number n but different total angular momentum j are split.

3.2.5 Classical Hydrogen Atom

Consider a classical hydrogen atom. A charged particle moves in an attractive Coulomb field according to the laws of classical mechanics. We focus on a circular orbit and argue non-relativistically. Denote by r the radius and by ω the angular velocity of the electron which has mass m and charge $-e$. The centrifugal force $mr\omega^2$ must be equal to the electrostatic attraction $e^2/4\pi r^2$. We conclude that radius and angular velocity are related by

$$\omega^2 = \frac{e^2}{4\pi\epsilon_0}\frac{1}{mr^3}. \tag{3.65}$$

This is Kepler's third law.

The orbiting electron causes an oscillating dipole moment $d = -er$. It radiates. The energy loss per unit time is given by

$$P = \frac{1}{3c}\frac{\mu_0}{4\pi}|\ddot{d}|^2 = \frac{1}{3c^3}\left(\frac{e^2}{4\pi\epsilon_0}\right)^3\frac{1}{m^2r^4}. \tag{3.66}$$

The total energy E of the orbiting electron is

$$E = \frac{m}{2}r^2\omega^2 - \frac{e^2}{4\pi\epsilon_0}\frac{1}{r} = -\frac{1}{2}\frac{e^2}{4\pi\epsilon_0}\frac{1}{r}. \tag{3.67}$$

Eliminating r we find

$$\dot{B} = \frac{16}{3m^2c^3}\frac{4\pi\epsilon_0}{e^2}B^4. \tag{3.68}$$

Here $B = -E$ is the positive binding energy and $\dot{B} = P$.

We abbreviate the constant in front of B^4 by K and apply the method of separation of variables. The result is

$$B_t = \frac{B_0}{\sqrt[3]{1 - 3KB_0^3\, t}}. \tag{3.69}$$

After the finite time

$$t_\infty = \frac{1}{3KB_0^3} \tag{3.70}$$

the binding energy becomes infinite. This time is 3.1×10^{-11} s for $B_0 = 13.6$ eV. The electron has spiraled into the center of the Coulomb force field thereby liberating an infinite amount of energy. In short: classical mechanics is not able to describe the hydrogen atom.

3.2.6 Summary

We have studied first a non-relativistic electron in an attractive electrostatic Coulomb potential. The decisive quantities are the electron's mass m_e, the unit charge e, the Coulomb force constant $1/4\pi\epsilon_0$ and $\hbar$, Planck's constant of action. Instead of the SI system of units (meter, kilogram, second and Ampère) we best employ atomic units. This amounts to setting $m_e = e = \hbar = 4\pi\epsilon_0 = 1$. The atomic unit of length a_* is called Bohr's radius, the unit of energy E_* is the Hartree.

Since the problem is spherically symmetric, a factorization into spherical harmonics and a radial wave function is appropriate. The latter is factorized again into a term for the behavior at small distances, another term for the decrease at infinity, and an interpolating power series. We derive a recursion relation for its coefficients. In general, the power series explodes and does not solve the problem. Only if one of the coefficients vanishes, we deal with a polynomial, and the solution is square-integrable. Although the energies of such bound state solutions might depend on a main quantum number n and the angular momentum quantum number l, they turn out to be $E_{nl} = -1/2n^2$; they are degenerate with respect to l. Here, $n = 1, 2, \dots$ and $l = 0, 1, \dots, n-1$. We also have discussed the corresponding normalized wave functions. The non-relativistic hydrogen atom can be solved analytically.

We sketched how to treat the problem relativistically. An electron, as a particle of spin $1/2$, should obey the Dirac equation. The quadratic energy-momentum relation is linearized by introducing matrices with certain anti-commutation relations. They live in a four-dimensional spinor space (which must not be confused with the four-dimensional time-space continuum). We have shown that there is a conserved probability current with positive probability density. Spin, orbital angular momentum and total angular momentum are discussed as well as the principle of minimal coupling to an electromagnetic external field.

These are the ingredients for solving the relativistic hydrogen atom by similar methods as above. The analytical solution for the bound states E_{nj} can be expanded with respect to the fine structure constant which is the inverse speed of light in atomic units. The non-relativistic limit for $E_{nj} - mc^2$ coincides with our previous result. It does not depend on the total angular momentum quantum number j which assumes values $j = 1/2$ for $l = 0$ and $j = l \pm 1/2$ for $l = 1, 2, \ldots$. The next term, suppressed by α^2, describes the fine splitting of otherwise degenerate levels. The non-relativistic level degeneracy is lifted.

Finally, the hydrogen atom has been modeled classically. The electron moves on Kepler orbits around the central proton. Since the system is an accelerated dipole, it radiates off more and more energy. After a finite time the electron will have emitted an infinite amount of energy and crashes into the proton. This simple calculation reemphasizes that atoms and molecules cannot exist in stable states if the laws of classical physics are applied.

3.3 Helium Atom

The helium atom is made of a helium nucleus and two electrons. The nucleus consists of two protons and two neutrons.[6] Again we may assume that the nucleus is fixed and the origin of a spherically symmetric Coulomb field with charge $2e$. However, there are two light particles moving in this attractive field which repel each other, a new and complicated situation which defies any analytic solution. Moreover, the two particles are identical. If the particles are in a state where both spins point upwards or downwards, there is no means to tell who is who. There are just two particles, one of them here, the other one there. The wave function must change sign if the two particles, location and spin, are interchanged. In order not to encounter all complications at the same time we treat the system in the non-relativistic limit.

3.3.1 *Wave Functions*

For a spin one half particle, like the electron, the spin angular momentum is represented by $\boldsymbol{S} = \hbar\boldsymbol{\sigma}/2$. The Pauli matrices are

$$\sigma^1 = \begin{pmatrix} 0 & 1 \\ 1 & 0 \end{pmatrix} \quad \sigma^2 = \begin{pmatrix} 0 & -\mathrm{i} \\ \mathrm{i} & 0 \end{pmatrix} \quad \sigma^3 = \begin{pmatrix} 1 & 0 \\ 0 & -1 \end{pmatrix}. \tag{3.71}$$

We arbitrarily choose the projection of spin onto the 3-axis. The wave function does not only depend on location $\boldsymbol{x}$, but also on spin $\uparrow$ or $\downarrow$. Such wave functions are eigenfunctions of S_3:

$$S_3\psi(\uparrow \boldsymbol{x}) = \frac{\hbar}{2}\psi(\uparrow \boldsymbol{x}) \ \text{ and } \ S_3\psi(\downarrow \boldsymbol{x}) = -\frac{\hbar}{2}\psi(\downarrow \boldsymbol{x}). \tag{3.72}$$

[6] The stable isotope Helium-3 with only one neutron is very rare.

The wave functions are normalized according to

$$\int \mathrm{d}^3x \sum_{\sigma=\downarrow,\uparrow} |\,\psi(\sigma\boldsymbol{x})\,|^2 = 1. \tag{3.73}$$

Equation (3.73) is easily generalized to the expression for the scalar product:

$$(\psi_2, \psi_1) = \int \mathrm{d}^3x \sum_{\sigma=\downarrow,\uparrow} \psi_2{}^*(\sigma, \boldsymbol{x})\, \psi_1(\sigma, \boldsymbol{x}). \tag{3.74}$$

Pauli Principle

Now assume there are two identical particles. One is found at $\boldsymbol{x}_a$ with spin σ_1, the other at $\boldsymbol{x}_b$ with spin σ_2. The probability to detect the pair is

$$\mathrm{d}W = |\,\psi(\sigma_a\boldsymbol{x}_a, \sigma_b\boldsymbol{x}_b)\,|^2\, \mathrm{d}^3x_a\, \mathrm{d}^3x_b. \tag{3.75}$$

There is a detector with volume d^3x_a at $\boldsymbol{x}_a$ such that it is triggered if the particle has polarization σ_a, i.e. down or up. Another such counter is positioned at $\boldsymbol{x}_b$ being sensitive for a particle with polarization σ_b. Equation (3.75) is the probability that both counters are triggered simultaneously. If so, is it the first particle which has triggered counter A while the second particle has triggered counter B, ore vice versa? This would be a sensible question if the first were a blue and the second a red particle, say, if they could be distinguished. But they are identical, interchanging the arguments in (3.75) cannot make a difference,

$$|\,\psi(\sigma_a\boldsymbol{x}_a, \sigma_b\boldsymbol{x}_b)\,|^2 = |\,\psi(\sigma_b\boldsymbol{x}_b, \sigma_a\boldsymbol{x}_a)\,|^2. \tag{3.76}$$

This equation says that interchanging the arguments must produce a factor α with $|\,\alpha\,| = 1$, a phase factor.

Under quite general assumptions it can be shown that $\alpha = 1$ for identical particles with integer spin and $\alpha = -1$ if the spin is half-integer. There are more possibilities, but they are not realized by nature.

In particular, a many-electron wave function changes sign if two groups of arguments $\sigma_a\boldsymbol{x}_a$ and $\sigma_b\boldsymbol{x}_b$ are interchanged. This fact is known as the Pauli principle.

The Pauli principle says that $\psi(\sigma\boldsymbol{x}, \sigma\boldsymbol{x}) = 0$. Two particles with the same polarization cannot be at the same place. In other words, electrons tend to avoid each other.

Spin-Singlet and Triplet

The total spin of a pair of spin = 1/2 particles is either 0 or 1. There are four different polarization states, either two times two or one (spin = 0) plus three (spin = 1). The states $\psi_{s\sigma}$ with total spin s and polarization σ are

$$\begin{aligned}\psi_{11}(\boldsymbol{x}_a,\boldsymbol{x}_b) &= \psi(\uparrow \boldsymbol{x}_a, \uparrow \boldsymbol{x}_b)\\ \psi_{10}(\boldsymbol{x}_a,\boldsymbol{x}_b) &= \frac{\psi(\uparrow \boldsymbol{x}_a, \downarrow \boldsymbol{x}_b) + \psi(\downarrow \boldsymbol{x}_a, \uparrow \boldsymbol{x}_b)}{\sqrt{2}}\\ \psi_{1-1}(\boldsymbol{x}_a,\boldsymbol{x}_b) &= \psi(\uparrow \boldsymbol{x}_a, \uparrow \boldsymbol{x}_b)\end{aligned} \tag{3.77}$$

and

$$\psi_{00}(\boldsymbol{x}_a,\boldsymbol{x}_b) = \frac{\psi(\uparrow \boldsymbol{x}_a, \downarrow \boldsymbol{x}_b) - \psi(\downarrow \boldsymbol{x}_a, \uparrow \boldsymbol{x}_b)}{\sqrt{2}}. \tag{3.78}$$

$\boldsymbol{S}_a$ and $\boldsymbol{S}_b$ are spin operators for particle a and b, respectively. The total spin is its sum $\boldsymbol{S} = \boldsymbol{S}_a + \boldsymbol{S}_b$. $S^{\pm} = S^1 \pm \mathrm{i}S^2$ are the up and down ladder operators, respectively. $S^+\psi_{11} = 0$ characterizes a spin $= 1$ state. The square root factors guarantee correct normalization. Note that ψ_{10} and ψ_{00} are orthogonal. The three states with total spin $= 1$ form a spin triplet, the spin $= 0$ state is a spin singlet.

The spin triplet wave functions $\psi_{1\sigma}(\boldsymbol{x}_a,\boldsymbol{x}_b) = -\psi_{1\sigma}(\boldsymbol{x}_b,\boldsymbol{x}_a)$ are anti-symmetric with respect to interchanging $\boldsymbol{x}_a$ and $\boldsymbol{x}_b$. This is a consequence of the Pauli principle and the fact that the spin part is symmetric. The spin singlet state $\psi_{00}(\boldsymbol{x}_a,\boldsymbol{x}_b) = \psi_{00}(\boldsymbol{x}_b,\boldsymbol{x}_a)$ must be symmetric.

Helium Atom Hamiltonian

The energy of a helium atom in non-relativistic approximation consists of five terms: the kinetic energies of the two electrons, their electrostatic attraction by the helium nucleus, and the electrostatic repulsion of the two electrons. We write

$$H = \frac{\boldsymbol{P}_a^2}{2} + \frac{\boldsymbol{P}_b^2}{2} - \frac{2}{|\,\boldsymbol{X}_a\,|} - \frac{2}{|\,\boldsymbol{X}_b\,|} + \frac{1}{|\,\boldsymbol{X}_b - \boldsymbol{X}_a\,|}, \tag{3.79}$$

in obvious notation and in atomic units. This Hamiltonian has to be applied to even or odd wave functions $\psi = \psi(\boldsymbol{x}_a,\boldsymbol{x}_b)$ if spin triplet or spin singlet states are to be described, respectively. Although the spin does not appear in expression (3.79), it indirectly affects the bound state energies via the admissible wave functions, symmetric or antisymmetric.

It is hopeless to strive for an analytic solution, we must resort to approximations or to numerical methods.

3.3.2 Minimal Ground State Energy

The ground state of a system is characterized by the lowest energy eigenvalue. The corresponding state is stationary as every energy eigenstate. However, transitions by perturbations to states with even lower energy are impossible. The ground state of a system is assumed if the system is left to itself. This statement is equivalent to the following, as we shall show later. A system in equilibrium with an environment of zero temperature is in its ground states. These remarks should have made clear why calculating the ground states is an important task of quantum theory.

Upper Bound

Assume that the eigenvalues of the self adjoint operator H are ordered increasingly, $E_0 \le E_1 \le \ldots$ Its eigenfunctions are $f_0, f_1, \ldots$. We assume that they form a complete set of orthonormal functions. The latter means

$$(f_j, f_k) = \delta_{jk}, \tag{3.80}$$

the former says that an arbitrary wave function may be expanded as

$$f = c_0 f_0 + c_1 f_1 + \ldots. \tag{3.81}$$

We then may write

$$(f, Hf) = |\, c_0\,|^2 E_0 + |\, c_1\,|^2 E_1 + \cdots \ge \left(|\, c_0\,|^2 + |\, c_1\,|^2 + \ldots\right) E_0. \tag{3.82}$$

Since the bracket on the right hand side equals $\|\, f\,\|^2$, the following inequality holds true:

$$E_0 \le \frac{(f, Hf)}{(f, f)} \quad \text{for all} \quad f \ne 0. \tag{3.83}$$

Any expectation value of H is an upper bound to the ground state energy. The equality sign holds if $f = f_0$ is inserted, the ground state wave function. We conclude, that a good approximation of the ground state wave function will deliver an expectation value close to the ground state energy.

There are also lower bounds for the ground state energy.[7] However, while an upper bound for an atom or molecule can be calculated with relative little effort and high precision, a lower bound is usually more difficult to obtain and generally rather crude.

If we could try out all wave functions, the ground state energy would be

$$E_0 = \inf_{\substack{f \in \mathcal{H} \\ f \ne 0}} \frac{(f, Hf)}{(f, f)}, \tag{3.84}$$

where $\mathcal{H}$ is the entire Hilbert space. For a subset $\mathcal{D} \subset \mathcal{H}$, we obtain an upper bound only:

$$E_0 \le \inf_{\substack{f \in \mathcal{D} \\ f \ne 0}} \frac{(f, Hf)}{(f, f)}, \tag{3.85}$$

[7] See for instance P. Hertel, E.H. Lieb, W. Thirring; *Lower bound to the energy of complex atoms*; J. Chem. Phys. 62, 3355–3356, 1975.

which is optimal for the subset $\mathcal{D}$. The larger and the better adapted the subset is chosen, the better the approximation of E_0 by the right hand side of (3.85). If the subset is a finite dimensional linear subspace, the task boils down to finding the smalles eigenvalue of a matrix. This is a standard problem of numerical linear algebra.

It is also possible to employ a set of trial functions which depend on a certain number of parameters in a non-linear way. Finding the best approximation to the ground state energy then becomes a problem of minimizing a non-linear functional, namely the energy expectation value. We shall try this now.

3.3.3 Sample Calculation

The first electron will be bound by the Helium nucleus in the hydrogen ground state wave function $\exp(-\kappa' r)$. Since the nuclear charge is $2e$, the parameter κ' would have the value 2 (atomic units). This tightly bound electron effectively screens off one nuclear charge, so the second electron sees a small charge ball with effective charge e, and it will settle down in a hydrogen atom like ground state $\exp(-\kappa'' r)$ with κ'' approximately 1.

Guided by this heuristic consideration we propose the following trial functions:

$$\psi_{\pm}(\boldsymbol{x}_a, \boldsymbol{x}_b) \propto \left\{ \mathrm{e}^{-\kappa' r_a}\,\mathrm{e}^{-\kappa'' r_b} \pm \mathrm{e}^{-\kappa' r_b}\,\mathrm{e}^{-\kappa'' r_a} \right\}. \tag{3.86}$$

ψ_+ is symmetric upon interchanging particles a and b, ψ_- is antisymmetric. The former describes the spin singlet, the latter the triplet. Both trial functions, symmetric and antisymmetric, depend on the two adjustable parameters κ' and κ''. We must calculate the energy expectation value as a function of the two parameters and minimize the result.

A helium atom might dissociate into a bound state of the helium nucleus and one electron, the remaining moving at infinity. The energy of the former is –2 a.u., the energy of the free electron is positive. Therefore, the dissociation threshold is –2 a.u. If the upper bound of the helium nucleus plus two electrons is below this threshold, the nucleus has bound two electrons; there is a stable helium atom.

Energy of Trial Configuration

One has to calculate the normalization $(\psi_{\pm}, \psi_{\pm})$ of the two test functions (3.86) and the five contributions to (3.79). Although tiresome, the integrals are straightforward to calculate. Only the Coulomb repulsion term is somewhat tricky. We encounter

$$\frac{1}{|\boldsymbol{x}_b - \boldsymbol{x}_a|} = \frac{1}{\sqrt{r_b^2 + r_a^2 - 2z r_b r_a}} = -\frac{1}{r_b r_a}\frac{\mathrm{d}}{\mathrm{d}z}\sqrt{r_b^2 + r_a^2 - 2z r_b r_a}, \tag{3.87}$$

which can be integrated for $r_b < r_a$ and $r_b > r_a$.

Altogether one obtains the expectation value

$$E_+(\kappa', \kappa'') = \frac{A\gamma + B\gamma^2}{2 - 3\delta + 3\delta^2 - \delta^3}, \tag{3.88}$$

where

$$\begin{aligned} \gamma &= \frac{\kappa' + \kappa''}{2} \\ \delta &= \left(\frac{\kappa' - \kappa''}{\kappa' + \kappa''}\right)^2 \\ A &= -\frac{27}{4} + \frac{75}{8}\delta - 10\delta^2 + \frac{27}{8}\delta^3 \\ B &= 2 - 3\delta + 6\delta^2 - 4\delta^3 + \delta^4 \end{aligned} \tag{3.89}$$

for the singlet state. A similar expression results for the triplet state.

The crudest approximation is to simply insert hydrogen atom wave functions. One finds for helium in the singlet spin state

$$E_0 \approx E_+(1, 1) = -2.3750. \tag{3.90}$$

This is an upper bound below the ionization threshold. We conclude that the helium atom exists.

A better approximation would be to use a hydrogen like ground state with effective charge κ for both electrons. We find

$$E_0 \approx \min_{\kappa} E_+(\kappa, \kappa) = -2.8477. \tag{3.91}$$

The effective charge of the best configuration is $\kappa = 1.69$.

This is only slightly better than an anti-symmetric product of one electron seeing the nuclear charge $\kappa = 2$, the other a hydrogen atom:

$$E_0 \approx E_+(2, 1) = -2.8409. \tag{3.92}$$

The best approximation within the class (3.86) is

$$E_0 \approx \min_{\kappa', \kappa''} E_+(\kappa', \kappa'') = -2.8757. \tag{3.93}$$

The optimal effective charges are $\kappa = 2.18$ and $\kappa = 1.19$.

The measured value is $E_0 = -78.975\,\text{eV} = -2.9024$ a.u. With (3.93) we are rather close.

The counterpart to (3.93) for the spin singlet is $E_0 \approx -2.160$ which must be compared with the experimental value $E_0 = -2.175$. Recall that the ground state energy of the He$^+$ ion is –2.000 E_*. Hence, also the spin-singlet forms a stable ion.

It is stable because ordinary radiative transitions to the triplet state are forbidden by symmetry rules. Only magnetic dipole transitions or collisions between particles may achieve such a singlet decay.

3.3.4 The Negative Hydrogen Ion

Can a proton bind not only one, but two electrons? Is there a bound state of the proton-electron-electron system? The energy of such a state should be below –1/2, the ground state energy of a hydrogen atom.

Well, the tools for answering this questions are here. One simply has to replace in (3.79) the nuclear charge 2e of helium by e for the proton. There is an expression for the energy expectation value rather similar to (3.88), one for the spin singlet and another for the triplet. Minimizing the triplet energy gives an upper bound of $E_0 \leq -0.513$ while the experimental value is $E_0 = -0.528$. There seems to be no singlet bound state.

H^- ions, or hydrogen anions, have been detected first in the atmosphere of the sun.

3.3.5 Summary

A theoretical treatment of the helium atom poses two challenges. First, how to describe a pair of identical particles? And second, how to solve eigenvalue equations which cannot be reduced to ordinary differential equations?

Particles of the same species, like electrons, are not only difficult to keep apart, they cannot be identified. One particle is described by a spin-location argument, the other particle by another spin-location argument. Interchanging such arguments must leave the wave function unchanged, possibly up to a phase factor. Nature realizes this by a plus sign for particles with integer spin and a minus sign for half-integer spin, like electrons. The Pauli exclusion principle is a consequence: two equally polarized electrons cannot be at the same location, they seem to avoid each other.

The wave function for the electron cloud of a helium atom must change sign if spin and location of the two electrons are interchanged. We construct wave functions of total spin = 1 (three states, triplet) and spin = 0 (one state, singlet). As a consequence, the spatial part of the electron configuration must be antisymmetric for the spin triplet and symmetric for the singlet.

The Hamiltonian of the helium atom is the sum of two hydrogen like atoms with charge two instead of one and an electron repulsion contribution. Without the latter, the problem would decouple into hydrogen atom problems. The repulsion energy, however, prohibits that.

Of the many approximation schemes we pick just one. We exploit the fact that the true solution minimizes the energy expectation value. The hope is that a plausible trial function will give a good approximation to the ground state energy. In particular, we propose a two-parameter family of manageable wave function such that the expectation value may be calculated analytically. Finding the best approximation then is an easy task. The spin triplet is more tightly bound than the singlet.

The results, compared with the experimental values, are surprisingly good in spite of such a simple computation.

The method also allows to establish an upper bound to the ground stated energy of the negative hydrogen ion.

3.4 Hydrogen Molecule

We discuss the hydrogen molecule H_2 as a prototype. There are two identical electrons and two identical protons, and for both the electrons and the protons the Pauli principle has to be observed. However, the mass of an electron is small as compared with the nucleon. Therefore, the electron cloud will adjust almost immediately to the current configuration of nuclei; they move slowly in the potential generated by the electron cloud. This Born–Oppenheimer approximation is the conceptual base of molecular and solid state physics.

We will carefully describe the Hamiltonian and the Born–Oppenheimer approximation. A simple model for the electron cloud is discussed. We also present plausibility arguments for the functional form of the potential. Finally, the motion of the two protons in their potential is analyzed.

3.4.1 *Wave Functions and Hamiltonian*

As before, we here understand by polarization σ an up or down value, or $\sigma = \pm 1$. The internal spin is characterized by an eigenvalue of S_3 with eigenvalues $\sigma\hbar/2$. $\boldsymbol{S}$ are the three spin operators and $\boldsymbol{S}^2 = s(s+1)\hbar^2$ with $s = 1/2$ for electrons or protons. Both are spin $=$ 1/2 particles and obey the Pauli principle.

Assume two proton counters A and B and two electron counters a and b. They are located at $\boldsymbol{x}_A, \boldsymbol{x}_B, \boldsymbol{x}_a$ and $\boldsymbol{x}_b$, respectively. Counter A detects protons with polarization σ_A, and so on. The amplitude for the event that all four counters are triggered simultaneously is denoted by $\psi(\sigma_A\boldsymbol{x}_A, \dots)$. The probability for this to happen is

$$\mathrm{d}W = \mathrm{d}V_A\,\mathrm{d}V_B\,\mathrm{d}V_a\,\mathrm{d}V_b\,|\,\psi(\sigma_A\boldsymbol{x}_A, \sigma_B\boldsymbol{x}_B, \sigma_a\boldsymbol{x}_a, \sigma_b\boldsymbol{x}_b)\,|^2. \tag{3.94}$$

Here $\mathrm{d}V_A$ is the active volume of counter A, and so on. If there is a hydrogen molecule, the total probability is

$$\int_A\int_B\int_a\int_b |\,\psi(\sigma_A\boldsymbol{x}_A, \sigma_B\boldsymbol{x}_B, \sigma_a\boldsymbol{x}_a, \sigma_b\boldsymbol{x}_b)\,|^2 = 1, \tag{3.95}$$

where the integral symbol is short for

$$\int_A = \sum_{\sigma_A=\downarrow,\uparrow} \int \mathrm{d}^3 x_A, \tag{3.96}$$

say. Recall that A and B label the two protons, a and b the two electrons.

Not all wave functions are allowed, they must respect the Pauli principle. Upon interchanging A with B, the wave function acquires a minus sign, and the same is true for the replacement of a by b. Equally polarized electrons or equally polarized protons cannot be at the same location.

In the sense of (3.95) square normalizable wave functions which obey the Pauli principle form a Hilbert space $\mathcal{H}$. In it we have to look for the ground state of the hydrogen molecule.

In non-relativistic approximation the Hamiltonian is

$$\begin{aligned} H =& T_A + T_B + T_a + T_b \\ &+ \frac{1}{R_{AB}} + \frac{1}{R_{ab}} \\ &- \frac{1}{R_{Aa}} - \frac{1}{R_{Ab}} - \frac{1}{R_{Ba}} - \frac{1}{R_{Bb}}. \end{aligned} \tag{3.97}$$

The kinetic andergy of the protons is

$$T_A = \frac{\boldsymbol{P}_A^2}{2M} \text{ and } T_B = \frac{\boldsymbol{P}_B^2}{2M} \text{ where } M = \frac{m_\mathrm{p}}{m_\mathrm{e}}, \tag{3.98}$$

and a similar expression, with M replaced by 1, for the electrons. $\boldsymbol{P}_A$ and so on are the linear momentum operators of the respective particles.

$$\frac{1}{R_{AB}} = \frac{1}{|\,\boldsymbol{X}_A - \boldsymbol{X}_B\,|} \tag{3.99}$$

describes the Coulomb repulsion between the protons. A similar expression accounts for the Coulomb repulsion between the two electrons. The remaining four negative contributions include the Coulomb attraction between differently charged particles, i.e. between protons and electrons. This Hamiltonian does not depend explicitly on the particle spins. It is symmetric with respect to interchanging the two protons and the two electrons. It is therefore a mapping of the Hilbert space $\mathcal{H}$ into itself. Here we do not bother about the domain of definition here. Test functions respecting the Pauli principle will do.

3.4.2 Born–Oppenheimer Approximation

The idea is to decouple the electron cloud from the nuclei. Instead of trying all functions ψ we only employ products,

$$\psi(\sigma_A \boldsymbol{x}_A, \sigma_B \boldsymbol{x}_B, \sigma_a \boldsymbol{x}_a, \sigma_b \boldsymbol{x}_b) = \psi_n(\sigma_A \boldsymbol{x}_A, \sigma_B \boldsymbol{x}_B) \cdot \psi_e(\sigma_a \boldsymbol{x}_a, \sigma_b \boldsymbol{x}_b). \quad (3.100)$$

Both wave functions ψ_n for the nuclei and ψ_e for the electrons shall respect Pauli's principle.

Denote by

$$W = T_a + T_b + \frac{1}{R_{ab}} - \frac{1}{R_{Aa}} - \frac{1}{R_{Ab}} - \frac{1}{R_{Ba}} - \frac{1}{R_{Bb}} \quad (3.101)$$

the energy of the electron cloud. Only X_A, X_B and electron operators are involved. For electronic wave functions, the nuclei's locations appear as mere constants. The ground state energy W_0 of the electron cloud is given by

$$W_0 = \inf_{\psi_e} \frac{(\psi_e, W\psi_e)}{(\psi_e, \psi_e)}. \quad (3.102)$$

Recall $\psi_e = \psi_e(\sigma_a \boldsymbol{x}_a, \sigma_b \boldsymbol{x}_b)$ with an obvious scalar product. W_0 may depend on X_A and X_B, but for symmetry reasons only the distance $R = | X_A - X_B |$ will remain: $W_0 = W_0(R)$.

The system of nuclei and a common electron cloud in its ground state is described by

$$H_{\mathrm{m}} = \frac{\boldsymbol{P}_A^2}{2M} + \frac{\boldsymbol{P}_B^2}{2M} + V(R), \quad (3.103)$$

with

$$V(R) = \frac{1}{R} + W_0(R). \quad (3.104)$$

We emphasize that this expression for the molecule's energy H_{m} is the result of the approximation (3.100) first formulated by Max Born and Robert Oppenheimer. The electrons, because they are so light, assume their ground state practically instantaneously.

3.4.3 The Molecular Potential

For large proton separation the lowest energy state of the system is that of two hydrogen atoms, $-1/2$ a.u. each. Therefore $W_\infty = W_0(\infty) = -1$ a.u.

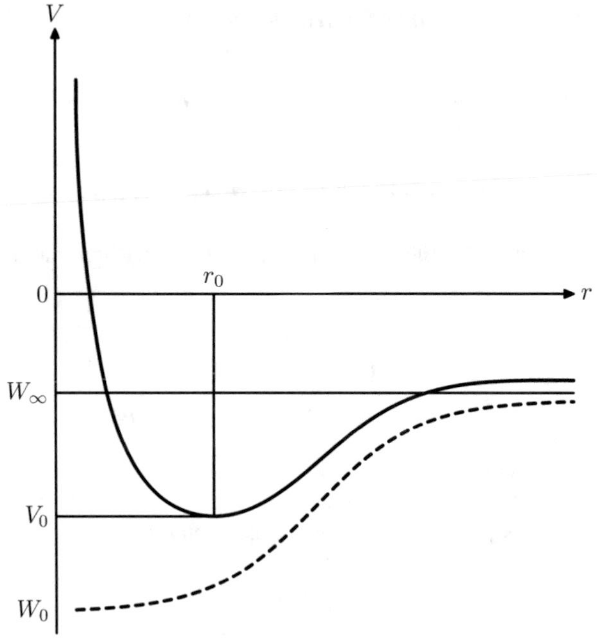

Fig. 3.2 Hydrogen molecular potential. The ground state energy $W_0(r)$ of the electron cloud is plotted versus nuclear distance r (*dashed*, schematic). W_0 and $W_\infty = -1$ are its values at zero distance and for $r \to \infty$. The potential energy $V(r)$ (*full line*) for the interacting two protons has an additional $1/r$ contribution describing their Coulomb repulsion. This potential energy is minimal at $r = r_0$ where it assumes the value V_0

For vanishing proton separation we encounter a point charge +2 and two electrons: the helium situation. The helium ground state energy is $W_0 = W_0(0) = -2.902$ a.u., in the electron spin singlet state.

$V(R)$, the entire potential, is minimal at $R_0 = 1.401$ a u. and has a value of $V_0 = V(R_0) = -1.166$ a.u. there.[8] These minimum data are experimental values.

Figure 3.2 visualizes the potential. For the spin singlet configuration of the electron cloud the interaction energy $W_0 = W_0(r)$ is strong enough so that the total potential $V(r) = 1/r + W_0(r)$ has a minimum at a finite distance $r = r_0$. It can therefore bind the two hydrogen atoms to a molecule.

In the case of the spin triplet configuration the electron binding contribution $W_0(r)$ is much weaker. Consequently, the total potential is never smaller than its value for infinite separation. Binding to a hydrogen molecule is not possible. There is no hydrogen molecule with parallel electron spins.

Clearly, $W_0(r)$ is the electronic ground state energy. If the cloud is in an excited state, there may be no binding. The molecule will dissociate upon electron excitation.

[8]Note that $V(\infty) = -1$ a.u.

3.4.4 *Molecular Vibrations*

We still have not yet completed the description of the hydrogen molecule. In Born–Oppenheimer approximation—the light electrons assume instantaneously their common ground state in the electrostatic field of two protons—we have still to solve for the eigenvalues of (3.103).

Center of Mass and Relative Motion

Let us denote the center of mass momentum by $\boldsymbol{P}_M$ and the relative momentum by $\boldsymbol{P}$. They are defined as

$$\boldsymbol{P}_M = \boldsymbol{P}_A + \boldsymbol{P}_B \text{ and } \boldsymbol{P} = \frac{\boldsymbol{P}_A - \boldsymbol{P}_B}{2}, \tag{3.105}$$

respectively. Both momenta commute with each other. The kinetic energy may be expressed as

$$T_A + T_B = \frac{\boldsymbol{P}_m^2}{4M} + \frac{\boldsymbol{P}^2}{2\mu}. \tag{3.106}$$

The first term is the kinetic energy of the entire molecule with mass $2M$. The second contribution describes the relative motion of the nuclei with respect to the center of mass. $\mu = M/2$ is the reduced mass. Both kinetic energy terms commute.

Note that the center of mass momentum $\boldsymbol{P}_m$ commutes with $\boldsymbol{X}_A - \boldsymbol{X}_B$, therefore also with $R = \mid \boldsymbol{X}_A - \boldsymbol{X}_B \mid$. It follows that the molecule as a whole moves freely, i.e. unaccelerated.

The following considerations refer to the center of mass reference system. With respect to it, the kinetic energy of the molecule as a whole vanishes. Therefore, the Hamiltonian of the system is

$$H_{\mathrm{m}} = \frac{\boldsymbol{P}^2}{2\mu} + V(R). \tag{3.107}$$

Vibrations

As we have explained in Sect. 3.1 on The Radial Schrödinger Equation, the kinetic energy is a sum of two terms, one for the radial motion and one for the angular momentum degree of freedom. In what follows, we only discuss the non-rotating hydrogen molecule, its vibrations. The orbital angular momentum of the molecule shall vanish, $l = 0$. Note that rotations will increase the energy by $l(l+1)\hbar^2/2\mu R^2$.

The relevant radial wave functions are $w = w(r)$ with

$$\int_0^\infty \mathrm{d}r \mid w(r) \mid^2 = 1. \tag{3.108}$$

The Schrödinger equation to be solved is

$$-\frac{1}{2\mu}w''(r) + V(r)\,w(r) = E\,w(r). \tag{3.109}$$

This equation refers to atomic units: masses are multiples of the electron mass.

The radial momentum is[9]

$$P = \frac{1}{\mathrm{i}}\frac{\mathrm{d}}{\mathrm{d}r}. \tag{3.110}$$

It commutes with the distance R between the protons according to

$$[R, P] = \mathrm{i}I, \tag{3.111}$$

with I as the unit operator. Put differently, R and P obey the canonical commutation relation.

Small Vibrations

The ground state of the molecule is characterized by the smallest possible energy. The internuclear distance will therefore be close to R_0 where the potential energy attains its minimum. In the vicinity of the minimum it can be approximated by

$$V = V_0 + \frac{1}{2}\,V_0''\,X^2 + \dots. \tag{3.112}$$

The term linear in the deviation $X = r - R_0$ is missing because the potential is minimal at R_0. With it, the Hamiltonian for molecular vibrations reads

$$H_\mathrm{m} = \frac{1}{2\mu}\,P^2 + \frac{\mu\Omega^2}{2}\,X^2 + V_0. \tag{3.113}$$

We could set $V_0'' = \mu\Omega^2$ because the second derivative is positive at the minimum. Note that the deviation X and the radial momentum P obey the canonical commutation rule.

We have calculated the energy eigenvalues before, in Sect. 2.5 on Small Oscillations. Equation (3.113) describes a harmonic oscillator the eigenvalues of which are

$$E_n = V_0 + \hbar\Omega\left(n + \frac{1}{2}\right), \tag{3.114}$$

for $n = 1, 2 \dots$.

Experimentally, the excitation energy is $\hbar\Omega = 0.54$ eV $= 0.0199E_*$. This is small if compared with $W_\infty - V_0 = 0.166E_*$, but not very small. Equation (3.114) is a moderately good approximation for a few eigenvalues only.

[9]Recall $\hbar = 1$ here.

3.4.5 Molecular Rotations

The kinetic energy of relative motion is

$$T_{\mathrm{m}} = \frac{\boldsymbol{P}^2}{2\mu} = \frac{P^2}{2\mu} + \frac{\boldsymbol{L}^2}{2\mu R^2}. \tag{3.115}$$

μ is the reduced mass of the molecule, $\boldsymbol{P}$ its linear momentum vector and P the linear momentum for radial motion. $\boldsymbol{L}$ stands for the angular momentum and R denotes the variable distance between the two nuclei. So far we have done away with the rotational energy term by assuming that the molecule has zero angular momentum, $l = 0$. We now want to relax this restriction.

The rotational energy should be approximated by replacing R with R_0, the equilibrium distance. Put otherwise, we assume a rigid rotator. Then the last term decouples from the vibrational degree of freedom since $\boldsymbol{L}$ commutes with P^2 and $V(R)$ or X^2. It follows that each vibrational energy level is increased by the eigenvalues of $\boldsymbol{L}^2/2\mu R_0^2$.

The molecule's energy levels are now

$$E_{nl} = V_0 + \hbar\Omega\left(n + \frac{1}{2}\right) + \frac{\hbar^2}{2\mu R_0^2}\, l(l+1). \tag{3.116}$$

$n = 0, 1, \dots$ is the harmonic oscillator quantum number, and $l = 0, 1, \dots$ denotes the angular momentum quantum number. Each level is $2l + 1$ fold degenerate because the eigenvalue of L_3 does not appear in the energy expression.

For the hydrogen molecule we obtain $E_r = \hbar^2/2\mu R_0^2 = 2.774 \times 10^{-4} E_*$. The corresponding wave length is $\lambda_r = 2\pi\hbar c/E_r = 164\,\mu\mathrm{m}$. Thus the transition from an $l = 1$ to an $l = 0$ state occurs at a wave length of $82.1\,\mu\mathrm{m}$, i.e. in the far infrared.

3.4.6 Summary

We have discussed the framework for the hydrogen molecule H_2, the simplest of all. The Hilbert space consists of square-integrable wave function of four groups of arguments: polarization and localization of the two protons and of the two electrons. For describing the molecule, a distinction is made between the electron cloud and the two heavy and practically immobile nuclei. The molecule's wave function is assumed to factorize into electronic and nuclear wave functions of two argument groups each. The Pauli principle must be taken into account.

For fixed positions of the nuclei, a minimum energy configuration describes the electron cloud and gives rise to a potential energy $V(r)$ which is a sum of Coulomb repulsion $1/r$ and interaction energy $W_0(r)$. For the electron singlet state, the potential energy at its minimum $r = R_0$ is below the potential energy at infinity. The latter corresponds to separate hydrogen atoms, or a dissociated molecule. Because of $V(R_0) < V(\infty)$ there exists a bound state for the electronic singlet spin state. This is not true for the triplet state.

The remaining task is to solve a two body problem with a certain potential energy which depends on the particle distance only. The molecule as a whole moves freely as a compound particle. We assume it to be at rest. The relative motion is described by the kinetic energy of a particle with reduced mass in the potential $V(r)$. For small deviations X from the equilibrium distance R_0 we may approximate the potential by a parabola. The resulting harmonic oscillator problem has been studied before.

The above applies to the relative radial motion of the two nuclei in absence of angular motion. With the approximation $R = R_0$, however, we encounter a rigid rotator the energy of which commutes with the rest. Each vibrational level then splits into a series of rotational energy levels $l(l+1)\hbar^2/2I$ with a momentum of inertia $I = \mu R_0^2$. The rotational levels are $(2l+1)$-fold degenerate. Transitions between such energy levels correspond to far infrared radiation while transitions between different vibrational levels are in the near infrared.

3.5 More on Approximations

In the preceding text we often formulated "models" by restricting the discussion to a finite-dimensional subspace. There are other cases where a sum over all vectors of a complete orthonormal base is involved, such as in the expression for the trace. To tackle such a problem often the sum over an infinite set has to be approximated by a finite sum. Another situation shows up when the minimum or maximum is to be calculated. Instead of considering all members of a set, we are content with a sufficiently large subset. There are many such approximation schemes all of which have one feature in common: replacing the Hilbert space by a suitable subspace. Choosing the proper subspace is a matter of physical intuition and is guided by considerations of ease of computation, either analytically or numerically.

Another situation is of interest as well. The Hamiltonian depends analytically on a parameter λ, in very many cases linearly. We then have to study the Hamiltonian $H_\lambda = H_0 + \lambda V$. The first term H_0 is supposed to be manageable, the second is proportional to an operator V regarded as a perturbation. Eigenvalues and eigenstates are expanded into a power serious in λ. For small perturbations the lowest order contribution hopefully provides a good approximation. The weaker the perturbation, the better the approximation.

We here discuss two methods for calculating approximations to the energy eigenvalues. The first one is based on the minimax principle of operator theory, the second features stationary perturbation theory.

3.5.1 The Minimax Theorem

Let us assume that the Hamiltonian $H : \mathcal{H} \rightarrow \mathcal{H}$ of a system is bounded from below and that at least the lower part of its spectrum consists of isolated eigenvalues:

$$H = E_1 \Pi_1 + E_2 \Pi_2 + \dots \tag{3.117}$$

where the Π_j are one-dimensional projectors on normalized and orthogonal wave functions e_j. The obey $(e_j, f_k) = \delta_{jk}$ and $\Pi_j e_k = \delta_{jk} e_k$. For simplicity of notation we assume the eigenvalues to be ordered, $E_1 \leq E_2 \leq \dots$.

Denote by $\mathcal{D} \subset \mathcal{H}$ an r-dimensional linear subspace. Courant's minimax theorem states that

$$E_r = \inf_{\substack{\mathcal{D} \subset \mathcal{H} \\ \dim \mathcal{D} = r}} \sup_{\substack{f \in \mathcal{D} \\ \| f \| = 1}} (f, Hf). \tag{3.118}$$

In words: find the largest expectation value in an r-dimensional subspace and calculate the smallest such result: this coincides with the eigenvalue E_r.

The minimax theorem was formulated by Richard Courant for an n-dimensional Hilbert space. However, since the dimension of $\mathcal{H}$ does not occur in (3.118), the generalization to infinite dimensions is rather plausible. The proof runs as follows.

Consider the linear subspace $\mathcal{F} = \text{span}\,\{e_1, e_2, \dots, e_r\}$ spanned by the lowest r eigenvectors. The expectation value of H in $\mathcal{F}$ obviously is a weighted average of eigenvalues $E_1, E_2, \dots, E_r$, the maximum therefore is attained for $f = e_r$. Since an infimum value over a set is smaller than or equal to a particular term, we conclude

$$\inf_{\substack{\mathcal{D} \subset \mathcal{H} \\ \dim \mathcal{D} = r}} \sup_{\substack{f \in \mathcal{D} \\ \| f \| = 1}} (f, Hf) \leq \sup_{\substack{f \in \mathcal{F} \\ \| f \| = 1}} (f, Hf) = E_r. \tag{3.119}$$

Now pick up an r-dimensional linear space $\mathcal{D}$. Define the following linear subspace $\mathcal{G} = \text{span}\,\{e_r, e_{r+1}, \dots, e_n\}$. For dimensional reasons the intersection of $\mathcal{D}$ with $\mathcal{G}$ cannot be the null space.

We chose a vector $g \in \mathcal{D} \cap \mathcal{G}$ with $\| g \| = 1$. Since $g \in \mathcal{G}$ we may represent it by $g = \alpha_r e_r + \alpha_{r+1} e_{r+1} + \dots + \alpha_n e_n$, and we estimate

$$(g, Hg) = \sum_{i=r}^{n} | \alpha_i |^2 E_i \geq E_r, \tag{3.120}$$

since $\sum | \alpha_i |^2 = 1$ and $E_i \geq E_r$ for $i = r, r+1, \dots, n$. Because of $g \in \mathcal{D}$ we conclude

$$\sup_{\substack{f \in \mathcal{D} \\ \| f \| = 1}} (f, Hf) \geq E_r, \tag{3.121}$$

and this remains true if we take the smallest such upper bound:

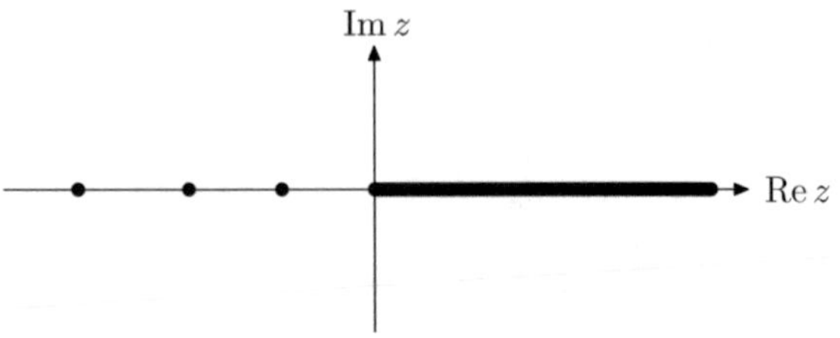

Fig. 3.3 Spectrum of a typical Hamiltonian. The spectrum is sketched in the complex plane. The lower part consists of isolated eigenvalues, the upper part is a continuum usually extending to infinity. Eigenvalues correspond to bound states, continuously distributed quasi-eigenvalues represent the dissociated system where the kinetic energies may assume arbitrary real values

$$\inf_{\substack{\mathcal{D}\subset\mathcal{H}\\ \dim\mathcal{D}=r}} \sup_{\substack{f\in\mathcal{D}\\ \| f \|=1}} (f, Hf) \geq E_r. \tag{3.122}$$

Equation (3.119) together with (3.122) prove the Courant minimax theorem (3.118).

3.5.2 Remarks

Finite Number of Eigenvalues

The Courant minimax theorem applies to self-adjoint operators H which are bounded from below, $H \geq B$. We also have assumed that the low part of the spectrum consist of isolated eigenvalues. Recall that the spectrum of a linear operator A is the set of complex numbers z for which $A - zI$ does not have a bounded inverse. This set $\sigma(A)$ consists of real numbers if $A = A^\star$ is self-adjoint. A typical situation is plotted in Fig. 3.3.

Assume we have three bound states only, but calculate E_4 with the aid of the minimax principle. It turns out that one gets stuck with the ionization threshold. All eigenvalues beyond the largest true value accumulate at the lower end of the continuous spectrum.

This is a sensible result. If we confine the system to a finite region of space the entire spectrum consists of eigenvalues. If the confining regions grows, a few eigenvalues remain isolated, but those above a certain threshold become closer and closer. In this sense the continuous spectrum is the limit of a set of closely spaced eigenvalues.

Rayleigh-Ritz Principle

The ground state energy is given by the infimum over all one-dimensional subspaces. This is so because the expectation value of H with respect to a normalized vector is the same as the corresponding supremum. An arbitrary one-dimensional subspace is spanned by a normalized vector such that we may write

$$E_1 = \inf_{\substack{f \in \mathcal{H} \\ \| f \| = 1}} (f, Hf). \tag{3.123}$$

This is the well-know Rayleigh-Ritz principle which we have come across with earlier in this chapter. In fact, the infimum may be replaced by a minimum because the ground state wave function in fact minimizes the functional.

Larger Operator Has Larger Eigenvalues

Assume that two self-adjoint linear operators $A, B : \mathcal{H} \to \mathcal{H}$ obey the relation

$$(f, Af) \leq (f, Bf) \text{ for all } f \in \mathcal{H}, \tag{3.124}$$

which is expressed as $A \leq B$. The eigenvalues $E(A)_r$ of A and $E(B)_r$ of B then obey

$$E(A)_r \leq E(B)_r, \tag{3.125}$$

provided both sets are labeled in increasing order.

From (3.124) follows

$$(f, Af) \leq \sup_{\substack{f \in \mathcal{D} \\ \| f \| = 1}} (f, Bf), \tag{3.126}$$

for all normalized $f \in \mathcal{D}$, and then

$$\sup_{\substack{f \in \mathcal{D} \\ \| f \| = 1}} (f, A, f) \leq \sup_{\substack{f \in \mathcal{D} \\ \| f \| = 1}} (f, Bf), \tag{3.127}$$

for all subsets $\mathcal{D} \subset \mathcal{H}$. Now take the infimum of the left hand side, over all r-dimensional subspaces $\mathcal{D}$:

$$\inf_{\substack{\mathcal{D} \subset \mathcal{H} \\ \dim \mathcal{D} = r}} \sup_{\substack{f \in \mathcal{D} \\ \| f \| = 1}} (f, A, f) \leq \sup_{\substack{f \in \mathcal{D} \\ \| f \| = 1}} (f, Bf), \tag{3.128}$$

which holds true for any r-dimensional subspace $\mathcal{D}$. The right hand side may be replaced by its minimal value:

$$\inf_{\substack{\mathcal{D} \subset \mathcal{H} \\ \dim \mathcal{D} = r}} \sup_{\substack{f \in \mathcal{D} \\ \| f \| = 1}} (f, A, f) \leq \inf_{\substack{\mathcal{D} \subset \mathcal{H} \\ \dim \mathcal{D} = r}} \sup_{\substack{f \in \mathcal{D} \\ \| f \| = 1}} (f, Bf). \tag{3.129}$$

This is the same as (3.125). We have presented the argument in detail in order to show that the infimum or supremum operations are to be applied in the right order. Note that (3.125) holds true even if A and B do not commute.

There are more consequences of the minimax theorem to be mentioned, and we shall do so at the appropriate place.

3.5.3 Stationary Perturbations

There are many situations where the Hamiltonian is of the form $H = H_0 + \lambda V$, the sum of a manageable term H_0 and a perturbation λV. The energy eigenstates are $f_r = f_r(\lambda)$, the eigenvalues will be denoted by $E_r = E_r(\lambda)$. Perturbation theory is about expanding both the eigenfunctions and the eigenstates into a power series in the perturbation strength λ.

We must solve

$$\big(H_0 + \lambda V\big)\,\big(f_r + \lambda f_r' + \dots\big) = \big(E_r + \lambda E_r' + \dots\big)\,\big(f_r + \lambda f_r' + \dots\big). \tag{3.130}$$

Note that we suppress the argument (λ) in f_r and E_r and that the prime here denotes differentiation with respect to λ.

Since eigenvectors can be scaled and multiplied by phase factors, there is a certain freedom in the choice of f_r. We restrict this freedom by insisting on normalized eigenvectors,

$$(f_s + \lambda f_s' + \dots,\, f_r + \lambda f_r' + \dots) = \delta_{rs}. \tag{3.131}$$

Now only the phases of f_r can be adjusted.

In zeroth order one finds

$$(f_s,\, f_r) = \delta_{sr} \;\text{ and }\; H_0 f_r = E_r f_r, \tag{3.132}$$

as expected.

In first order one has to solve

$$H_0 f_r' + V f_r = E_r f_r' + E_r' f_r. \tag{3.133}$$

The scalar product with f_r yields

$$E_r' = (f_r,\, V f_r). \tag{3.134}$$

The scalar product with f_s for $r \neq s$ gives

$$(f_s,\, f_r') = \frac{(f_s,\, V f_r)}{E_r - E_s} \tag{3.135}$$

or

$$f_r' = \sum_{s \neq r} \frac{(f_s,\, V f_r)}{E_r - E_s}\, f_s. \tag{3.136}$$

We have made use of $f = \sum_s (f_s,\, f) f_s$ and of $(f_r,\, f_r') = 0$.

Now we want to work out higher order effects. One should differentiate $H(\lambda) f_r(\lambda) = E_r(\lambda) f_r(\lambda)$ various times with respect to λ and evaluate at $\lambda = 0$.

The calculation follows the above pattern, but becomes more and more tedious. Here we just show the expression for the energy up to second order:

$$E_r(\lambda) = E_r + \lambda\,(f_r, V f_r) + \lambda^2 \sum_{s \neq r} \frac{|\,(f_s, V f_r)\,|^2}{E_r - E_s} + \dots . \tag{3.137}$$

The first order correction is just the average of the perturbation λV. Also the second order contribution is rather plausible. The closer a different level s or the larger the matrix element $(f_s, V f_r)$, the larger the energy shift. Higher levels decrease, lower levels increase the energy, irrespective of the sign of the perturbation.

3.5.4 Coping with Degeneracies

It might be that two different eigenstates f_s and f_{s+1} have the same energy, i.e. they are degenerate. In this case, we would have to divide by zero when evaluating (3.136) of (3.137). To remedy this fault one has to make sure that the matrix elements $(f_{s+1}, V f_s) = (f_s, V f_{s+1})$ vanish[10] New eigenvectors $\bar{f}_s$ and $\bar{f}_{s+1}$ have to be introduced as suitable linear combinations of the old eigenvectors f_s and f_{s+1}. The off-diagonal matrix elements $(\bar{f}_s, V \bar{f}_{s+1}) = (\bar{f}_{s+1}, V \bar{f}_s) = 0$ now vanish. A perturbation theoretical treatment with $\bar{f}_r$ instead of f_r will exclude terms with zero denominator.

We shall illustrate the role of degeneracy when studying the energy shifts caused by an external electro-static field. The well-known $2s$-$2p$ degeneracy of the non-relativistic hydrogen atom must be taken into account properly.

3.5.5 Summary

In this section we have presented two different procedures for approximating bound state energies. The first one is based on Courant's minimax principle. The eigenvalues are characterized by an infimum over finite dimensional subspaces. The approximation consists of regarding only some instead of all such subspaces. The Rayleigh-Ritz principle is a special case. As a by-product we learn that a less-or-equal relation between two self-adjoint operators extends to the eigenvalues.

The second scheme of approximating energy eigenvalues is perturbation theory. The addition of a small term λV to the Hamiltonian H_0 of the unperturbed system is studied by expanding the effects in a power series of the perturbation. Degenerate eigenvalues may be handled easily.

[10] It can be arranged that both are real.

3.6 Stark and Zeeman Effect

The bound states and consequently the energy levels of atoms or molecules will change if the system is exposed to an external static electric or magnetic field. Such fields interact with the electrons, and the level changes may be observed by spectroscopic methods. We will study the interaction of static electric and magnetic fields here by investigating the simplest case, the hydrogen atom.

We recapitulate the interaction of a charge distribution with a slowly varying field. The rigid charge distribution may be expanded into point-like multipoles, and normally only the lowest non-vanishing term will be seen by the field. These are the electric dipole moment (since the atom is neutral) and the magnetic dipole moment (because magnetic monopoles do not exist).

The external static field is always very small on the atomic scale and may be treated as a perturbation. We here present the most common approximation method, the truncation of the entire Hilbert space to a suitable finite-dimensional space of manageable functions.

In the case of the hydrogen atom and for the lowest energy levels a set energy eigenfunctions is to be recommended because analytical expressions are known which allow to study the relevant matrix elements. In this context we shall encounter the notion of parity.

Two degenerate levels of different parity show the linear Stark effect: an energetic splitting which depends linearly on the electric field strength. Two non-degenerate levels split by an amount which is quadratic in the field strength.

We also discuss the corresponding findings for the Zeeman effect, the dependency of energy levels on an external magnetic induction.

3.6.1 Multipoles

The electromagnetic field is defined by its action on charged particles. If a particle at location $\boldsymbol{x}$, velocity $\dot{\boldsymbol{x}}$, momentum $\boldsymbol{p}$ and charge q moves in an electromagnetic field, it is exposed to a force

$$\dot{\boldsymbol{p}} = q\,\{\boldsymbol{E} + \dot{\boldsymbol{x}} \times \boldsymbol{B}\}\,. \tag{3.138}$$

$\boldsymbol{E} = \boldsymbol{E}(t, \boldsymbol{x})$ is the electric and $\boldsymbol{B} = \boldsymbol{B}(t, \boldsymbol{x})$ the magnetic induction field.

The fields are generated by the electric charge density ρ^{e} and the current density $\boldsymbol{j}^{\mathrm{e}}$. Both are fields, i.e. they depend on time and location. Maxwell's equations describe this:

$$\begin{aligned} \epsilon_0 \nabla \cdot \boldsymbol{E} &= \rho^{\mathrm{e}} \\ -\epsilon_0 \dot{\boldsymbol{E}} + \frac{1}{\mu_0} \nabla \times \boldsymbol{B} &= \boldsymbol{j}^{\mathrm{e}} \\ \nabla \cdot \boldsymbol{B} &= 0 \\ \dot{\boldsymbol{B}} + \nabla \times \boldsymbol{E} &= 0. \end{aligned} \tag{3.139}$$

They are relativistically covariant, just as well as (3.138) and the continuity equation

$$\dot{\rho}^{\mathrm{e}} + \nabla \cdot \boldsymbol{j}^{\mathrm{e}} = 0 \tag{3.140}$$

which may be deduced from the set of Maxwell equations.

Potentials

We speak of an external field because the generating charge- and current distribution is not part of the system. The field is static if it does not change with time. Note that Maxwell's equations for a static field decouple into

$$\epsilon_0 \nabla \cdot \boldsymbol{E} = \rho^{\mathrm{e}} \text{ and } \nabla \times \boldsymbol{E} = 0 \tag{3.141}$$

for the electric field and in

$$\nabla \cdot \boldsymbol{B} = 0 \text{ and } \frac{1}{\mu_0} \nabla \times \boldsymbol{B} = \boldsymbol{j}^{\mathrm{e}} \tag{3.142}$$

for the magnetic induction.

The second equation of (3.141) is automatically fulfilled if the electrostatic field is the gradient of a scalar potential ϕ:

$$\boldsymbol{E} = -\nabla \phi \text{ where } -\epsilon_0 \Delta \phi = \rho^{\mathrm{e}}. \tag{3.143}$$

Likewise, a rotation free induction field may be viewed as the curl of a vector potential $\boldsymbol{A}$:

$$\boldsymbol{B} = \nabla \times \boldsymbol{A} \text{ where } \nabla \cdot \boldsymbol{A} = 0 \text{ and } \frac{1}{\mu_0} \Delta \boldsymbol{A} = \boldsymbol{j}^{\mathrm{e}}. \tag{3.144}$$

We have made use of the fact that the representation of a rotation free field by the rotation of a vector potential is far from unique. Another condition, namely that the vector field may be divergence free, may safely be imposed. $\nabla \cdot \boldsymbol{A} = 0$ is called the Coulomb gauge.

The spherically symmetric Greens function for $-\Delta G(\boldsymbol{x}) = \delta^3(\boldsymbol{x})$ is the Coulomb potential. The general solution of electro- and magnetostatics therefore read

$$\phi(\boldsymbol{x}) = \frac{1}{4\pi\epsilon_0} \int \mathrm{d}^3 y \, \frac{\rho^{\mathrm{e}}(\boldsymbol{y})}{|\, \boldsymbol{y} - \boldsymbol{x} \,|} \tag{3.145}$$

and

$$A(\boldsymbol{x}) = -\frac{\mu_0}{4\pi}\int \mathrm{d}^3 y \, \frac{\boldsymbol{j}^{\mathrm{e}}(\boldsymbol{y})}{|\,\boldsymbol{y}-\boldsymbol{x}\,|}. \tag{3.146}$$

Far Field

Assume the charges are contained in a region $|\,\boldsymbol{y}\,| < R$ and consider the far field $R \ll r = |\,\boldsymbol{x}\,|$. A power series expansion in $1/r$ is appropriate. We find

$$\phi(r\boldsymbol{n}) = \frac{1}{4\pi\epsilon_0}\left\{\frac{Q}{r} + \frac{\boldsymbol{n}\cdot\boldsymbol{d}}{r^2} + \dots\right\} \tag{3.147}$$

for the electric potential. $\boldsymbol{n}$ is a unit vector indicating direction.

$$Q = \int \mathrm{d}^3 y \, \rho^{\mathrm{e}}(\boldsymbol{y}) \tag{3.148}$$

is the total charge. If it does not vanish, it will dominate the far field.

$$\boldsymbol{d} = \int \mathrm{d}^3 y \, \rho^{\mathrm{e}}(\boldsymbol{y})\, \boldsymbol{y} \tag{3.149}$$

denotes the electric dipole moment. It comes into play only if the system's total charge vanishes. Moreover, only then the electric dipole moment is well defined such that its value does not depend on the coordinate system.

Let us repeat this for the magnetic field. There is no $1/r$ contribution because the magnetic monopole

$$\int \mathrm{d}^3 y \, j^{\mathrm{e}}_i = \int \mathrm{d}^3 y \, j^{\mathrm{e}}_k \, \nabla_k y_i = -\int \mathrm{d}^3 y \, y_i \, \nabla_k j^{\mathrm{e}}_k = 0 \tag{3.150}$$

vanishes. Charge is conserved, $\boldsymbol{\nabla}\cdot\boldsymbol{j}^{\mathrm{e}} = 0$. The least rapidly decreasing contribution is

$$A(r\boldsymbol{n}) = \frac{\mu_0}{4\pi}\left\{\frac{\boldsymbol{m}\times\boldsymbol{n}}{r^2} + \dots\right\}, \tag{3.151}$$

it is caused by a magnetic dipole moment

$$\boldsymbol{m} = \frac{1}{2}\int \mathrm{d}^3 y \, \boldsymbol{y}\times\boldsymbol{j}^{\mathrm{e}}. \tag{3.152}$$

Only if the magnetic dipole moment vanishes do we need the magnetic quadrupole moment, and so on.

An external electric field exerts a force $\boldsymbol{F} = \boldsymbol{\nabla}(\boldsymbol{d}\cdot\boldsymbol{E})$ on the electric dipole, and a torque $\boldsymbol{N} = \boldsymbol{d}\times\boldsymbol{E}$. Similar expressions hold true for the action of a magnetic induction on a magnetic dipole. Therefore the interaction energy between an external static electric or magnetic field and a charge-current distribution is

$$V(\boldsymbol{x}) = -\boldsymbol{d} \cdot \boldsymbol{E} - \boldsymbol{m} \cdot \boldsymbol{B}, \tag{3.153}$$

in dipole approximation.

Parity

We have to take selection rules into account. Denote by Π the parity operator which sends $\boldsymbol{X}$ to $-\boldsymbol{X}$ and $\boldsymbol{P}$ to $-\boldsymbol{P}$. Π is self-adjoint, obeys $\Pi^2 = I$, and is therefore unitary as well. It describes a symmetry and is compatible with the canonical commutation relations. Its eigenvalues are real and lie on the unit circle, i.e. they are $\eta = \pm 1$. A pure state f with $\Pi f = +f$ has even parity, a state with $\Pi f = -f$ is odd. The dipole moment obeys $\Pi \boldsymbol{d} \Pi = -\boldsymbol{d}$, it is of odd parity. Because of

$$(f_k, \boldsymbol{d} f_j) = -(\Pi f_k, \boldsymbol{d} \Pi f_j) = -\eta_k \eta_j \, (f_k, \boldsymbol{d} f_j) \tag{3.154}$$

the matrix element vanishes unless the parities η_j and η_k differ.

A spherical harmonic function transforms as $\Pi Y(\theta, \phi) = Y(\pi - \theta, \phi + \pi)$. It follows that the spherical harmonics have even or odd parity if the angular momentum quantum number l is an even or odd integer, respectively:

$$\Pi \, Y_{lm} = (-1)^l \, Y_{lm}. \tag{3.155}$$

3.6.2 Electric Dipole Moment

Consider a neutral system of charged particles. Its dielectric dipole moment is

$$\boldsymbol{d} = \sum_a q_a \boldsymbol{X}_a \tag{3.156}$$

where q_a is the charge of particle a at $\boldsymbol{X}_a$. The dipole moment is well defined if $\sum q_a = 0$. For a hydrogen atom this amounts to

$$\boldsymbol{d} = -e\boldsymbol{X}_{\mathrm{e}} + e\boldsymbol{X}_{\mathrm{p}} = -e\boldsymbol{X} \tag{3.157}$$

where the subscripts e and p refer to the electron and the proton, respectively. $\boldsymbol{X}$ is the relative location.

Recall that the eigenstates $f_{nlm} = f_{nlm}(r)$ of a hydrogen atom are labeled by a principal quantum number $n = 1, 2, \dots$ and an orbital angular momentum quantum number $l = 0, 1, \dots, n-1$, and in addition by a magnetic quantum number $m = -l, -l+1, \dots, l$. We look for matrix elements $(f_s, \boldsymbol{d} f_r)$ where r and s are groups of quantum numbers.

The parity selection rule requires that the angular momentum quantum numbers of f_r and f_s differ by 1, 3, and so on. Otherwise the matrix element vanishes.

We assume that the electrostatic field $\boldsymbol{E} = (0, 0, \mathcal{E})$ points in 3-direction. Therefore only the dipole moment component $d = d_3 = -eX_3$ is of interest. We obtain, with the wave functions as calculated in Sect. 3.2 on the Hydrogen Atom, the following matrix elements for low-lying states:

$$d(2p, 1s) = -(f_{2p0}, X_3\, f_{1s0}) = \sqrt{2}2^7 3^{-5} = 0.74494, \tag{3.158}$$

in atomic units. Likewise

$$d(2p, 2s) = -(f_{2p0}, X_3\, f_{2s0}) = -3. \tag{3.159}$$

The Hamiltonian for an atom or molecule exposed to a static external electric field $\boldsymbol{E}$ is given by

$$H(E) = H_0 - \mathcal{E}\, d_3. \tag{3.160}$$

H_0 is the energy without field and $\boldsymbol{d}$ the dipole moment. The electric field strength is assumed to vary but slowly as compared with the atom's extension. Therefor in (3.154) the value at the center of mass location is understood. Because of rotational symmetry, we may choose the coordinate system such the field points in 3-direction, $\boldsymbol{E} = (0, 0, \mathcal{E})$. Consequently $\boldsymbol{E} \cdot \boldsymbol{d} = \mathcal{E}\, d_3$.

The atomic unit of electric field strength $E_* = 0.5142\ \mathrm{GV\,m^{-1}}$ is many orders of magnitude bigger than what can be realized in a laboratory. Put otherwise, $\mathcal{E}$ is always tiny as compared with its atomic unit. We therefore regard $\mathcal{E}\, d_3$ as a small perturbation.

3.6.3 Stark Effect

Enumerate by $1 = (1s)$, $2 = (2s)$, $3 = (2p)$, $4 = (3s)$, $5 = (3p)$ and $6 = (3d)$ the low-lying hydrogen eigenstates with $m = 0$. We set up the Hamiltonian as restricted to the six-dimensional subspace $\mathcal{D}$ spanned by $f_1, f_2, \ldots, f_6$. The dipole moment is represented by the following matrix:

$$\begin{pmatrix} 0 & 0 & 0.74494 & 0 & 0.29831 & 0 \\ 0 & 0 & -3.00000 & 0 & 1.76947 & 0.00000 \\ 0.74494 & -3.00000 & 0.00000 & 0.54179 & 0 & 2.45185 \\ 0 & 0 & 0.54179 & 0 & -7.34847 & 0 \\ 0.29831 & 1.76947 & 0 & -7.34847 & 0 & -5.19615 \\ 0 & 0 & 2.45185 & 0 & -5.19615 & 0 \end{pmatrix} \tag{3.161}$$

In Fig. 3.4 we have plotted the energy shifts caused by an external field $\mathcal{E}$.

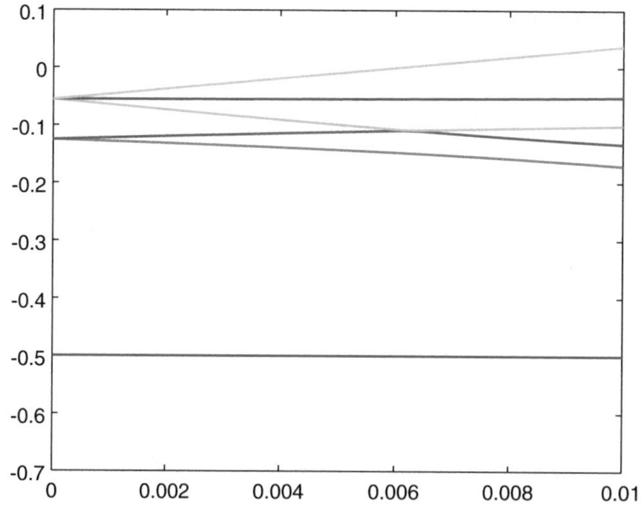

Fig. 3.4 Hydrogen atom Stark effect. The hydrogen energy levels for $n = 1, 2, 3$ are plotted versus the electric field strength $\mathcal{E}$, both in atomic units

The upper levels are not reliable, they would be repelled by $n = 4$ and higher states. Because of degeneracy, the Stark effect is predominantly linear in the strength of an external electrostatic field. Quadratic effects are barely visible.

Polarizability

The hydrogen atom ground state is not degenerate and does not experience the linear Stark effect. Its energy varies as $E_1(E) = E_1(0) + d(\mathcal{E})\mathcal{E}/2$ where $d(\mathcal{E}) = \alpha\mathcal{E}$. The induced dipole moment is proportional to the electric field strength; the polarizability α being the constant of proportionality.

Second order perturbation theory, as discussed in the previous section, gives the following expression:

$$\alpha = 2 \sum_{n=2}^{\infty} \frac{\mid d(np, 1s) \mid^2}{\epsilon_n - \epsilon_1}. \tag{3.162}$$

The first contribution by the $2p$ state gives 2.96 atomic units. This has to be compared with the analytic result $\alpha = 4.5$. $3p$ contributes with 0.40, and so on.

The problematic point is i.e. olated eigenvalues, although there are infinitely many, do not exhaust the spectrum. Instead of a sum over all p-eigenstates, a sum over the entire spectrum is required. Put otherwise, the system of proper eigenvectors is not complete. All contributions to (3.156) are positive, therefore we systematically underestimate the polarizability. Since the non-relativistic hydrogen atom in an electric field can be diagonalized analytically, the exact value $\alpha = 9/2a_*^3$ is known.

3.6.4 Magnetic Dipole Moment

Consider a single particle with charge q. Its state shall be described by a normalized wave function $f = f(\boldsymbol{x})$. We define its current density by

$$\boldsymbol{j}^{\rm e} = \frac{q}{2m}\frac{\hbar}{\rm i}\left\{f^*(\nabla f) - (\nabla f)^* f\right\}. \tag{3.163}$$

The divergence of this current is

$$\begin{aligned}\nabla\cdot\boldsymbol{j}^{\rm e} &= \frac{q}{2m}\frac{\hbar}{\rm i}\{f^*(\Delta f) - (\Delta f)^* f\}\\ &= -q\frac{1}{{\rm i}\hbar}\left\{f^*(Hf) - (Hf)^* f\right\}\\ &= -q\left\{f^*\dot{f} + \dot{f}^* f\right\} = -\dot{\rho}^{\rm e}.\end{aligned} \tag{3.164}$$

$\rho^{\rm e} = q \mid f \mid^2$ is the charge density. The continuity equation $\dot{\rho}^{\rm e} + \nabla\cdot\boldsymbol{j}^{\rm e} = 0$ proves that expression (3.163) in fact describes the electric density of a particle with charge q. Note that the potential energy is absent in (3.164).

The magnetic moment (3.152) turns out to be

$$\boldsymbol{m} = \frac{q}{m}\int {\rm d}^3x\, f^*(\boldsymbol{x})(\boldsymbol{X}\times\boldsymbol{P})f(\boldsymbol{x}) = \frac{q}{m}(f, \boldsymbol{L}f). \tag{3.165}$$

We have made use of $-{\rm i}\nabla$ being self-adjoint. $\boldsymbol{L}$ is the orbital angular momentum operator, $\boldsymbol{m}$ an expectation value.

The the total angular momentum of the electron is a sum $\boldsymbol{J} = \boldsymbol{L} + \boldsymbol{S}$ of orbital and spin angular momentum. Related to it are magnetic momenta

$$\boldsymbol{m}_L = \frac{-e}{m_{\rm e}}\boldsymbol{L} \tag{3.166}$$

and

$$\boldsymbol{m}_S = 2\frac{-e}{m_{\rm e}}\boldsymbol{S}, \tag{3.167}$$

respectively. The factor $g = 2$ in front of the spin magnetic moment is one of the spectacular results of Dirac's relativistic theory. We have discussed it briefly in Sect. 3.2 on the Hydrogen Atom.

Table 3.2 Angular momenta of some hydrogen atom configurations. Orbital momentum $l = 1$ and spin $s = 1/2$ combine to total angular momentum $j = 1/2$ and $j = 3/2$. The table shows their composition of spherical harmonics and spin-up and spin-down states. The expectation value $\langle \hat{L}_3 + 2\hat{S}_3 \rangle = \gamma j_3$ is given in the last row

j_3	$s_{1/2}$	$p_{1/2}$	$p_{3/2}$
$3/2$			$Y_{1,1} \uparrow$
$1/2$	$Y_{0,0} \uparrow$	$(Y_{1,0} \uparrow - Y_{1,1} \downarrow)/\sqrt{2}$	$(Y_{1,0} \uparrow + Y_{1,1} \downarrow)/\sqrt{2}$
$-1/2$	$Y_{0,0} \downarrow$	$(Y_{1,-1} \uparrow - Y_{1,0} \downarrow)/\sqrt{2}$	$(Y_{1,-1} \uparrow + Y_{1,0} \downarrow)/\sqrt{2}$
$-3/2$			$Y_{1,-1} \downarrow$
γ	2	4/3	2/3

3.6.5 Zeeman Effect

The lowest levels of a hydrogen atom are denoted by $1s_{1/2}, 2s_{1/2}, 2p_{1/2}$, and $2p_{3/2}$, in nlj-notation. In non-relativistic approximation all states with the same main quantum number n are degenerate. Relativistically, $2s_{1/2}$ and $2p_{1/2}$ are still degenerate, but the state $2p_{3/2}$ differs in energy by the fine structure correction $E(2p_{3/2}) - E(2p_{1/2}) = 1.7 \times 10^{-6}$.

Suppose a hydrogen atom is exposed to an external induction field $\mathcal{B}$ pointing in 3-direction. Then a certain energy level is shifted by

$$\Delta E = \frac{e\mathcal{B}}{m_e} \langle L_3 + 2S_3 \rangle. \tag{3.168}$$

$\langle \dots \rangle$ is the expectation value in the respective state.

Some angular momentum configurations are summarized in Table 3.2.

$$\mu_B = \frac{e\hbar}{2m_e} = 9.79 \times 10^{-9}\,\text{eV}\,\text{T}^{-1} \tag{3.169}$$

is Bohr's magneton. A magnetic induction field much higher than one tesla (T) is difficult to produce in the laboratory. So the Zeeman effect is usually rather small. Here are the values for the hydrogen atom $n = 2$ states:

- The two $2s_{1/2}$ levels are shifted by $\pm\mu_B\mathcal{B}$
- The two $2p_{1/2}$ levels are shifted by $\pm(2/3)\mu_B\mathcal{B}$.
- The four $2p_{3/2}$ energy levels split into four equidistant levels with a shift of $(2/3)\mu_B\mathcal{B}$ between two adjacent levels.

3.6.6 Summary

We have recapitulated how to characterize fixed charge- and current distributions by multipoles. They are expectation values, i.e. depend on the state of the system. There are no matrix elements between states of equal parity, a concept we have discussed in passing.

Next, we focused on level shifts brought about by an external static electric field. Generally this energy shift depends quadratically on the field strength, and we speak of the quadratic Stark effect. However, if there are degenerate levels, there is a linear Stark effect which facilitates to identify the parity of energy levels. Examples for the hydrogen atom have been worked out.

When discussing the electric polarizability we encounter the problem that the set of bound states does not span the entire Hilbert space. The contributions of scattering states (continuous spectrum) are required as well.

The magnetic moment of a system has two contributions, one proportional to orbital angular momentum, the other to spin. The proportionality constant for the latter is twice the classical value. This follows from Dirac's equation for a relativistically correct description. We have discussed in some detail the degenerate $2s_{1/2}$ and $2p_{1/2}$ pair as well as the $2p_{3/2}$ state. Both groups are degenerate if the hydrogen atom is dealt with in non-relativistic approximation. The magnetic moment of a system is responsible for energy shifts which are linear in the magnetic induction strength B. This is known as the Zeeman effect; it helps in interpreting spectral lines.

Chapter 4
Decay and Scattering

So far we have mostly dealt with bound states. They are stationary, therefore eigenstates of a system's Hamiltonian. They are also pure states to be described by one-dimensional projections which in turn are characterized by normalized wave functions.

We now turn to processes instead of stationary situations. The state of a system changes with time.

There may be a time-dependent external force field which prohibits the state to be stationary. The system without the field may attain one of various eigenstates, but the presence of the field causes transitions between them. The field normally is a rapidly oscillating electric or magnetic field which we treat here according to the laws of classical electrodynamics.

A system in an excited state may decay without external stimulation into one of its lower levels thereby emitting a photon. This phenomenon again is caused by a coupling of the material part to an electromagnetic field. The latter, however, will not be a classical field but is of quantum nature. The topic falls into the branch of quantum field theory, in particular quantum electrodynamics. We here will elaborate Einstein's trick to derive the decay probability rate for a spontaneous transition.

Interacting particles may form bound states. However, for higher energies this is not possible, the interaction manifests itself by one of them deflecting the other one. In classical thinking, the trajectories of interacting particles are no more straight trajectories with constant velocities. More appropriate is a formulation in terms of scattering. Initially, at large separation, the particles move as plane waves with sharply defined momenta. Long after a collision, when the particles had been close enough to interact via a potential, they will again move as free particles. However, with a certain probability, the momenta might have changed, the particles have been deflected. There is a certain probability that a particle has been removed from the incoming beam. Such experiments are described by a cross section.

P. Hertel, *Quantum Theory and Statistical Thermodynamics*,
Graduate Texts in Physics, DOI 10.1007/978-3-319-58595-6_4

We shall discuss the scattering amplitude for colliding particles. Its absolute square is the differential cross section for the deflection by a certain angle. The scattering amplitude is the far-field of a wave function which solves an integral form of an appropriate Schrödinger equation. In a natural way this integral equation lends itself to an approximation of which we here study the first order only. It turns out that the scattering amplitude is the Fourier transform of the interaction potential, in first order, or Born approximation.

After these preparations the Coulomb scattering problem can easily be solved giving rise to Rutherford's famous expression. The latter happens to coincide with the exact solution and is identical to the classical result. It can easily be generalized to a fully relativistic description of Coulomb scattering.

We finally treat the scattering of charged point particles on a charge distribution, such as the ground state configuration of a hydrogen atom. Remarks on revealing the structure of composed particles conclude this chapter on decay and scattering.

4.1 Forced Transitions

The energy levels of a system, if coupled to a static external field, will be shifted. If, however, the system is coupled to an oscillating field, transitions between otherwise stationary states might occur. Such transitions may take place from lower to higher levels, and we speak of excitations. They may also occur from a higher to a lower level (de-excitation). Both types are enforced by a rapidly oscillating field in resonance. The condition for this is $E_1 = E_2 \pm \hbar\omega$ where E_1 and E_2 are two energy levels and ω is the angular frequency of the oscillating perturbation.

There are also transitions $E_2 = E_1 + \hbar\omega$ from the upper level E_2 to the lower level without an external field. We call them spontaneous, they are dealt with in the next section.

Here we study forced transitions by an oscillating external electric field in dipole approximation. We restrict ourselves to a model system with only two energy levels. The latter is not really a restriction since close to a resonance all other states barely contribute.

4.1.1 Time Dependent External Field

Assume two energy eigenstates f_1 and f_2 with energies $E_1 = \hbar\omega_1$ and $E_2 = \hbar\omega_2$. The Hamiltonian of the system is characterized by

$$H_0 f_j = \hbar\omega_j f_j \quad \text{for} \quad j = 1, 2. \tag{4.1}$$

This Hamiltonian is to be augmented by a term which describes the interaction with an external electric field

$$H(t) = H_0 - \boldsymbol{d} \cdot \boldsymbol{\mathcal{E}}(t). \tag{4.2}$$

Here $\boldsymbol{\mathcal{E}}(t)$ is the oscillating external electric field at the systems location and $\boldsymbol{d}$ stands for the dipole moment operator. See Sect. 3.6 on the Stark and Zeeman Effect. We assume the field to point in 3-direction and will write $\mathcal{E}$ instead of $\mathcal{E}_3$ and d instead of d_3.

Equation of Motion

Assume an arbitrary state f described by a normalized vector. It may be expanded as

$$f(t) = c_1(t)\, \mathrm{e}^{-\mathrm{i}\omega_1 t} f_1 + c_2(t)\, \mathrm{e}^{-\mathrm{i}\omega_2 t} f_2. \tag{4.3}$$

This family of states has to obey the Schrödinger equation

$$-\frac{\hbar}{\mathrm{i}} \dot{f}(t) = \{H_0 - d\, \mathcal{E}(t)\}\, f(t). \tag{4.4}$$

Note that the time-dependence caused by H_0 is known and has already been incorporated.

By projecting the Schrödinger equation onto f_1 and f_2 we obtain the following coupled system of differential equations:

$$\begin{aligned} -\frac{\hbar}{\mathrm{i}} \dot{c}_1(t) &= \mathcal{E}(t)\, d_{11}\, c_1(t) + \mathcal{E}(t)\, d_{12}\, \mathrm{e}^{\mathrm{i}\omega_0 t}\, c_2(t) \\ -\frac{\hbar}{\mathrm{i}} \dot{c}_2(t) &= \mathcal{E}(t)\, d_{21}\, \mathrm{e}^{-\mathrm{i}\omega_0 t}\, c_1(t) + \mathcal{E}(t)\, d_{22}\, c_2(t). \end{aligned} \tag{4.5}$$

Here d_{jk} stands for the dipole moment matrix element $(f_j, d\, f_k)$. It is a property of the unperturbed system and does not change with time. We have abbreviated the energy difference by $\hbar\omega_0 = \hbar\omega_2 - \hbar\omega_1 = E_2 - E_1$.

Normally, the eigenstates of our system have a definite parity, either positive or negative. Dipole matrix elements which refer to the same parity vanish. We therefore drop the terms with d_{11} and d_{22}. Since d is self-adjoint the off-diagonal elements are related by $d_{12}^* = d_{21}$. By properly choosing the phases of f_1 and f_2, we may make $d_{12} = d_{21} = d$ a real number. Thus (4.5) simplifies to

$$\begin{aligned} -\frac{\hbar}{\mathrm{i}} \dot{c}_1(t) &= d\, \mathcal{E}(t)\, \mathrm{e}^{+\mathrm{i}\omega_0 t}\, c_2(t) \\ -\frac{\hbar}{\mathrm{i}} \dot{c}_2(t) &= d\, \mathcal{E}(t)\, \mathrm{e}^{-\mathrm{i}\omega_0 t}\, c_1(t). \end{aligned} \tag{4.6}$$

Solution by Iteration

Assume that the system initially, at $t = 0$, was in state f_1. We then have to integrate (4.6) with the initial condition $c_1(0) = 1$ and $c_2(0) = 0$.

The differential equations and the initial conditions may be combined. The resulting set of integral equations read

$$\begin{aligned} c_1(t) &= 1 - \frac{\mathrm{i}d}{\hbar} \int_0^t \mathrm{d}s\, \mathcal{E}(s)\, \mathrm{e}^{+\mathrm{i}\omega_0 s}\, c_2(s) \\ c_2(t) &= 0 - \frac{\mathrm{i}d}{\hbar} \int_0^t \mathrm{d}s\, \mathcal{E}(s)\, \mathrm{e}^{-\mathrm{i}\omega_0 s}\, c_1(s). \end{aligned} \tag{4.7}$$

This form of the equations of motion lends itself to a solution by iteration.

The zeroth order is $c_1(t) = 1$ and $c_2(t) = 0$. Inserting this into the right-hand side of (4.7) gives the first order approximation $c_1(t) = 1$ and

$$c_2(t) = -\frac{\mathrm{i}d}{\hbar} \int_0^t \mathrm{d}s\, \mathcal{E}(s)\, \mathrm{e}^{-\mathrm{i}\omega_0 s}\,. \tag{4.8}$$

The next order approximation does not modify $c_2(t)$, but

$$c_1(t) = 1 - \left(\frac{\mathrm{i}d}{\hbar}\right)^2 \int_0^t \mathrm{d}s' \int_0^{s'} \mathrm{d}s''\, \mathcal{E}(s')\, \mathcal{E}(s'')\, \mathrm{e}^{\mathrm{i}\omega_0 (s' - s'')}\,. \tag{4.9}$$

Continuing this way we obviously obtain a power series expansion in the perturbing field $\mathcal{E}$. Whether it converges towards the searched for solution is quite another question.

Anyhow, as long as the transition amplitude $c_2(t)$ is small, the approximation (4.8) is good.

Oscillating Perturbation

Let us now specialize to a harmonically oscillating external field,

$$\mathcal{E}(t) = \mathcal{E} \cos(\omega t). \tag{4.10}$$

We have to calculate

$$\int_0^t \mathrm{d}s\, \mathrm{e}^{\mathrm{i}\omega s}\, \mathrm{e}^{-\mathrm{i}\omega_0 s} = \frac{\mathrm{e}^{\mathrm{i}(\omega - \omega_0)t} - 1}{\mathrm{i}(\omega - \omega_0)}, \tag{4.11}$$

and the same with ω replaced by $-\omega$. One obtains for the linear response (4.8) the following expression:

$$c_2(t) = \frac{d\,\mathcal{E}}{2\hbar}\left\{\frac{1 - \mathrm{e}^{\mathrm{i}(\omega-\omega_0)t}}{\omega-\omega_0} + \frac{1 - \mathrm{e}^{\mathrm{i}(\omega+\omega_0)t}}{\omega+\omega_0}\right\}. \tag{4.12}$$

For $\omega \approx \omega_0$ the first term dominates. Its absolute square reads

$$W_{1\to 2}(t) = |\, c_2(t)\,|^2 \approx \frac{|\, d\,\mathcal{E}\,|^2}{\hbar^2(\omega-\omega_0)^2}\sin^2\frac{\omega-\omega_0}{2}t. \tag{4.13}$$

$|\, c_2(t)\,|^2$ is the probability that the system transits from state f_1 to f_2, after being perturbed for a time span t. This statement is true close to resonance $\omega \approx \omega_0$ and as long as the transition probability is small. Otherwise the approximation $c_1(t) = 1$ would not be justified.

4.1.2 Detailed Balance

Equation (4.13) is a remarkable result. It says that the transition probabilities $W_{1\to 2}(t)$ and $W_{2\to 1}(t)$ are equal, for the same perturbing field $\mathcal{E} = \mathcal{E}(t)$.

Throughout we assume E_2 the upper and E_1 the lower energy level such that ω_0 is positive.

Solving the equations of motion with $c_1(t) \approx 1$ yields to (4.12), the consequence of which is (4.13). Doing the same for $c_2(t) \approx 1$ likewise gives (4.12) with ω_0 replaced by $-\omega_0$—the same expression. It follows that $W_{1\to 2}(t)$ is the same as $W_{2\to 1}(t)$.

The probabilities of excitation $f_1 \to f_2$ and de-excitation $f_2 \to f_1$ by a forced transition are the same. In fact, the reason for this is the relation $(f_2, d\, f_1) = (f_1, d\, f_2)^*$ and the fact that probabilities are absolute squares of amplitudes.

Although (4.13) is the result of an approximation the statement $W_{1\to 2} = W_{2\to 1}$ is correct in general provided time reversal symmetry applies. In particular, the principle of detailed balance is an important prerequisite for Boltzmann's equation which describes the statistical behavior of a classical many-particle system.

4.1.3 Incoherent Radiation

We digress on the intensity of electromagnetic radiation, be it coherent or incoherent. The result is applied to a two level system exposed to such radiation. Our result will be a transition probability rate.

Energy Transport by the Electromagnetic Field

The classical electromagnetic field is defined by its action on charged particles. A trajectory $t \to \boldsymbol{x}_t$ of a slow particle with mass m and charge q deviates from free motion by

$$m\ddot{\boldsymbol{x}}_t = q\,\{\boldsymbol{E}(t,\boldsymbol{x}_t) + \dot{\boldsymbol{x}}_t \times \boldsymbol{B}(t,\boldsymbol{x}_t)\}\,. \tag{4.14}$$

The six components of the electromagnetic field are real-valued functions of one time and three location arguments. This lengthy remark is to clarify that electric and magnetic field strengths are always real.

It is a consequence of Maxwell's equations that charge is conserved, as expressed by

$$\dot{\varrho}^{\mathrm{e}} + \nabla \cdot \boldsymbol{J}^{\mathrm{e}} = 0. \tag{4.15}$$

The charge within a region $\mathcal{V}$ increases only if there is a net influx of charge across the surface $\partial\mathcal{V}$,

$$\int\limits_{\mathcal{V}} \mathrm{d}V\,\varrho^{\mathrm{e}} = \int\limits_{\partial\mathcal{V}} \mathrm{d}\boldsymbol{A}\cdot\boldsymbol{j}^{\mathrm{e}}. \tag{4.16}$$

There is a similar balance equation for energy, namely

$$\dot{u} + \nabla \cdot S = \boldsymbol{j}^{\mathrm{e}} \cdot \boldsymbol{E}. \tag{4.17}$$

The energy density of an electromagnetic field is given by

$$u = \frac{\epsilon_0}{2}\boldsymbol{E}^2 + \frac{1}{2\mu_0}\boldsymbol{B}^2, \tag{4.18}$$

the energy current density, also called Poynting's vector, by

$$\boldsymbol{S} = \frac{1}{\mu_0}\boldsymbol{E}\times\boldsymbol{B}. \tag{4.19}$$

The right hand side of (4.17) describes an increase or decrease of field energy. Whenever a current flows counter to the direction of the electric field, the latter's energy decreases; it is transferred to the system of charges.

Plane Waves

Let us study a plane wave traveling in x-direction which is polarized along the z-axis. This situation is described by

$$\boldsymbol{E} = \begin{pmatrix} 0 \\ 0 \\ \mathcal{E}\,\cos(\omega t - kx) \end{pmatrix}. \tag{4.20}$$

$\mathcal{E}$ is the field amplitude, ω the angular frequency and k the wave number. The wave shall propagate in vacuum, therefore the charge density ϱ^{e} and the current density $\boldsymbol{j}^{\mathrm{e}}$ vanish identically.

$\epsilon_0 \nabla \cdot \boldsymbol{E} = 0$ is already satisfied. Maxwell's equation $\dot{\boldsymbol{B}} = \nabla \times \boldsymbol{E}$ gives

$$\boldsymbol{B} = \begin{pmatrix} 0 \\ -\mathcal{B} \cos(\omega t - kx) \\ 0 \end{pmatrix}, \tag{4.21}$$

with $\mathcal{B} = k\mathcal{E}/\omega$. Again, $\nabla \cdot \boldsymbol{B} = 0$ is automatically fulfilled.

The remaining Maxwell equation $\mu_0 \epsilon_0 \dot{\boldsymbol{E}} = -\nabla \times \boldsymbol{B}$ gives $\omega = kc$ where $c = 1/\sqrt{\epsilon_0 \mu_0}$, the speed of light in vacuum. Equations (4.20) and (4.21) describe in fact a solution of Maxwell's equations.

This plane wave transports field energy in 3-direction:

$$\boldsymbol{S} = \frac{\mathcal{E}^2}{Z_0} \begin{pmatrix} \cos^2(\omega t - kx) \\ 0 \\ 0 \end{pmatrix} \quad \text{where } Z_0 = \sqrt{\frac{\mu_0}{\epsilon_0}}. \tag{4.22}$$

$Z_0 = 376.73$ Ohm is the vacuum impedance. As expected, the wave and energy propagation directions coincide.

Spectral Intensity

A constant beam of light is a mixture of plane waves with no correlation between different frequencies. It follows that the electric fields of such plane waves will not add, in the average, but their intensities do so. There will be a more thorough discussion in Chap. 6 on Fluctuations and Dissipation. However, we touched the subject already when explaining Feynman's heuristic rules for probability amplitudes.

The beam's intensity therefore may be written as

$$S = \int_0^\infty \mathrm{d}\omega \, I(\omega), \tag{4.23}$$

as an integral over the spectral intensity. Likewise, (4.13) should be integrated over all angular frequencies. Because $\sin^2 x/x^2$ is peaked at $x \approx 0$, we obtain

$$\int_0^\infty \mathrm{d}\omega \, \frac{\sin^2(\omega - \omega_0)t/2}{(\omega - \omega_0)^2} \approx \frac{\pi}{2} \, t \, I(\omega_0). \tag{4.24}$$

If summed over frequency, the transition probability $W_{1\to 2}$ grows linear with the time t of illumination, at least initially. Hence the probability rate for a transition between two stationary states is

$$\Gamma_{1\to 2} = \left.\frac{\mathrm{d}W_{1\to 2}}{\mathrm{d}t}\right|_{t=0}. \tag{4.25}$$

Before presenting our final result, we should mend a deficiency. So far the calculation was made in a coordinate system where the atom or molecule was oriented parallel to the field direction. However, the target particle will be oriented at random. This requires that we replace

$$|\,(f_2, d_3\, f_1)\,|^2 \text{ by } \frac{1}{3} \sum_{j=1,2,3} |\,(f_2, d_j\, f_1)\,|^2. \tag{4.26}$$

This then gives

$$\Gamma_{1\to 2} = \frac{\pi Z_0}{3\hbar^2}\, I(\omega_0) \sum_{j=1,2,3} |\,(f_2, d_j f_1)\,|^2. \tag{4.27}$$

f_1 and f_2 are two energetically different stationary states, the energy difference being $\hbar\omega_0$. $I(\omega_0)$ is the spectral intensity of the illumination with this angular frequency. It is multiplied by a constant and describes the transition rate from state 1 to state 2 or vice versa.

An Example

Assume a gas of hydrogen atoms, the most abundant chemical substance in the universe. The atom shall be in its ground state $f_1 = 1s$. If this atom is illuminated with ultraviolet light of wave length $\lambda_0 = 0.121\,\mu\text{m}$, transitions to $f_2 = 2p$ are energetically possible. We calculate

$$(2p0, d_1 1s) = (2p0, d_2 1s) = 0 \text{ and } (2p0, d_3 1s) = 2^{15/2}\, 3^{-5} \tag{4.28}$$

in atomic units ea_*. Therefore

$$\Gamma_{1\to 2} = \frac{4.2 \times 10^{12}}{\text{s}} \frac{I(\omega_0)}{\text{J m}^{-2}}. \tag{4.29}$$

The states $2pm$ describe different orientations of the atom. However, since we averaged over them, the transition rates do not depend on the magnetic quantum number m of the angular momentum.

To give a concrete number, assume a beam of intensity $\bar{S} = 10\ \text{W cm}^{-2}$ equally uniformly distributed between $1.0 \times 10^{16}\,\text{s}^{-1}$ and $2.0 \times 10^{16}\,\text{s}^{-1}$. The corresponding spectral intensity is $I = 1.0 \times 10^{-11}\,\text{J m}^{-2}$. With these numbers the transition probability rate is $\Gamma_{1s\to 2p} = 42\ \text{s}^{-1}$. By a flash of one millisecond, forty two out of thousand atoms will be excited. For a longer flash, the spontaneous decay to the ground state has to be taken into account.

The latter will be the subject of the following section.

4.1.4 Summary

We established the equations of motion of a quantum system in an external, rapidly oscillating electric field. We focus on two energetically different stationary states. The external field, if static, would affect the energy and the wave function of the levels while an oscillating field, if in resonance, forces transitions from one to another. It turns out that the upward transition probability (from the lower to the higher level) is the same as for the downward transition. This finding has far reaching consequence for the equilibrium state of matter.

We next studied the energy transport by electromagnetic radiation, in particular by plane waves. Incoherent light is characterized by the fact that the phases between different plane waves are random. Thus, when averaging, the superposition of field contributions with different wave vectors cancel.

In our case we arrived at an expression for the transition probability which grows linearly with illumination time, and the proper notion is the transition probability rate. This expression is adapted for the case that the atoms or molecules of the probe are randomly oriented.

A simple example, the transition between the $1s$ and the $2p$ states of atomic hydrogen demonstrates the orders of magnitude for a typical laboratory experiment. It is out of the scope of this book to discuss how a probe of hydrogen atoms is prepared and how the incoherent ultraviolet light beam is set up.

Although we focused on a two-state system, this is not really a loss of generality. In case of resonance—the light energy $\hbar\omega_0$ just fits the energy difference between a pair of states—just these two states determine the transition rate.

4.2 Spontaneous Transitions

An atom or molecule has a number of bound states. They are eigenfunctions of the system's Hamiltonian, the eigenvalues being the discrete energy levels. As such they are stationary; no expectation value depends on time. Therefore, if the system has been prepared to be in an energy eigenstate, it will remain so forever.

If the constituents of the system are charged, there is a coupling to the electromagnetic field. In the previous section we have discussed the action of an external electromagnetic field, one i.e. produced by external means: a thermal radiation source or a laser. Spontaneous transitions do not fit into this framework.

4.2.1 Einstein's Argument

Assume an atom or a molecule with two levels, the second one being higher: $E_2 - E_1 = \hbar\omega_0 > 0$. This particle shall be located in a cavity the walls of which are kept at a constant temperature T. The atom will be in equilibrium with a photon gas the spectral intensity of which is given by Planck's formula,

$$I(\omega) = \frac{\hbar\omega^3}{\pi^2 c^3} \frac{1}{e^{\hbar\omega/k_B T} - 1}. \tag{4.30}$$

We shall derive this spectral intensity distribution for black-body radiation in Sect. 5.5 on Gases.

Being in equilibrium with a photon gas of temperature T, the atom is in an equilibrium state for temperature T as well. This means that its energy levels E_j are occupied with probability

$$w_j \propto e^{-E_j/k_B T}. \tag{4.31}$$

We shall discuss this as well in Chap. 5 on Thermal Equilibrium.

Let us denote the previously calculated probability transition rates by $\Gamma^{\text{ind}}_{1\to 2} = \Gamma^{\text{ind}}_{2\to 1}$, for *induced*. There is also a rate $\Gamma^{\text{sp}}_{2\to 1}$ for a *spontaneous* transition. With this we may write down the following balance equations:

$$\begin{aligned} \dot{w}_1 &= w_2\{\Gamma^{\text{ind}}_{2\to 1} + \Gamma^{\text{sp}}_{2\to 1}\} - w_1 \Gamma^{\text{ind}}_{1\to 2} \\ \dot{w}_2 &= w_1 \Gamma^{\text{ind}}_{1\to 2} - w_2\{\Gamma^{\text{ind}}_{2\to 1} + \Gamma^{\text{sp}}_{2\to 1}\}. \end{aligned} \tag{4.32}$$

Obviously, $\dot{w}_1 + \dot{w}_2 = 0$. Energy is only exchanged. In the equilibrium state, however, there is no change. From $\dot{w}_1 = \dot{w}_2 = 0$ we conclude

$$\Gamma^{\text{sp}}_{2\to 1} = \frac{4}{3} \frac{\omega_0^3}{4\pi\epsilon_0 \hbar c^3} \sum_{j=1,2,3} |\,(f_1, d_j f_2)\,|^2. \tag{4.33}$$

We have inserted (4.30), (4.31) and (4.27). Note that f_2 and f_1 are normalized energy eigenstates and remember that $\boldsymbol{d}$ stands for the electric dipole moment as an observable.

Equation (4.33) describes the probability rate for the spontaneous transition from an energetically higher stationary state f_2 to a lower state f_1. There must be no external field to stimulate the transition, it occurs spontaneously.

The electric dipole contribution should be augmented by other multipoles. However, even the next important term—caused by the magnetic dipole moment—is suppressed by an additional power of the fine structure constant $\alpha \approx 1/137$. Magnetic dipoles, electric quadrupoles and so on play a role only if the electric dipole transition is forbidden by symmetry.

4.2.2 *Lifetime of an Excited State*

The total decay rate of an excited state f_i is given by

$$\frac{1}{\tau_i} = \Gamma_i = \sum_{E_k < E_i} \Gamma^{\text{sp}}_{i \to k}. \tag{4.34}$$

One has to sum over all final states f_k with energy below E_i. By the way, a trivial consequence is that the ground state cannot decay.

If the system was at time $t = 0$ with certainty in a state f_i, it will still be in this excited state with probability

$$w_i = \mathrm{e}^{-\Gamma_i t} = \mathrm{e}^{-t/\tau_i}. \tag{4.35}$$

This explains why τ_i is called the lifetime of the excited state f_i.

The exponential decay law has been thoroughly checked. If the excited state f_i has survived the time t' and if one waits for another time t'', the exponential law (4.35) says $w = w'w''$ where

- w' is the probability to have survived a time span t',
- w'' denotes the probability to survive another time span t'', and
- $w = w'w''$ gives the probability for not having decayed during a time span $t' + t''$.

If a state f_i has not yet decayed up to now, its future is completely independent of this fact. An old excited state is just as good as a freshly generated. Roughly speaking, excited states have no memory; they forget their past—when have they been generated?—immediately.

4.2.3 *Comment on Einstein's Reasoning*

Let us reflect for a minute on Einstein's reasoning. He imagined an atom or a molecule in thermal contact with black body radiation of a certain temperature. Then he set up a balance equation for the population and de-population, for an arbitrary pair of states. In equilibrium, just as many excited states must decay as there are generated. This led to his expression for the spontaneous decay rate Γ_i^{sp}. In this expression, neither the temperature nor the Boltzmann constant show up; the expression is true for arbitrary situations. If A is true and B as well, then there is not contradiction if C holds true as well.

His expression for the diffusion constant follow this pattern.

Also the discovery of the theory of Special relativity. On the one hand, there is Galilei's principle of relativity, namely that there is no absolute definition of rest.

On the other hand, there is the consequence of Maxwell's equations, namely that the speed of light is always the same. Both seemingly contradicting findings are not in conflict if the transformation from one inertial frame to another is special. The most important consequence is that every body has its own proper time.

Even the theory of General relativity follows this pattern. Pseudo forces, such as the Coriolis force, show up if the frame of reference is not inertial. Pseudo forces on a body are always proportional to its mass. Just as gravitational forces. Therefore, gravitational forces are pseudo forces. They may be transformed away by choosing a better suited frame of reference. In fact, the metric tensor of the space-time manifold is the gravitational field.

4.2.4 Classical Limit

The energy loss per unit time of a system in state f_i is given by

$$\sum_{E_k<E_i}(E_i-E_k)\Gamma^{\mathrm{sp}}_{i\to k}=\frac{4}{3}\sum_{E_k<E_i}\frac{\omega_{ki}^4}{4\pi\epsilon_0 c^3}\sum_{j=1,2,3}(f_i,d_j f_k)(f_k,d_j f_i), \tag{4.36}$$

where $E_i-E_k=\hbar\omega_{ki}$. We are going to rewrite this expression.

Each observable A can be associated with a corresponding rate $\dot{A}$,

$$\dot{A}=\frac{\mathrm{i}}{\hbar}[H,A]. \tag{4.37}$$

For energy eigenstates in particular we find

$$(f_k,\dot{A}f_i)=-\frac{\mathrm{i}}{\hbar}(E_i-E_k)(f_k,Af_i). \tag{4.38}$$

With this (4.36) reads

$$\sum_{E_k<E_i}(E_i-E_k)\Gamma^{\mathrm{sp}}_{i\to k}=\frac{4}{3}\sum_{E_k<E_i}\frac{1}{4\pi\epsilon_0 c^3}\sum_{j=1,2,3}(f_i,\ddot{d}_j f_k)(f_k,\ddot{d}_j f_i). \tag{4.39}$$

The sum extends over all energy eigenstates with energy E_k below E_i. It may be shown that for highly excited states the sum over <u>all</u> states is twice as large. One may easily verify this for a harmonic oscillator. With it we obtain

$$P\approx\frac{2}{3}\frac{1}{4\pi\epsilon_0 c^3}(f_i,\ddot{\boldsymbol{d}}^2 f_i) \tag{4.40}$$

as power loss.

Equation (4.40) coincides with the classical expression for the energy loss of a charged particle, namely

$$P = \frac{2}{3}\frac{1}{4\pi\epsilon_0 c^3} q^2 \mid \ddot{x}_i \ddot{x}^i \mid . \tag{4.41}$$

$x_i(s) = \{ct(s), \boldsymbol{x}(s)\}$ denotes the trajectory of a classical particle with s as proper time, and q is its electric charge. For slow particles, time and proper time are the same, and only the spatial part $\ddot{\boldsymbol{d}}^2$ will contribute. Quantum theory says: its expectation value.

Note that the classical result is valid for highly excited states only. Also note that the classical variables, like the position of a charged particle, turn out to be expectation values.

We spoke of power loss. This refers to the energy of matter; the missing energy is transferred to the radiation field. Each decay $f_i \to f_k$ is accompanied by the creation of a photon with energy $\hbar\omega_{ki} = E_i - E_f$. Therefore, the energy of matter plus radiation is still a conserved quantity. We refrain from discussing time reversal invariance here since the arguments brought forward are somewhat weak.

4.2.5 *Summary*

An external electromagnetic field induces transition between otherwise stationary states of a system. But also without a stimulating field, the system may undergo transitions from an energetically higher stationary state f_i to a stationary state f_k with lower energy. Although the problem belongs to the field of quantum electrodynamics, a derivation of the spontaneous transition probability rate $\Gamma^{\text{sp}}_{i\to k}$ is possible, as shown by Albert Einstein. He considered the atom or molecule to be in thermal equilibrium with the black body radiation of a certain temperature. Only with a particular expression for the spontaneous transition rate a consistent description is possible. The final result does not depend on temperature and other details of the argument.

We reflect briefly on the methodological approach Albert Einstein's in this and other of his basic discoveries.

We also discuss the exponential decay law and its significance. An excited state which has not yet decayed does not know of its age.

We finally discuss the power loss, in particular of highly excited states. We are able to verify the classical expression.

4.3 Scattering Amplitude

In Chap. 2 with Simple Examples we have introduced already the concepts of scattering and the scattering amplitude. Feynman's heuristic rules for amplitudes and their combinations were sufficient to explain how, with the aid of neutron scattering experiments, the structure of a molecule may be determined. Here we generalize the discussion. The forces between particles directly determine the scattering cross section. In fact, most of our knowledge on elementary and not so elementary particles is derived from scattering experiments.

4.3.1 Cross Section

A typical scattering experiment looks like follows. A beam of particles, electrons say, all with the same momentum, is guided to hit a thin foil of gold, say. The latter, the target, contains n scattering centers per unit volume, its thickness shall be denoted by $\mathrm{d}x$. Since the foil is supposed to be thin, there are $n\,\mathrm{d}x$ scatterers per unit area. The probability that a particle of the beam passes the foil unhindered can be expressed as

$$w(\mathrm{d}x) = 1 - \mathrm{d}x\, n\, \sigma. \tag{4.42}$$

The proportionality constant σ, an area, is called the cross section. It may be interpreted as the area which the scattering center presents to a particle in the beam. Integrating (4.42) gives

$$w(x) = \mathrm{e}^{-x\, n\, \sigma}. \tag{4.43}$$

x denotes the depth under the foil's surface. w is the probability of the incoming particle to be still in the incident beam.

The particle is no more in the direct beam because it was deflected by some angle θ. With probability

$$\mathrm{d}W(\theta) = \mathrm{d}x\, n\, \mathrm{d}\Omega\, \sigma_{\text{diff}} \tag{4.44}$$

it is missing in the direct beam and scattered into the solid angle $\mathrm{d}\Omega$ around θ. σ_{diff} is the differential cross section. Clearly, a full solid angle integration of the differential cross section gives the entire cross section,

$$\sigma = \int \mathrm{d}\Omega\, \sigma_{\text{diff}}(\theta) = 4\pi\, \frac{1}{2} \int_{-1}^{1} \mathrm{d}\cos\theta\, \sigma_{\text{diff}}(\theta). \tag{4.45}$$

We explicitly assume that the beam or the target is not polarized and the azimuthal angle ϕ does not show up.

There is another view on scattering. We do not focus on a single particle of the beam, but on a particular scatterer. This scatterer at $\boldsymbol{x} = 0$ is illuminated with a beam of particles of direction $\boldsymbol{n}_{\text{in}}$ the current density of which is j_{in} particles per unit area. The suffix stands for <u>in</u>coming or <u>in</u>cident.

Per unit time, $\sigma\, j_{\text{in}}$ particles are scattered in the sense that they are removed from the incident beam. At a distance r in direction $\boldsymbol{n}_{\text{out}}$ we position a counter and measure the current density j_{out} of scattered particles, namely

$$j_{\text{out}} = \frac{\sigma_{\text{diff}}(\theta)}{r^2}\, j_{\text{in}}. \tag{4.46}$$

$\boldsymbol{n}_{\text{in}} \cdot \boldsymbol{n}_{\text{out}} = \cos\theta$ defines the scattering angle θ. If (4.46) is integrated over a sphere of radius r one in fact obtains the flux of scattered particles $\sigma\, j_{\text{in}}$.

The first point of view is closer to experiment, the second is better suited for theoretical considerations, as we shall see now.

4.3.2 *Scattering Amplitude*

We describe the current of incoming particles by a plane wave, a not-normalizable wave function

$$\psi_{\text{in}}(\boldsymbol{x}) = \mathrm{e}^{\,ik\boldsymbol{n}_{\text{in}} \cdot \boldsymbol{x}} \tag{4.47}$$

with momentum $p = \hbar k$. This wave function describes free, non-interacting particles. There is a scatterer at $\boldsymbol{x} = 0$ which interacts with the incoming particles via a potential $V = V(\boldsymbol{x})$. This situation is inadequately described by (4.47), there must be an addition so that $\psi = \psi_{\text{in}} + \psi_{\text{out}}$. This wave function has to obey the Schrödinger equation

$$\frac{\hbar^2}{2m}(\Delta + k^2)\psi(\boldsymbol{x}) = V(\boldsymbol{x})\,\psi(\boldsymbol{x}). \tag{4.48}$$

Note that ψ may be replaced by ψ_{out} on the left-hand side.

Let us define the inverse of the operator on the left hand side of (4.48). We introduce a Green's function G as a spherically symmetric solution of

$$\frac{\hbar^2}{2m}(\Delta + k^2)\, G(\boldsymbol{x}) = \delta^3(\boldsymbol{x}). \tag{4.49}$$

The solution is

$$G(\boldsymbol{x}) = -\frac{2m}{\hbar^2}\,\frac{\mathrm{e}^{\,ikr}}{4\pi r}, \tag{4.50}$$

and we may rewrite the Schrödinger differential equation (4.48) into its integral form

$$\psi(\boldsymbol{x}) = \mathrm{e}^{\mathrm{i}k\boldsymbol{n}_{\text{in}}\cdot\boldsymbol{x}} - \frac{2m}{\hbar^2}\int \mathrm{d}^3y\, \frac{\mathrm{e}^{\mathrm{i}k|\boldsymbol{x}-\boldsymbol{y}|}}{4\pi|\boldsymbol{x}-\boldsymbol{y}|}\, V(\boldsymbol{y})\,\psi(\boldsymbol{y}). \tag{4.51}$$

For large distance r in direction $\boldsymbol{n}_{\text{out}}$ we calculate

$$\psi_{\text{out}} \approx f(\theta)\,\frac{\mathrm{e}^{\mathrm{i}kr}}{r}. \tag{4.52}$$

The scattering amplitude $f = f(\theta)$ may be read off as

$$f(\theta) = -\frac{2m}{\hbar^2}\frac{1}{4\pi}\int \mathrm{d}^3y\; \mathrm{e}^{-\mathrm{i}k\,\boldsymbol{n}_{\text{out}}\cdot\boldsymbol{y}}\; V(\boldsymbol{y})\,\psi(\boldsymbol{y}). \tag{4.53}$$

Note that dimensionally the scattering amplitude is a length. f depends on the momentum $p = \hbar k$ of the particles, via ψ implicitly on the incoming direction $\boldsymbol{n}_{\text{in}}$ and explicitly on the outgoing direction $\boldsymbol{n}_{\text{out}}$. For a spherical potential, however, only the angle between $\boldsymbol{n}_{\text{in}}$ and $\boldsymbol{n}_{\text{out}}$ shows up, namely θ.

What is the significance of the scattering amplitude, why is it called so? Now, comparing (4.46) with (4.52) tells us

$$\sigma_{\text{diff}}(\theta) = |\, f(\theta)\,|^2. \tag{4.54}$$

The differential cross section describes the probability for an incoming particle do be deflected by the scatterer at an angle θ. The probability is an absolute square of a complex amplitude f. The scattering amplitude gives the amount of outgoing spherical wave to be associated with an incoming plane wave.

4.3.3 Center of Mass and Laboratory Frame

Note that so far we assumed the scatterer to be fixed at $\boldsymbol{x} = 0$. This is true if it is very heavy. The incoming and the scattered outgoing particles have the same momentum in this case.

In general, if the masses of the scattering and the scattered particle are comparable, we must state whether we refer to the laboratory frame of reference or to the center of mass inertial system.

In the center of mass reference system the incoming beam particle has momentum $\boldsymbol{p}_1 = p(0,0,1)$ and the incoming target particle $\boldsymbol{p}_2 = p(0,0,-1)$. Their energies are $E_1 = \sqrt{m_1^2c^4 + p^2c^2}$ and $E_2 = \sqrt{m_2^2c^4 + p^2c^2}$, respectively. The outgoing beam particle's momentum is $\boldsymbol{p}_3 = p(\sin\theta\cos\phi, \sin\theta\sin\phi, \cos\theta)$, and the outgoing target particle must have momentum $\boldsymbol{p}_4 = -\boldsymbol{p}_3$. The latter follows from momentum conservation, $\boldsymbol{p}_1 + \boldsymbol{p}_2 = \boldsymbol{p}_3 + \boldsymbol{p}_4 = 0$. For elastic scattering processes—$m_3 = m_1$ and $m_4 = m_2$—the momenta p of incoming and outgoing particles are the same. This is a consequence of energy conservation. The differential cross section in general does not depend on the azimuthal angle ϕ, but on the deflection angle θ only.

Theoretical considerations as described above refer to a two-body system where the center of mass motion of the total mass $M = m_1 + m_2$ has been split off. The remaining relative motion, as described by the distance $\boldsymbol{x}$ is with respect to an inertial frame where the center of mass is at rest. This will make the two particles appear as one particle with relative mass $m = m_1 m_2/(m_1 + m_2)$ located at position $\boldsymbol{x}$. Therefore, (4.54) pertains to the center of mass reference frame. The differential cross section σ_{diff}, the absolute square of the scattering amplitude, is understood with respect to a center of mass description.

A simple Galilei or Lorentz transformation applied to each of the four momentum vectors will make $\boldsymbol{p}_2$ to vanish. Then the incoming target particle, the scatterer, is at rest, and we speak of the Laboratory systems. For slow particles, where velocities simply add as vectors, each of the four particle velocities has to be incremented by the velocity $\boldsymbol{u} = \boldsymbol{p}_2/m_2$. This lets the momentum of the scatterer vanish, $\boldsymbol{p}_2^{\text{lab}} = 0$, while the incoming beam particle has momentum

$$\boldsymbol{p}_1^{\text{lab}} = \frac{m_1 + m_2}{m_2}\boldsymbol{p}. \tag{4.55}$$

Its kinetic energy in non-relativistic approximation, is

$$T_1^{\text{lab}} = \frac{m_1 + m_2}{m_2}\frac{\boldsymbol{p}^2}{2m}. \tag{4.56}$$

Note that

$$T^{\text{cm}} = \frac{\boldsymbol{p}^2}{2m} \tag{4.57}$$

is the total kinetic energy in the center of mass frame.

A simple calculation shows that the deflection angle in the laboratory frame is given by

$$\cos\theta^{\text{lab}} = \frac{m_1 + m_2\cos\theta}{\sqrt{m_1^2 + m_2^2 + 2m_1 m_2\cos\theta}}. \tag{4.58}$$

$\theta = \theta^{\text{cm}}$ is the deflection angle in the center of mass reference system. Again, for a very massive scatterer there is no difference between the deflection angles in the center of mass and laboratory system.

In most cases the deflection angle enters the differential cross section via the momentum transfer $\boldsymbol{\Delta} = \boldsymbol{p}_3 - \boldsymbol{p}_1$. It is invariant under Galilei transformations. Its magnitude is best expressed in terms of center of mass quantities:

$$\Delta = 2p \sin \frac{\theta}{2}. \tag{4.59}$$

4.3.4 Relativistic Description

The following remarks pertain to a situation where there are two incoming particles with masses m_1 and m_2 which leave the reaction zone as particles with masses m_3 and m_4. They are described by an energy-momentum four-vector each:

$$p = (p_0, \boldsymbol{p}) = (E/c, \boldsymbol{p}), \tag{4.60}$$

the zeroth components being the energy up to a factor c. Such four vectors have a pseudo scalar product

$$p \cdot q = p_0 q_0 - p_1 q_1 - p_2 q_2 - p_3 q_3 \tag{4.61}$$

which is invariant if p and q are transformed by a Poincaré transformation. If p is the energy-momentum vector of a particle with (rest) mass m, the expression

$$p \cdot p = m^2 c^2 \tag{4.62}$$

is in fact invariant. It characterizes a property of the particle which does not depend on the state of motion of that particle, namely its mass.

We consider a scattering process $p_1 + p_2 \to p_3 + p_4$ where the four-momenta stand for the respective particles. Energy-momentum conservation requires

$$p_1 + p_2 = p_3 + p_4. \tag{4.63}$$

The process is invariantly described by two variables

$$s = (p_1 + p_2)^2 = (p_3 + p_4)^2 \tag{4.64}$$

and

$$t = (p_3 - p_1)^2 = (p_4 - p_2)^2. \tag{4.65}$$

p^2 here stands for $p \cdot p$.

The variable s is the square of the total energy of the incoming particles in the center of mass frame. Likewise, t is the square of the momentum transfer from particle one to three or from two to four.

There is a third invariant, namely

$$u = (p_3 - p_2)^2 = (p_4 - p_1)^2. \tag{4.66}$$

However, because of

$$s + t + u = (m_1^2 + m_2^2 + m_3^2 + m_4^2)c^2 \tag{4.67}$$

only two of the tree so-called Mandelstam variables s, t, u are independent.

4.3.5 *Summary*

We have discussed the scattering of particles in a beam which hits a target. One viewpoint is that a particular incoming particle is scattered by a target with n scatterers per unit volume. The probability for that particle to be removed from the beam of incoming particles is described by a cross section. If the deflected particles are sorted by their deflection angle we arrive at the notion of a differential cross section.

Another viewpoint is to focus on a particular scatterer which is bombarded by a stream of incoming particles. Again the experiment is to be described by a differential cross section.

We describe the stream of incoming particles with definite momentum by a plane wave. However, because the Hamiltonian includes a potential energy describing the scattering center, the plane wave of the incoming particle must be augmented by an addition which far away from the scattering center behaves as an outgoing spherical wave, the amplitude of which depends on the direction $\boldsymbol{n}_{\text{out}}$.

We rewrite the Schrödinger equation—by incorporating the incoming wave—into an integral form such that the scattering amplitude may be read off. Its absolute square is the differential cross section.

In the remainder of this section we discuss the kinematics of the scattering process, in the center of mass reference frame as well as in the laboratory setup. In the former, the center of mass of the two interacting particles is at rest, in the latter setup the scatterer is at rest initially. A fully relativistic description by Mandelstam variables s and t is also mentioned.

4.4 Coulomb Scattering

We want to calculate the differential cross section for the ever-present Coulomb interaction between two charged particles. Although this case is the only known example of an analytic solution, we explain a widely used approximation scheme due to

Max Born. Firstly, because the exact and the approximate expressions for the differential cross section happen to be the same. Secondly, because Born's approximation scheme can be applied to more complicated situations.

As an example we shall investigate the scattering of electrons by hydrogen atoms. Although the scattering on electrons and on protons is known, what if they form a bound state? We derive the differential cross section and discuss how—with certain reservation—the wave function of a bound state can be reconstructed.

In a final subsection we address the intimate relation between differential cross section and the structure of composite particles in general. As we know by now, most so called elementary particles turn out to be not elementary at all, if probed by fast particles transferring a large amount of momentum.

4.4.1 Scattering Schrödinger Equation

Recall: Two particles with mass m_1 and m_2 interact via a potential $V = V(\boldsymbol{x})$. The center of mass is supposed to be at rest. We may treat this situation as the motion of one particle with relative mass $m = m_1 m_2/(m_1 + m_2)$ in the force field of the potential.

Scattering of the two particles is described by a solution of the Schrödinger equation

$$\psi(\boldsymbol{x}) = \mathrm{e}^{ik\boldsymbol{n}_{\text{in}}\cdot\boldsymbol{x}} - \frac{2m}{\hbar^2}\int \mathrm{d}^3y\, \frac{\mathrm{e}^{ik|\boldsymbol{x}-\boldsymbol{y}|}}{4\pi|\boldsymbol{x}-\boldsymbol{y}|}\, V(\boldsymbol{y})\,\psi(\boldsymbol{y}), \tag{4.68}$$

here in integral form. See the preceding section. Note that ψ appears on the left and on the right hand side. $\boldsymbol{n}_{\text{in}}$ is a unit vector in the direction of the incoming beam, and $\hbar k$ the momentum of either of the two particles. The solution is an incoming plane wave and an outgoing spherical wave in the far-zone, namely

$$f(\theta)\frac{\mathrm{e}^{ikr}}{4\pi r}, \tag{4.69}$$

at $\boldsymbol{x} = r\boldsymbol{n}_{\text{out}}$ for large particle separation r. The scattering amplitude f depends on the angle θ via $\boldsymbol{n}_{\text{in}}\cdot\boldsymbol{n}_{\text{out}} = \cos\theta$. After all, the potential shall be spherically symmetric. The scattering amplitude itself is given by

$$f(p,\theta) = -\frac{2m}{\hbar^2}\frac{1}{4\pi}\int \mathrm{d}^3y\, \mathrm{e}^{-ik\,\boldsymbol{n}_{\text{out}}\cdot\boldsymbol{y}}\, V(\boldsymbol{y})\,\psi(\boldsymbol{y}). \tag{4.70}$$

$p = \hbar k$ is the momentum of either of the two particles.

The differential cross section in the center of mass frame is the absolute square of the scattering amplitude,

$$\sigma_{\text{diff}}(p,\theta) = |\, f(p,\theta)\,|, \tag{4.71}$$

the cross section itself is the integral

$$\sigma(p) = \int \mathrm{d}\Omega \; \sigma_{\text{diff}}(p,\theta) = 2\pi \int_0^{\pi} \mathrm{d}\theta \; \sin\theta \, \sigma_{\text{diff}}(p,\theta). \tag{4.72}$$

4.4.2 Born Approximation

The Schrödinger scattering equation (4.68) is a formal solution of the Schrödinger eigenvalue equation where the auxiliary condition—incoming plane wave—has been incorporated as well as spherical symmetry. Formally, it may be regarded as a fixed point problem. Max Born did just this and obtained a first approximation. The right hand side of (4.68) defines an operator $\psi \to T(\psi)$, and we look for a fixed point $\psi = T(\psi)$.

Fixed point problems may be solved as follows. One begins with a plausible zeroth approximation ψ_0. The next approximation is $\psi_1 = T(\psi_0)$, then comes $\psi_2 = T(\psi_1)$, and so on. The mapping T is contracting if $\| T(\psi'') - T(\psi') \| \le K \| \psi'' - \psi' \|$ holds true for all ψ', ψ'' in a certain domain of definition and with a suitable norm, with $K < 1$. In this case the sequence of approximations ψ_n converges towards a solution of $\psi = T(\psi)$.

If there were no interaction, the incoming plane wave ψ_0 of (4.47) were already a solution. Max Born worked out the improved solution as

$$\psi_1(\boldsymbol{x}) = \mathrm{e}^{ik\boldsymbol{n}_{\text{in}} \cdot \boldsymbol{x}} - \frac{2m}{\hbar^2} \int \mathrm{d}^3y \, \frac{\mathrm{e}^{ik|\,\boldsymbol{x}-\boldsymbol{y}\,|}}{4\pi|\,\boldsymbol{x}-\boldsymbol{y}\,|} \, V(\boldsymbol{y}) \, \mathrm{e}^{ik\boldsymbol{n}_{\text{in}} \cdot \boldsymbol{y}} . \tag{4.73}$$

This gives the following scattering amplitude:

$$f_1(p,\theta) = -\frac{2m}{\hbar^2}\frac{1}{4\pi} \int \mathrm{d}^3y \; \mathrm{e}^{-ik\,\boldsymbol{n}_{\text{out}} \cdot \boldsymbol{y}} \, V(\boldsymbol{y}) \, \mathrm{e}^{ik\boldsymbol{n}_{\text{in}} \cdot \boldsymbol{y}} , \tag{4.74}$$

in first order approximation, i.e. linear in the potential V.

Denote the wave vector transfer by

$$\boldsymbol{\Delta} = k\,(\boldsymbol{n}_{\text{out}} - \boldsymbol{n}_{\text{in}}). \tag{4.75}$$

With this (4.74) turns out to be

$$f_1(p,\theta) = -\frac{2m}{\hbar^2}\,\hat{V}(\boldsymbol{\Delta}). \tag{4.76}$$

In first order Born approximation, the scattering amplitude is proportional to the Fourier transform of the potential. A remarkable result.

Note that a spherically symmetric potential has a spherically symmetric Fourier transform. Therefore, $\hat{V}$ is spherically symmetric as well and depends on $\Delta = 2k \sin\theta/2$ only. This is compatible with the dependency of the scattering amplitude on $p = \hbar k$ and θ.

4.4.3 Scattering of Point Charges

Point charges interact via the Coulomb potential

$$V^{\mathrm{C}}(\boldsymbol{x}) = \frac{1}{4\pi\epsilon_0} \frac{q_1 q_2}{|\,\boldsymbol{x}\,|}. \tag{4.77}$$

Its Fourier transform is not well defined. We here treat this potential as belonging to a photon of arbitrarily small mass $\kappa\hbar/c$:

$$\int \mathrm{d}^3 x \; \mathrm{e}^{-\mathrm{i}\boldsymbol{k}\cdot\boldsymbol{x}} \frac{\mathrm{e}^{-\kappa|\,\boldsymbol{x}\,|}}{|\,\boldsymbol{x}\,|} = \frac{4\pi}{k} \int_0^\infty \mathrm{d}r \; \mathrm{e}^{-\kappa r} \sin kr = \frac{4\pi}{k^2 + \kappa^2}. \tag{4.78}$$

With $\kappa \to 0$ we obtain

$$f^{\mathrm{C}} = -2m \frac{q_1 q_2}{4\pi\epsilon_0} \frac{1}{|\,\boldsymbol{\Delta p}\,|^2}. \tag{4.79}$$

This is the Coulomb scattering amplitude which is inversely proportional to the square of the momentum transfer. We dropped the suffix 1 for the first order Born approximation because (4.79) happens to be exact, up to an irrelevant phase factor.

By squaring (4.79) we obtain Rutherford's famous formula for the differential cross section of electric point charges:

$$\sigma^{\mathrm{C}}_{\mathrm{diff}} = \frac{m^2}{(4\pi\epsilon_0)^2} \frac{q_1^2 q_2^2}{4p^2 \sin^4\theta/2}. \tag{4.80}$$

There is no distinction between attraction and repulsion. Remarkably, Planck's constant of action $\hbar$ does not show up. No wonder that the same result is found in the framework of classical mechanics.

By the way, if in (4.79) we replace $|\,\boldsymbol{\Delta p}\,|^2$ by the Mandelstam variable t, the correct, i.e. fully relativistic expression for Coulomb scattering is obtained.

Note that the cross section for Coulomb scattering is infinite. No wonder, the electrostatic force between to charged particles has an infinite range.

4.4.4 Electron-Hydrogen Atom Scattering

Let us now scatter electrons on a charge distribution, such as a hydrogen atom in its ground state. The normalized wave function for the latter is

$$f_{1s0}(\boldsymbol{x}) = \frac{1}{\sqrt{\pi}} \, \mathrm{e}^{-|\,\boldsymbol{x}\,|}, \tag{4.81}$$

in atomic units. The charge density of a hydrogen atom at $\boldsymbol{x} = 0$ therefore is

$$\varrho^{\mathrm{e}}(\boldsymbol{x}) = \delta^3(\boldsymbol{x}) - |\, f_{1s0}(\boldsymbol{x})\,|^2. \tag{4.82}$$

The corresponding electrostatic potential V is the convolution of the Coulomb potential V^{C} and the charge density ϱ^{e}. Therefor, its Fourier transform is the product of the Fourier transforms of the factors of the convolution:

$$\hat{V}(\boldsymbol{\Delta}) = \hat{V}^{\mathrm{C}}(\boldsymbol{\Delta})\, g(\boldsymbol{\Delta}). \tag{4.83}$$

$g = g(\boldsymbol{\Delta})$, the Fourier transformed charge density, is called a form factor. It is to be calculated as

$$g(\boldsymbol{\Delta}) = 1 - \int \mathrm{d}^3 y\, |\, f_{1s0}(\boldsymbol{y})\,|^2 \, \mathrm{e}^{-\mathrm{i}\boldsymbol{\Delta}\cdot\boldsymbol{y}}. \tag{4.84}$$

In our case—ground state of the hydrogen atom—the result is

$$\begin{aligned} g(\Delta) &= 1 - 2\int_0^\infty \mathrm{d}r\, r^2\, \mathrm{e}^{-2r} \int_{-1}^{+1} \mathrm{d}z\, \mathrm{e}^{-\mathrm{i}\Delta r z} \\ &= 1 - \frac{2}{\Delta}\int_0^\infty \mathrm{d}r\, r\, \mathrm{e}^{-2r}\, \sin \Delta r \\ &= 1 - \left\{ \frac{1}{1 + (a_* \Delta/2)^2} \right\}^2 . \end{aligned} \tag{4.85}$$

In the last expression we have insert the atomic unit a_* of length, or Bohr's radius. As it should be, the form factor is a dimensionless quantity depending on Δ only, the modulus of the wave vector transfer.

The amplitude for scattering electrons on an hydrogen atom in its ground state is

$$f = \frac{2m}{\hbar^2}\frac{e^2}{4\pi\epsilon_0}\frac{g(\Delta)}{\Delta^2} \tag{4.86}$$

with the form factor (4.85).

4.4.5 Form Factor and Structure

The form factor can be measured, at least up to a common phase. The differential cross section is

$$\sigma_{\text{diff}}(p,\theta) = \frac{m^2}{(4\pi\epsilon_0)^2}\frac{1}{\hbar^4\Delta^4} \mid g(\Delta)\mid^2 \tag{4.87}$$

with

$$\Delta = \Delta(p,\theta) = 2\,\frac{p}{\hbar}\sin^2\frac{\theta}{2} \tag{4.88}$$

the wave vector transfer. The dependency on momentum p and scattering angle θ only via the wave number transfer Δ reflects an interaction which is mediated by a potential.

By measuring the differential cross section, the absolute square of the form factor can be determined. Being a Fourier transform of a smooth potential $V(\boldsymbol{x}) = V(-\boldsymbol{x})$, the form factor and its derivative are real and continuous, and can easily be reconstructed. Hence, with some caveats, the Fourier transform of the wave function can be measured in a scattering experiment.

Scattering of point charges, here electrons, on a composite particle allows to unveil the latter's structure. We have discussed this before in Sect. 2.3 on Neutron Scattering on Molecules. Here we shall reword and deepen our previous findings.

Scattering of Neutrons on Molecules

Assume an atom which is illuminated by a monochromatic beam of slow neutrons. At low energies, the neutrons and the electrons of the atom ignore each other: both are neutral. There is, however, a strong interaction between the beam neutrons and the protons and neutrons which form the nucleus. Such nuclear forces have a short range, and we may approximate the neutron-nucleus potential by a delta function:

$$V(\boldsymbol{x}) = A\delta^3(\boldsymbol{x}). \tag{4.89}$$

The scattering amplitude is proportional to its Fourier transform,

$$f^{\text{A}} = -\frac{2m}{\hbar^2}\,\hat{V} = -\frac{2m}{\hbar^2}\,A, \tag{4.90}$$

where m is the neutron-nucleus reduced mass. There is no dependency on the momentum vector transfer. For low neutron energies the differential cross section is isotropic and also independent on momentum.

Let us now replace the target by a diatomic molecule of the same two atoms. One is sitting at $\boldsymbol{x}_1 = -\boldsymbol{a}/2$, the other one at $\boldsymbol{x}_2 = \boldsymbol{a}/2$. The potential, as seen by the incoming neutron, is

$$V(\boldsymbol{x}) = A\left\{\delta^3(\boldsymbol{x}_1) + \delta^3(\boldsymbol{x}_2)\right\}. \tag{4.91}$$

The corresponding scattering amplitude is

$$f^{\mathrm{M}} = -\frac{2m}{\hbar^2}\, 2\cos\frac{\boldsymbol{\Delta}\cdot\boldsymbol{a}}{2}, \tag{4.92}$$

it depends on the wave vector transfer.

Since the target molecules will be randomly oriented, we have to average the differential cross section over all orientations. Another correction concerns the reduced masses in (4.90) and (4.92), they are slightly different. For simplicity we assume a sufficiently massive nucleus where the reduced mass is the same as the neutron mass. With this we obtain

$$\sigma^{\mathrm{M}}_{\mathrm{diff}}(\Delta) = 2\left\{1 + \frac{\sin\Delta a}{\Delta a}\right\}\sigma^{\mathrm{A}}_{\mathrm{diff}}. \tag{4.93}$$

The superscripts M and A stand for molecule and atom, respectively. Δ denotes the wave vector transfer, it depends on the scattering energy and the deflection angle. Only the distance a between the two nuclei matters since we have averaged over orientations. For very small wave vector transfer, $a\Delta \ll 1$, the ratio of differential cross sections is four indicating full interference. At large momentum transfer $a\Delta \gg 1$ the differential cross sections of the two nuclei simply add. They act as independent scatterers, there is no interference at all.

Both examples show that the structure of a composite particle may be reconstructed by measuring the form factor.

Consider, for instance, the scattering of electrons on protons or neutrons. For low momentum transfer, the proton appears to be a point particle with charge $q = 1$, and the neutron, since neutral, is not noticed at all.

With increasing Δp the form factor of the proton decreases from 1 and the form factor of the proton increases from 0. The particles reveal a structure, their charge distribution has an extension. For the hydrogen form factor, for example, we obtain

$$g(\Delta) = 0 + \frac{1}{2}a^2\Delta^2 + \ldots, \tag{4.94}$$

with a as Bohr's radius.

For very high momentum transfer more and more details of the structure can be discerned: protons and neutrons are composed of much tinier constituents which Richard Feynman called partons.

This hypothetical partons have been better and better characterized. All heavy particles (proton, neutron and other instable hadrons) are made up of three quarks which are held together by the exchange of glue particles, or gluons.

We stop here because the branch of particle physics, or high energy physics, has developed into a field of its own. To investigate the structure of such particles, fundamental or composite, requires scattering experiments with high momentum transfer which is possible only for high energies. These experiments must be guided by theory which is necessarily relativistic; it must be a quantum field theory since particles may be produced or they may vanish in scattering process.

In this book we only briefly touch in Sect. 5.4 the topic of Second Quantization, a framework where the number of particles in a system is an observable and not constant.

4.4.6 Summary

A scattering process is described by its amplitude. This scattering amplitude characterizes the far-field region of a solution of Schrödinger's equation. Schrödinger's equation with an incoming plane wave plus an additional contribution is best formulated as an integral equation where the searched-for solution appears on both sides. This situation defines a fixed-point problem which can be solved by iteration. We here discuss one iteration only, known as Born's approximation; it is of first order in the interaction potential. As it turns out, the scattering amplitude is proportional to the Fourier transform of the potential.

We apply the Born approximation method to the scattering of two point charges interacting by a Coulomb potential. The result is Rutherford's famous formula for the scattering of electrons on nuclei. By the way, a rigorous solution for the inverse distance potential is possible; up to an irrelevant phase it coincided with the Born approximation. We also mention that the corresponding problem has an identical solution in classical mechanics. The potential does not contain a length scale. This is the reason why classical physics and quantum physics, with and without Born's approximation, result in the same expression for the differential cross section.

We next turn to the scattering of point charges on a composite target, the hydrogen atom for example. Rutherford's formula has to be modified by a form factor. It corrects the scattering amplitude for unstructured targets by a factor which depends on momentum transfer only. The form factor, which can be measured in a scattering experiment, allows to determine the structure of the composite particle.

We finally reformulate our previous rather heuristic discussion in Sect. 2.3 on Neutron Scattering on Molecules in the terminology introduced above. Again, even if the target molecules are oriented randomly, the structure of this molecule may be unveiled by scattering experiments.

We conclude with a few remarks on the structure of heavy particles like proton and neutron. At low energies, the target particles are point masses carrying charge one or zero, respectively. At high momentum transfer, scattering reveals that they have size and structure. The are made up of three quarks which are kept together by the exchange of glue particles.

Chapter 5
Thermal Equilibrium

So far we have mainly discussed systems with a few particles only. Such systems can be well isolated from their respective environment. They are autonomous in so far that the Hamiltonian contains all dynamical variables which are involved. Although the environment may act on the system, this influence is described by external parameters. There is no feed-back to the environment.

The states of small system is usually the ground state—a pure state—or the result of an excitation, in general a pure state as well. If the system is ideally isolated from its environment, the equation of motion is such that a pure state will remain a pure state.

If the system is large it is practically impossible to isolate it for a reasonable time. Just think about a gas in an enclosing vessel. The walls of it vibrate and will transfer energy. The environment emits radiation, and the system absorbs it. Even if the system does not receive or deliver energy in the mean, the environment influences the system's state. In fact it mixes it more and more.

We speak of equilibrium if the state cannot be mixed any more because it is already mixed maximally. We shall discuss a measure for the degree of mixing, the entropy, and calculate the state with maximal entropy for a given energy content. This equilibrium, or Gibbs state, is characterized by a temperature which is the same for the system and the environment. We therefore speak of thermodynamic equilibrium represented by the Gibbs state.

Phenomenological thermodynamics has a long tradition. It describes with a few rules simple observations on ordinary matter in ordinary conditions. The first and second main law are examples. We set out to embed thermodynamics into the framework of quantum physics. In particular, the entire field of continuum physics is included.

We also present the method of second quantization. With it one may study systems where the number of particles of a certain species is not fixed but may change because the system is open to particle exchange or because of particle creation or absorption in chemical reactions or excitation of phonons and photons.

P. Hertel, *Quantum Theory and Statistical Thermodynamics*,
Graduate Texts in Physics, DOI 10.1007/978-3-319-58595-6_5

In a section on gases we study weakly interacting fermions, bosons and their common classical limit, the Boltzmann gas. The photon gas, or black body radiation, is treated as well. We describe additional degrees of freedom, molecule rotations and vibrations, which modify the expressions for the heat capacity.

Interactions between molecules affect the pressure law; we go one step beyond the ideal gas limit. Interestingly, by such studies one can get information about the intermolecular potential.

We also discuss the vibrations of a crystal lattice as a phonon gas. Likewise, the conduction electrons of a solid may be approximated as an ideal gas of quasi-electrons.

The interaction of an external electric or magnetic field will polarize matter. Besides the partial alignment of permanent electric or magnetic dipoles, we also discuss Heisenberg' model of a ferromagnet with its phase transition from the para-magnetic to the ferromagnetic state.

5.1 Entropy and the Gibbs State

Up to now we have focused on systems with a few particles only. We either looked for stationary states or discussed situations where a simple atom or molecule changed its state by excitation or relaxation, or where two particles interact by scattering. The system's states could always be described by wave functions, i.e. by pure, or unmixed states.

To prepare a pure states means to perform a measurement$M$ with possible outcomes f_j associated with measurement values m_j. Preparing the system in state, say f_1, is achieved by suppressing all other outcomes $f_2, f_3, \dots$

The larger a system, the more states are to be found in a certain energy interval. It soon becomes impossible to suppress all but one outcome of a preparation procedure. Moreover, perturbations caused by the environment can no more be neglect. The autonomous system which is totally isolated from its environment, turns out to be fiction, in particular so if the system is large. It is virtually impossible to prepare a pure state of a large systems. And even if the effort were successful, the state would not remain to be pure for long. The states of large systems are always mixed, more or less so.

We explain in this section why states tend to be mixed more and more and finally settle down as the appropriate equilibrium state which is stationary and stable against further mixing.

In order to be self-contained, we very briefly recapitulate the framework of quantum physics. Observables, states and expectation values are the essentials. The so-called first main law of thermodynamics is a simple consequence, it distinguishes between work and heat.

We than develop a measure of the amount of mixing of a state, the information gain per measurement. This entropy is never negative; it vanishes for pure states. The entropy functional is invariant under unitary transformations and it increases upon further mixing.

We give a precise meaning to the observation that, without doing something about it, structure decays and differences are wiped away. This is the second main law in full generality. We later indicate a proof of a much weaker formulation.

The way from here to an explicit form of the equilibrium, or Gibbs state, is routine. Notions like temperature, chemical potentials and free energy are given a precise meaning.

5.1.1 *Observables and States*

Let us recapitulate the conceptual framework of quantum physics. Although more abstract formulations are possible, we base the discussion on Hilbert spaces.

Hilbert Space

Any physical system is described by a Hilbert space $\mathcal{H}$. This is a linear vector space with a scalar product $(g, f) \in \mathbb{C}$ for all $g, f \in \mathcal{H}$. The Hilbert space is complete in the natural topology induced by the norm $\| f \|^2 = (f, f)$. Cauchy sequences of vectors always converge toward a vector within the Hilbert space. A linear subspace $\mathcal{L} \subset \mathcal{H}$ inherits the scalar product. However, it must not be complete.

A base is a set of mutually orthogonal normalized vectors $g_1, g_2, \dots$ which span the entire Hilbert space. In other words, each vector f can be expanded according to $f = \alpha_1 g_1 + \alpha_2 g_2 + \dots$

Linear Operators

Of particular interest are linear mappings $A : \mathcal{H} \to \mathcal{H}$ of the Hilbert space into itself. Such a linear mapping is also called a linear operator. Linear operators preserve linear combinations, i.e. $A(\alpha f + \beta g) = \alpha Af + \beta Ag$ for $f, g \in \mathcal{H}$ and $\alpha, \beta \in \mathbb{C}$.

Operators themselves can be multiplied by scalars, added, and multiplied. αA is defined by $(\alpha A)f = \alpha Af$ for a complex number α and a vector $f \in \mathcal{H}$. We likewise define $A + B$ by $(A + B)f = Af + Bf$. And finally, AB is declared by $(AB)f = A(Bf)$. The zero-operator 0 causes $0f = 0$ for any vector f, where the 0 on the right hand side is the zero vector defined by $f + 0 = f$. There is a one-operator, or unity I defined by $If = f$ for all vectors f. Note that AB and BA may differ, the algebra of operators is non-commutative.

For each operator A there is an adjoint operator $A^\star$ which is characterized by $(g, AF) = (A^\star g, f)$. Adjoining is similar to complex conjugating: $(\alpha A)^\star = \alpha^* A^\star$, $(A + B)^\star = A^\star + B^\star$, and $(AB)^\star = B^\star A^\star$. Note the order of multiplication.

Operators have a norm as defined by $\| Af \| \le \| A \| \| f \|$ for all $f \in \mathcal{H}$. The norm $\| A \|$ is the smallest such bound. The norm has the desired properties. $\| A \| = 0$ implies $A = 0$, and vice versa. It is multiplicative, $\| \alpha A \| = | \alpha | \| A \|$. The triangle inequality $\| A + B \| \leq \| A \| + \| B \|$ is fulfilled, and its multiplication counterpart as well, namely $\| AB \| \leq \| A \| \| B \|$. And finally, the operator norm is compatible with adjoining: $\| A^\star \| = \| A \|$. Operators with a finite norm are bounded. They behave like the matrices of linear algebra.

The trace TrA of an operator is defined as follows. Chose a base $g_1, g_2, \dots$ and work out $\sum (g_j, Ag_j)$. If this sum over the diagonal elements is absolutely convergent, another base will result in the same value for the trace. However, not all linear operators have a trace. Note that $A \to \text{Tr}A$ is a linear functional and that $\text{Tr}AB = \text{Tr}BA$ holds true, provided both TrA and $\| B \|$ exist, or vice versa.

A projection operator Π, or projector, is characterized by $\Pi = \Pi^\star$ and $\Pi^2 = \Pi$. A projector projects the Hilbert space onto the liner subspace $\Pi\mathcal{H}$, the set of all vectors Πf for $f \in \mathcal{H}$. Indeed, $\Pi f = f$ for $f \in \Pi\mathcal{H}$. We write $\Pi[g]$ for the projector onto a one-dimensional subspace spanned by a unit vector g. It is defined by $\Pi[g]f = (g, f)\, g$ for any vector f.

Normal Operators

Very often one has to do with a decomposition of unity into mutually orthogonal projectors. Two projectors Π' and Π'' are orthogonal if their product vanishes, $\Pi'\Pi'' = 0$, and consequently $\Pi''\Pi' = 0$. Consider a family of mutually orthogonal projectors Π_j with $\Pi_j \Pi_k = \delta_{jk}\Pi_j$. A decomposition of unity is given by $I = \Pi_1 + \Pi_2 + \dots$.

A base $g_1, g_2, \dots$ defines a decomposition of unity into mutually orthogonal one-dimensional projectors. Each $\Pi_j = \Pi[g_j]$ projects on the linear one-dimensional subspace spanned by g_j, as explained above.

We now are ready to introduce a particularly important class of operators.

An operator N is normal if it commutes with its adjoint, $NN^\star = N^\star N$. The following operators are normal:

- Unitary operator U. Defined by $UU^\star = U^\star U = I$ or $U^{-1} = U^\star$.
- Self-adjoint operator M. It is adjoint to itself, $M = M^\star$.
- Positive (non-negative) operator B. It is characterized by $(f, Bf) \geq 0$ for all $f \in \mathcal{H}$ or, equivalently, $B = AA^\star$ with some A. Clearly, B is self-adjoint.
- Probability operator ϱ. A positive operator with unit trace, $\varrho \geq 0$ and $\text{Tr}\varrho = 1$.

For each normal operator N there is a decomposition $I = \Pi_1 + \Pi_2 + \dots$ of unity and a set of complex numbers (eigenvalues) $\nu_1, \nu_2, \dots$ such that

$$N = \nu_1 \Pi_1 + \nu_2 \Pi_2 + \dots \tag{5.1}$$

holds true.

The eigenvalues of unitary operators lie on the unit circle, they are phase factors. The eigenvalues of self-adjoint operators are real numbers. They are even positive (non-negative) if the operator has this property. The eigenvalues of a probability operator are never negative and sum up to one (counting multiplicities).

Essentials

The essentials of quantum physics may be formulated as follows.

- Each observable property of a system, or observable, is associated with a self-adjoint operator M. Its eigenvalues are real, they are the possible outcomes of a measurement.
- A state of the system is described by a probability operator ϱ. Its eigenvalues are probabilities.
- If the system is in state ϱ and one measures an observable M, the value to be expected, or expectation value, is $\varrho(M) = \langle M \rangle_\varrho = \text{Tr}\varrho M$. The functional $M \to \langle M \rangle_\varrho$ is linear and normalized in the sense of $\langle I \rangle_\varrho = 1$.

Unfortunately, the most interesting observables like energy, linear momentum and angular momentum are unbounded; they cannot be defined on the entire Hilbert space. Fortunately, there is always a way how to cope with such inessential difficulties.

5.1.2 *First Main Law*

As usual, we denote by H the energy of the system, an observable. In this chapter we reserve the symbol U for its expectation value,

$$U = \langle H \rangle_\varrho = \text{Tr}\varrho\, H. \tag{5.2}$$

It is also called the internal energy. We silently assume that the system as a whole is at rest. Therefore its kinetic energy vanishes, and a possible potential energy plays no role because it is constant.

The system's Hamiltonian may depend on external parameters $\lambda_1, \lambda_2 \dots$ which we denote collectively by λ. Such parameters, like an external electric field, act on the system. However, there is no feedback; they are controlled from outside. External parameters describe interactions of the system's environment with the particles of the system, but the source of such forces is so big that the system may not change its state. If a mass at the surface of the earth moves under the influence of gravitation, the earth remains unaffected.

All expectation values are linear with respect to the observable and linear with respect to the state. In particular, the internal energy depends linearly on the state ϱ and on the Hamiltonian $H = H(\lambda)$. Therefore, small changes are related by

$$\text{d}U = \text{Tr}H(\lambda)\,\text{d}\varrho + \text{Tr}\varrho\,\text{d}H(\lambda) = \text{d}Q + \text{d}W. \tag{5.3}$$

This finding is known as the first law of thermodynamics: a change of the system's internal energy may be due to heating by

$$\text{d}Q = \text{Tr}H(\lambda)\,\text{d}\varrho \tag{5.4}$$

or by performing work on the system,

$$\mathrm{d}W = \mathrm{Tr}\,\varrho\,\mathrm{d}H(\lambda). \tag{5.5}$$

The latter expression for work can be written as

$$\mathrm{d}W = -\sum_r \Lambda_r \mathrm{d}\lambda_r \quad \text{with} \quad \Lambda_r = -\mathrm{Tr}\,\varrho\,\frac{\partial H(\lambda)}{\partial \lambda_r}. \tag{5.6}$$

The Λ_r are generalized forces. They would be forces if λ were positions.

Let us emphasize that the first main law is valid in general and does not refer to anything thermodynamic. In thermodynamics, however, the heat term in fact describes the increase of internal energy due to heat conduction, internal friction or Joule's heat.

5.1.3 Entropy

So far we have discussed closed systems only. The Hamiltonian—which governs the time evolution of a system—is a combination of its observables and of external parameters. We also say that such a system is totally isolated from its environment.

This is a fiction. It is impossible to isolate a system from its environment forever. There are weak electromagnetic waves, particles of cosmic radiation and so on. Although they might not transfer energy to the system, these perturbations, originating in the environment, may influence the system's time evolution. If a system is isolated against energy and particle transfer, but not totally, we speak of good isolation: $U = \mathrm{Tr}\,\varrho H$ will remain constant, but the systems's energy does not describe the time evolution completely. In other words, Schrödinger's equation

$$\dot{\varrho} = -\frac{\mathrm{i}}{\hbar}[H, \varrho]. \tag{5.7}$$

is not sufficient.

Obviously, this is the more true the larger the system. Large systems can be at best well isolated. Total isolation is possible only for a short time which becomes shorter and shorter the more particles belong to the system. The reason is that, with growing number of particles, the number of different states within a certain energy interval grows tremendously. The energy difference between two of them gets smaller and smaller. Consequently the rate of change increases because such energy differences appear in the denominator, as demonstrated in Sect. 4.1 on Forced Transitions.

We shall also discuss the situation where the system under discussion is part of a much larger system such that energy and particles may be exchanged freely. Such open systems are not isolated at all.

Shannon's Entropy

Consider a well isolated system in a certain state ϱ. In general, the latter is a mixture of pure states as described by one-dimensional projectors:

$$\varrho = \sum_j p_j \Pi_j, \tag{5.8}$$

with $\Pi_j = \Pi[g_j]$, the g_j forming a base. The probabilities p_j are subject to

$$p_j \geq 0 \text{ and } \sum_j p_j = 1. \tag{5.9}$$

Clearly, if one probability equals one, say $p_k = 1$, the remaining values must vanish. In this case the system's state is described by the wave function g_k, and no other different wave functions are possible. g_k is certain, all other states are impossible. Such a probability operator is said to describe a pure state.

In general, the system's state ϱ is a mixture of pure states described by g_j. Mixed states can be further mixed:

$$\varrho = (1 - \alpha)\,\varrho' + \alpha\,\varrho'' \text{ where } 0 \leq \alpha \leq 1. \tag{5.10}$$

ϱ is a probability operator if ϱ' and ϱ'' are. Note that ϱ' and ϱ'' may not commute. In this case (5.10) is more than mixing the probabilities.

We shall develop here a plausible answer to the question: how much mixed is a state? To this end we perform the following thought experiment.

Imagine we have prepared a system in a specific way as reflected by its probability operator $\varrho = \sum p_j \Pi[g_j]$. We want to measure the index of the pure state contained in ϱ, i.e. the observable $J = \sum j \Pi[g_j]$. For this we must repeat the measurement again and again under identical conditions. The results are recorded, $R = \{j_1, j_2, \ldots, j_N\}$. In the n'th repetition out of N the index j_n was measured.

Now, before performing the experiment one knows that the index j will occur $n_j \approx p_j N$ times. Afterwards, one knows the values as recorded in R in addition. There are

$$\Omega = \frac{N!}{n_1!n_2!\ldots} \tag{5.11}$$

such records,[1] and one of them is the result of the experiment. To learn which one, $\mathrm{ld}\,\Omega$ yes-or-no questions have to be answered.

The information gain in bits[2] therefore is $\mathrm{ld}\,\Omega$. Since we have repeated the measurement very often, the n_j are large numbers, so Stirling's formula $\lg x! = x \lg x + \ldots$ is applicable:

[1]Interchanging the positions of the same index will not result in a new record.

[2]That explains why we invoke the logarithmus dualis, i.e. for base 2.

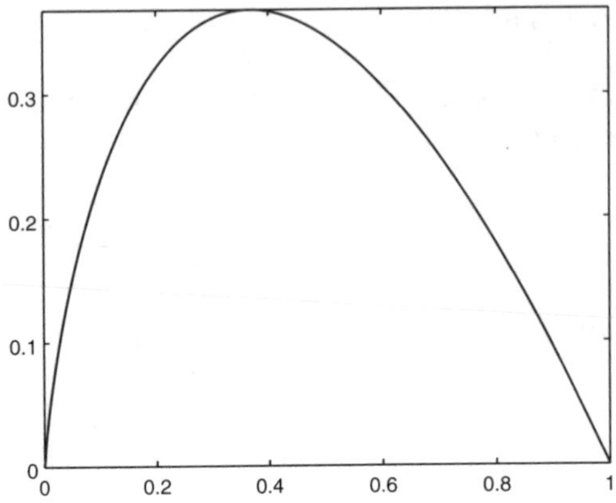

Fig. 5.1 Entropy contributions. The function $x \to -x \ln x$ is non-negative and concave in the interval [0, 1]. It vanishes at both ends

$$\mathrm{ld}\,\Omega = N\,\mathrm{ld}\,N - n_1\,\mathrm{ld}\,n_1 - n_2\,\mathrm{ld}\,n_2 - \cdots = -N \sum_j p_j\,\mathrm{ld}\,p_j. \tag{5.12}$$

As it should be, the information gain grows linear with N, for a large number N of repetitions. Hence,

$$\mathcal{S}_*(\varrho) = -\sum_j p_j\ \mathrm{ld}\,p_j = -\mathrm{Tr}\varrho\ \mathrm{ld}\,\varrho \tag{5.13}$$

is the searched for measure of the degree of mixing: the information gain per measurement.

The entropy functional $\mathcal{S}_* = \mathcal{S}_*(\varrho)$ was derived by Claude Shannon to provide a measure of uncertainty in the context of cryptology. We will use it as a measure for mixing. In fact, physicists prefer the natural logarithm and they measure entropy in units of the Boltzmann constant k_B, for historical reasons. Henceforth, we will call

$$\mathcal{S}(\varrho) = -k_B \mathrm{Tr}\varrho\ \ln\varrho \tag{5.14}$$

the entropy of the state described by ϱ. This entropy functional has the following properties:

- The entropy is never negative.
- Pure states are characterized by vanishing entropy.
- The entropy is invariant under unitary transformations.
- Mixing increases the entropy.

The function $x \to -x\ln(x)$ is positive in $x \in (0, 1)$ while it vanishes for $x = 0$ and for $x = 1$. Since the entropy is a sum of such terms, it cannot be negative. See Fig. 5.1.

If the entropy vanishes, the probabilities p_j must vanish or be equal to one. Since they add up to one, exactly one probability, say p_k, evaluates to one while all other vanish. Therefore, $\varrho = \Pi[g_k]$ is a one-dimensional projector describing a pure state.

Let U be unitary and $\varrho' = U\varrho U^\star$. If ϱ is a probability operator, $\varrho \geq 0$ and $\mathrm{Tr}\varrho = 1$, then ϱ' is a probability operator as well. $\varrho = AA^\star$ translates into $\varrho' = (UAU^\star)(UAU^\star)^\star$, and $\mathrm{Tr}\varrho' = \mathrm{Tr}U\varrho U^\star = \mathrm{Tr}U^\star U\varrho = \mathrm{Tr}\varrho$.

The most interesting property concerns mixing. If ϱ is a mixture of ϱ' and ϱ'' according to (5.10), the entropy of the mixed state is never smaller than the linearly interpolated entropy:

$$S((\varrho) \geq (1-\alpha)S(\varrho') + \alpha S(\varrho''). \tag{5.15}$$

Here $\varrho = (1-\alpha)\varrho' + \alpha\varrho''$ and $0 \leq \alpha \leq 1$. The entropy functional is concave.

5.1.4 Second Main Law

In the preceding chapters we have discussed closed systems. The exchange of internal energy and of particles is prevented. Such a system is ideally isolated if it is also autonomous in the sense that changes in the environment do not affect the system. In the Schrödinger picture the states of an autonomous system change with time according to

$$\varrho_t = \mathrm{e}^{-\frac{\mathrm{i}}{\hbar}tH}\,\varrho\,\mathrm{e}^{+\frac{\mathrm{i}}{\hbar}tH}, \tag{5.16}$$

i.e. by a unitary transformation. It follows that the entropy remains constant,

$$\frac{\mathrm{d}}{\mathrm{d}t}S(\varrho_t) = 0. \tag{5.17}$$

For a well isolated closed system the environment affects the time evolution. Since the system's environment behaves chaotically, at least unpredictable, it will affect the system in such a way that the latter becomes more and more mixed. In other words,

$$\frac{\mathrm{d}}{\mathrm{d}t}S(\varrho_t) \geq 0 \tag{5.18}$$

for a well isolated system.

The justification for the second law of thermodynamics might appear convincing, but it has some flaws. It may be true that a chaotic environment destroys order. But can disorder really be described by one number, namely the entropy of a state? Another criticism is that the second main law is rather encompassing. Although it predicts the approach to equilibrium and the properties of the equilibrium state, it never has been verified far away from equilibrium. We here consider the second law in the form of (5.18) as an axiom some consequences of which are supported by a plenitude of observations. It reflects a realistic view on life: if nothing is done to restore it, all order soon or later gets lost. Obviously, the second law violates a naive concept of time reversal invariance. We shall postpone a discussion of this paradoxon.

5.1.5 The Gibbs State

Assume a well isolated system described by its energy H. The number of particles is fixed, energy may fluctuate about a certain fixed vale $U = \mathrm{Tr}\varrho_t H$. What happens if we just wait?

Now, the environment will still act on the system by mixing its state more and more. Consequently, the state ϱ_t will creep towards equilibrium which is characterized by maximal entropy in accordance with (5.17).

The equilibrium, or Gibbs state G can no more increase its entropy because it is already at the summit:

$$\mathcal{S}(G) = \sup_{\varrho} \mathcal{S}(\varrho) \text{ where } \varrho \geq 0 \text{ , } \mathrm{Tr}\varrho = 1 \text{ , } \mathrm{Tr}\varrho H = U. \tag{5.19}$$

Since the entropy functional is concave, the supreme is attained at a unique probability operator G.

Now, a maximum is characterized by vanishing slope. Let $\mathrm{d}\varrho$ be an arbitrary small difference between two states. It follows that

$$\mathrm{d}\langle\, I \,\rangle = \mathrm{Tr}\,\mathrm{d}\varrho\, I = 0. \tag{5.20}$$

The entropy close to the searched for Gibbs state is

$$\mathrm{d}\mathcal{S} = \mathcal{S}(G + \mathrm{d}\varrho) - \mathcal{S}(G) = -k_{\mathrm{B}} \mathrm{Tr}\,\mathrm{d}\varrho \ln G. \tag{5.21}$$

There is another contribution from differentiating the logarithm,

$$\ln(G + \mathrm{d}\varrho) - \ln G = \int_0^1 \mathrm{d}x\; G^{-x} \mathrm{d}\varrho\, G^{x-1}, \tag{5.22}$$

the expectation value $\mathrm{Tr}G\dots$ of which is $\mathrm{Tr}\mathrm{d}\varrho$, which vanishes according to (5.20). Different trial states should have the same internal energy U which amounts to

$$\mathrm{d}\langle\, H \,\rangle = \mathrm{d}Q = \mathrm{Tr}\,\mathrm{d}\varrho\, H = 0. \tag{5.23}$$

Thus we have to solve $\mathrm{d}\mathcal{S} = 0$ with the auxiliary conditions (5.20) and (5.23). For this we multiply with Lagrange multipliers, (5.21) with $-T$ and (5.20) with $-F$, and add,

$$\mathrm{Tr}\,\mathrm{d}\varrho\; \{k_{\mathrm{B}} T \ln G - F I + H\} = 0, \tag{5.24}$$

now for unrestricted $\mathrm{d}\varrho$. The expression in curly brackets must vanish and we obtain

$$G = \mathrm{e}^{(FI - H)/k_{\mathrm{B}}T}. \tag{5.25}$$

Recall that F, T are numbers, I, H and G linear operators. The expression (5.25) for the equilibrium, or Gibbs state will be the starting point of many calculations.

The Gibbs state is stationary because it commutes with the energy, $[G, H] = 0$.

5.1.6 *Free Energy and Temperature*

The Lagrange multiplier F is the system's free energy, T its temperature. F serves to normalize the Gibbs state:

$$F = -k_B T \ln \mathrm{Tr}\, \mathrm{e}^{-H/k_B T} \tag{5.26}$$

guarantees $\mathrm{Tr} G = 1$. Since the energy observable will depend on external parameters, $H = H(\lambda)$, the free energy also depends on them and on T, i.e. $F = F(T, \lambda)$. The free energy as a function of its natural variables serves as a thermodynamic potential, its derivatives are related with generalized forces. We shall come back to this soon.

Let us now explain why T is the temperature. It shows up here as a Lagrangian parameter, but does it mean the same as in everyday life?

A first hint comes from the fact that T grows if a system is heated. This knowledge is embodied in the related English words hot versus heat, or heiß versus Hitze in German, or caliente versus calor in Spanish.

We calculate for $U = \mathrm{Tr} G H$ the derivative with respect to T,

$$\frac{\partial U}{\partial T} = \frac{\partial}{\partial T} \frac{\mathrm{Tr}\, \mathrm{e}^{-H/k_B T} H}{\mathrm{Tr}\, \mathrm{e}^{-H/k_B T}} = \frac{\langle H^2 \rangle_G - \langle H \rangle_G^2}{k_B T^2}. \tag{5.27}$$

The nominator on the right hand side is the variance of H and cannot be negative. We conclude that $T \to U$ is an increasing function. For given internal energy U it allows to determine the corresponding temperature T.

Let us now comment on another property of temperature. Consider a system Σ in thermal equilibrium. Its state is the Gibbs state

$$G \propto \mathrm{e}^{-\beta H}, \tag{5.28}$$

with β short for $1/k_B T$. Now introduce a fictitious border dividing the system into two subsystems Σ_1 and Σ_2 with energies H_1 and H_2, respectively. The total energy is $H = H_1 + H_2 + H_{12}$. The latter term describes the coupling of the subsystems; its contributions are interactions between pairs of particles where one sits in Σ_1, the other in Σ_2. The interaction is of short range, so only particles close to the border contribute. In other words, the interaction energy H_{12} is very small as compared with H_1 or H_2, provided the entire system is large in some sense. We conclude

$$G \propto e^{-\beta(H_1 + H_2 + H_{12})} \approx e^{-\beta H_1} e^{-\beta H_2}. \tag{5.29}$$

Replacing the exponential of a sum by the product of the exponentials, which is always valid for numbers, is possible here too because H_1 and H_2 commute.

What we just have explained is the following. If a system is in a Gibbs state with temperature T, its subsystems are also in Gibbs states with the same temperature. If we normalize the Gibbs states,

$$\begin{aligned} G_1 &= e^{(F_1 - H_1)/k_B T} \\ G_2 &= e^{(F_2 - H_1)/k_B T} \\ G &= e^{(F - H)/k_B T}, \end{aligned} \tag{5.30}$$

we find

$$F = F_1 + F_2. \tag{5.31}$$

The free energy of a system is an additive quantity.

Temperature is measured by bringing the system in thermal contact with a thermometer. Once equilibrium is attained, the temperature of the system is the same as the temperature of the thermometer.

5.1.7 Chemical Potentials

A system may not only exchange energy with its environment, but also particles. Particles come in species which we label by $a = 1, 2, \ldots$ N_a is the number of particles of species a. Its eigenvalues are $0, 1, \ldots$ which characterizes a number operator.

Denote by

$$\bar{N}_a = \langle N_a \rangle_\varrho = \mathrm{Tr}\varrho N_a \tag{5.32}$$

the average number of particles of species a. We generalize the second main law to systems which allow an exchange of particles as well. The equilibrium, or Gibbs state G^* is characterized by maximal entropy for fixed average energy $\bar{H} = U$ and for fixed average particle numbers $\bar{N}_a$. The asterisk indicates that energy and particle numbers may fluctuate around their average values. G^* is the grand canonical Gibbs state to be contrasted with the canonical Gibbs state G.

We have to augment (5.20) and (5.23) by the restriction

$$\mathrm{d}\bar{N}_a = \mathrm{Tr}\,\mathrm{d}\varrho\, N_a = 0 \tag{5.33}$$

and arrive at

$$G^* = \mathrm{e}^{(F^* I - H + \sum_a \mu_a N_a)/k_B T}\,. \tag{5.34}$$

F^* is the grand canonical free energy. The Lagrange multipliers μ_a are chemical potentials, one for each species. We will discuss their properties when required.

5.1.8 Minimal Free Energy

It follows from (5.26) that the free energy F, the internal energy $U = \mathrm{Tr} G H$ and the equilibrium entropy $S = \mathcal{S}(G)$ are related, namely by

$$F = U - TS. \tag{5.35}$$

Motivated by this we define the free energy functional

$$\mathcal{F}(\varrho) = \mathrm{Tr} \varrho \{H + k_B T \ln \varrho\}\,. \tag{5.36}$$

It depends on the temperature T which is kept constant.

Assume that this functional is extremal (minimum or maximum) at $\varrho = G'$. For a small variation $\mathrm{d}\varrho$,

$$\mathrm{Tr}\,\mathrm{d}\varrho\, I = 0, \tag{5.37}$$

we obtain

$$\mathrm{d}\mathcal{F}(G') = \mathrm{Tr}\,\mathrm{d}\varrho\, \left\{H + k_B T \ln G'\right\}. \tag{5.38}$$

We multiply (5.37) with a Lagrange multiplier F' and add to (5.38). Now the variation is unrestricted which results in

$$G' = \mathrm{e}^{(F' I - H)/k_B T}\,. \tag{5.39}$$

We conclude $G' = G$ and $F' = F$. Since the functional $\varrho \to \mathcal{F}(\varrho)$ is strictly convex, for $T > 0$, the Gibbs state minimizes the free energy functional.

A generalization to the grand canonical situation is straightforward:

$$\mathcal{F}^*(\varrho) = \mathrm{Tr} \varrho \left\{ H - \sum_a \mu_a N_a + k_B T \ln \varrho \right\}. \tag{5.40}$$

Note that $F = \mathcal{F}(G)$ or $F^* = \mathcal{F}(G^*)$ respectively are the free energies in the equilibrium state. The principal of minimal free energy is the starting point of many approximations. It simplifies to the Ritz variational principle for zero temperature.

5.1.9 *Summary*

We have recollected the essentials of quantum physics. Keywords are Hilbert space, base of mutually orthogonal and normalized vectors, linear mappings onto itself, or operators and their adjoints. An important class of diagonalizable, or normal operators is characterized by a decomposition of unity into mutually orthogonal projectors and corresponding complex eigenvalues. Unitary and self-adjoint operators have phases or real numbers as eigenvalues. The trace is a linear functional serving as an integral. Observables are described by self-adjoint operators, states are positive operators of unit trace. The expectation value is a linear functional of the observable which is linear in the state as well.

An expectation value changes if the state or the observable change. Applied to the energy observable, this defines heat and work and is known as the first main law of thermodynamics. A misnomer, because thermodynamics is not involved.

Then we distinguished between autonomous (or ideally isolated) systems, well isolated systems (may exchange energy with the environment) and open systems which are allowed to exchange particles. The entropy of an ideally isolated system remains constant. Since the environment's dynamics is unknown or even unpredictable, the state of a well isolated or open system becomes more and more mixed. The entropy, as a measure of mixing, increases until a maximum is reached. This statement is known as the second main law of thermodynamics.

We have calculated the state of maximal entropy, the Gibb state, which describes the equilibrium of a system with its environment. There are two Lagrange multipliers. One, the free energy, serves to normalize the Gibbs state. The second parameter, the temperature, is fixed by the internal energy of the system. We discussed its properties.

If the system may exchange particles with its environment the Gibbs state also depends on chemical potentials, one for each species.

Finally, we briefly mentioned that not only the entropy functional, but also a free energy functional may serve to define the Gibbs state; it is minimal there.

The section which you have just studied described the conceptual and mathematical framework of statistical thermodynamics. It can be reformulated so that Hilbert spaces, operators and other quantum physical object do not appear explicitly. The field of thermodynamics proper will be the subject of the following section.

5.2 Thermodynamics

Thermodynamics has a long tradition. Its rules were formulated before the invention of quantum theory. Today everybody knows that matter is composed of interacting particles. However, in earlier periods, this was an interesting hypothesis, one among others. Physicists and chemists succeeded in establishing a system of rules and laws without the notion of grainy matter. Atoms and molecules as well as photons were too small—as compared with humans—to be observed directly.

In this section we deduce some of the rules which form a branch of knowledge called phenomenological thermodynamics. It is about physical systems in equilibrium, about temperature changes and external parameters, volume say, and generalized forces such as pressure.

Originally developed for a deeper understanding of the steam engine (which initiated the industrial revolution) and of chemistry, thermodynamics became an encompassing science of equilibrium. The young Max Planck was convinced that thermodynamics provided the foundation for all branches of science. Not so Ludwig Boltzmann who strongly defended his view that the laws of thermodynamics derive from the random behavior of large collections of particles.

5.2.1 Reversible and Irreversible Processes

One talks about a system which is in a particular state. This state may be either in equilibrium or it is away from equilibrium.

Any state has an internal energy U and an entropy S. If the system is in an equilibrium state, it also has a temperature.

There are also external parameters. They appear in the Hamiltonian, but are not observables of the system. Instead they describe the action of the environment on the system disregarding any feedback.

The prototype of such an external parameter is the electric field strength E which may be kept constant by a power grid or by feedback control. Another such external parameter is the region $\mathcal{V}$ of space within which the particles are confined. For gases or liquids, the thermodynamic properties depend on the volume $V = \text{vol}(\mathcal{V})$ only; they are fluids which cannot support shear forced. For gases and liquids the volume V of the confining region is an external parameter. In thermodynamics, the volume is simply one of the thermodynamic variables. If the thermodynamic variables are kept fixed, the system will finally attain its equilibrium state.

If we consider a sequence of states being the result of very slowly changing the thermodynamic variables, the system will always have enough time for attaining its equilibrium state. It passes, in the course of time, through a sequence of equilibrium states, and we speak of a reversible process. Reversible, because the variable changes may be undone equally slowly.

Assume the external parameters change with time according to $\lambda = \lambda_t$ and the environment temperature as $T = T_t$. Such a process is reversible if

$$\varrho_t \approx G(T_t, \lambda_t). \tag{5.41}$$

On the other hand, if the thermodynamic variables are changing rapidly, there will not be sufficient time for attaining the equilibrium state. Before it has been reached, the variables have changed again.

How slow is slow? Well, the approach to equilibrium is an exponential relaxation with a certain time constant This time constant depends on how large the system is.

The smaller, the faster an equilibrium is attained. On the other hand, the thermodynamic functions which are calculated, usually are limits. For a certain accuracy, a minimal size is required. Therefore, one must find a compromise between reasonable relaxation time (small system) and negligible fluctuations (large systems).

5.2.2 Free Energy as Thermodynamic Potential

Consider a thermodynamic system where N identical particles are confined to move in a region of volume V. The Hamiltonian for it is

$$H = \sum_{r=1}^{N} \left\{ \frac{\boldsymbol{p}_r}{2m} + w\,\theta_{\mathcal{V}}(\boldsymbol{x}_r) \right\}. \tag{5.42}$$

There is a potential energy w required to bring a particle from inside $\mathcal{V}$ through the wall to the outside region. w is finally sent to infinity. More professional is to include $\mathcal{V}$ in the domain of definition: wave functions must vanish outside. Anyway, $H = H(\mathcal{V})$.

Associated with this Hamiltonian H is the Gibbs or equilibrium state

$$G = \mathrm{e}^{(F - H)/k_{\mathrm{B}}T} \tag{5.43}$$

with temperature T and free energy

$$F = -k_{\mathrm{B}}T \ln \mathrm{Tr}\,\mathrm{e}^{-H/k_{\mathrm{B}}T}. \tag{5.44}$$

The entropy S of this equilibrium state is

$$S = -k_{\mathrm{B}} \mathrm{Tr} G \ln G = \frac{U - F}{T}, \tag{5.45}$$

with

$$U = \mathrm{Tr} G H \tag{5.46}$$

the internal energy.

The partial derivative of the free energy with respect to temperature is

$$\frac{\partial F}{\partial T} = \frac{F}{T} - \frac{U}{T} = -S. \tag{5.47}$$

For a fluid, a gas or a liquid, the free energy depends on the external parameter $\mathcal{V}$ only via its volume. The derivative, at constant temperature, with respect to volume is the pressure p. We therefore may write

$$\mathrm{d}F(T, V) = -S\,\mathrm{d}T - p\,\mathrm{d}V. \tag{5.48}$$

The free energy here serves as a thermodynamic potential. Its partial derivatives define generalized, or thermodynamic forces. These are the entropy describing the reaction on temperature changes and the pressure as the generalized force associated with volume changes. Note that the right hand side of an equation like (5.48) indicates the independent variables. Also note that the equation refers to reversible state changes. After all, it has been derived from the Gibbs state (5.26). Clearly, entropy S and pressure p are functions of the independent thermodynamic variables T and V.

5.2.3 More Thermodynamic Potentials

One may interchange the role of variable and derivative by a Legendre transformation.

The internal energy U as a thermodynamic potential is defined by

$$U(S, V) = \inf_{T} \{F(T, V) + TS\}. \tag{5.49}$$

It follows that $S \to U(S, V)$ is concave. If there are no phase transitions, the thermodynamic potentials are differentiable, the infimum in (5.49) is a minimum, and the derivative with respect to T vanishes:

$$\frac{\partial F}{\partial T} + S = 0. \tag{5.50}$$

Hence the new variable is $S = -\partial F / \partial T$ and we may write

$$\mathrm{d}U(S, V) = T\mathrm{d}S - p\mathrm{d}V. \tag{5.51}$$

The independent variables are S and V, the corresponding derivatives T and $-p$. If the internal energy is known as a function of entropy and volume, temperature and pressure may be calculated.

While F is useful for isothermic[3] processes, the internal energy proves helpful for adiabatic processes. Such processes are slow, hence reversible, and at constant entropy as is the case for a rapidly expanding gases.

For thermodynamic processes encountered in biology, not only the temperature, but also the pressure is constant. They are best described by a thermodynamic potential depending on the temperature and pressure variables. So, V must be replaced by p. For this we define the Gibbs free energy, or Gibbs potential $G = F + pV$. With the normal meaning of the symbols, its differential is

[3] At constant temperature.

Table 5.1 Thermodynamic potentials. A gas or liquid of a fixed number of identical particles are confined within a region of volume V. T stands for temperature, S for entropy and p for pressure. Free energy and temperature are Lagrange multipliers

Name	Symbol	Definition	Variables	Derivatives
Free energy	F	$\mathrm{Tr}G = 1$	(T, V)	$\mathrm{d}F = -S\mathrm{d}T - p\mathrm{d}V$
Internal energy	U	$U = F + TS$	(S, V)	$\mathrm{d}U = T\mathrm{d}S - p\mathrm{d}V$
Gibbs potential	G	$G = F + pV$	(T, p)	$\mathrm{d}G = -S\mathrm{d}T + V\mathrm{d}p$
Enthalpy	H	$H = U + PV$	(S, p)	$\mathrm{d}H = T\mathrm{d}S + V\mathrm{d}p$

$$\mathrm{d}G(T, p) = -S\mathrm{d}T + V\mathrm{d}p. \tag{5.52}$$

A fourth potential is useful for streaming gases, the enthalpy. It is defined by $H = F + TS + pV$, and its differential is given by

$$\mathrm{d}H(S, p) = T\mathrm{d}S + V\mathrm{d}p. \tag{5.53}$$

We summarize these findings in Table 5.1.

The free energy is concave in T and convex in V. A Legendre transformation changes this. For example, the internal energy is a convex function of entropy and a convex function of volume. Such signs of the second derivative will play a role in the next subsection.

5.2.4 Heat Capacity and Compressibility

The heat capacity of a system is the amount of energy required to raise the temperature by one unit. We here assume N_A molecules in the system[4] which amounts to one mole.

The isochoric, or iso-volumetric heat capacity is defined as

$$C_V = \left.\frac{\mathrm{d}Q}{\mathrm{d}T}\right|_{\mathrm{d}V=0}. \tag{5.54}$$

We may express this as a second derivative of the free energy:

$$.C_V = -T\,\frac{\partial^2 F(T, V)}{\partial T^2} \tag{5.55}$$

For one mole, (5.55) is the molar heat capacity at constant volume.

[4] $N_\mathrm{A} = 6.022141 \times 10^{23}\,\mathrm{mol}^{-1}$ is Avogadro's number. N_A atoms of the C^{12} isotope weigh 12 g by definition.

The isothermal compressibility is defined as

$$\kappa_T = -\frac{1}{V}\left.\frac{\mathrm{d}V}{\mathrm{d}p}\right|_{\mathrm{d}T=0}. \tag{5.56}$$

With $\partial F(T, V)/\partial V = -p$ we find

$$\frac{1}{\kappa_T} = V\,\frac{\partial^2 F(T, V)}{\partial V^2}. \tag{5.57}$$

Just as the heat capacity at constant volume, the compressibility at constant temperature is always positive.

Some quantities grow proportional to the size of a system, and other quantities remain constant. The former are called extensive, the latter intensive. Particle numbers N, volume V, energy U and free energy F as well as entropy are extensive. Temperature and pressure are intensive. Heat capacity is an extensive quantity, compressibility is intensive. This can be read off from (5.55) and (5.57).

The heat capacity at constant pressure, i.e.

$$C_p = \left.\frac{\mathrm{d}Q}{\mathrm{d}T}\right|_{\mathrm{d}p=0}, \tag{5.58}$$

can be expressed with the aid of the free energy, although this is rather cumbersome. One may write

$$\mathrm{d}S = -\frac{\partial^2 F}{\partial T^2}\,\mathrm{d}T - \frac{\partial^2 F}{\partial V \partial T}\,\mathrm{d}V \tag{5.59}$$

for the entropy change and

$$\mathrm{d}p = -\frac{\partial^2 F}{\partial T \partial V}\,\mathrm{d}T - \frac{\partial^2 F}{\partial V^2}\,\mathrm{d}V \tag{5.60}$$

for the pressure change. The latter expression must vanish, therefore

$$C_p = T\left\{\frac{{F_{TV}}^2}{F_{VV}} - F_{TT}\right\}. \tag{5.61}$$

Here F_{TV} stands for $\partial^2 F/\partial T \partial V$, and so on. By the way, (5.61) says $C_p > C_V$. The heat capacity at constant pressure is larger than at constant volume.

The natural choice for calculating the isobaric heat capacity is the Gibbs potential $G = G(T, p)$. We find

$$C_p = -T\,\frac{\partial^2 G(T, p)}{\partial T^2}. \tag{5.62}$$

More elegant, but no indication why C_p should be larger than C_V, for all gases or liquids.

If a gas expands rapidly, but still reversibly, there is no time for heat exchange. The process occurs at constant entropy, because of $T\mathrm{d}Q = \mathrm{d}S = 0$. The corresponding compressibility is defined as

$$\kappa_S = -\frac{1}{V}\frac{\mathrm{d}V}{\mathrm{d}p}\bigg|_{\mathrm{d}S=0}. \tag{5.63}$$

By a similar chain of arguments as for the isobaric heat capacity, we arrive at

$$\frac{1}{\kappa_S} = V\left\{F_{VV} - \frac{{F_{TV}}^2}{F_{TT}}\right\}. \tag{5.64}$$

Clearly, the adiabatic compressibility is always smaller than its isothermal counterpart: $\kappa_S < \kappa_T$.

However, never forget that a mere reformulation does not generate new information. In our case, there is one primary thermodynamic potential, the free energy, depending on two variables. Since partial differentiations commute, there are three independent second derivatives. We stop this here because phenomenological thermodynamics is just a sideline of this small book on quantum physics.

5.2.5 *Chemical Potential*

Imagine a system which may exchange particles with its environment. A biological cell is an example, but also a block of ice swimming in water with an atmosphere above it. The well known phenomenon of osmotic pressure is another example where the small water molecules may freely pass a semi-permeable membrane, but not the large sugar particles.

Let us label the particle species by $a = 1, 2, \dots$ and by N_a the number of particles of species a within the system. As we have shown before, the thermodynamic equilibrium, or Gibbs state is characterized by maximal entropy which amounts to

$$G^* = \mathrm{e}^{(F^*I - H + \mu_1 N_1 + \mu_2 N_2 + \dots)/k_\mathrm{B}T}. \tag{5.65}$$

The Lagrange multiplier T is the temperature, the μ_a are called chemical potentials. The grand-canonical free energy, a Lagrange multiplier as well, is given by $\mathrm{Tr}G^* = 1$, i.e.

$$F^*(T, V, \mu_1, \mu_2, \dots) = -k_\mathrm{B}T \ln \mathrm{Tr}\,\mathrm{e}^{(-H + \mu_1 N_1 + \mu_2 N_2 + \dots)/k_\mathrm{B}T}. \tag{5.66}$$

The differential of this thermodynamic potential is

Table 5.2 Stoichiometric coefficients. A water molecule may dissociate either into a proton and OH ion (reaction 1) or into H_2 and O_2 molecules (reaction 2). The stoichiometric coefficients for these two reactions are listed

Reaction	H_2O	H^+	OH^-	H_2	O_2
$H_2O \to H^+ + OH^-$	-1	1	1	0	0
$2H_2O \to 2H_2 + O_2$	-2	0	0	2	1

$$\mathrm{d}F^*(T, V, \mu_1, \mu_2, \dots) = -S\mathrm{d}T + \mathrm{d}W - \bar{N}_1\mathrm{d}\mu_1 - \bar{N}_2\mathrm{d}\mu_2 - \dots. \tag{5.67}$$

Here $\mathrm{d}W$ denotes the work spent on the system by changing its external parameters, the barred numbers are averages:

$$\bar{N}_a = \langle N_a \rangle = \mathrm{Tr}G^* N_a. \tag{5.68}$$

For simplicity of notation let us specialize to one kind of particles only the chemical potential of which is denoted by μ, its number by N. Note that N is a number operator with eigenvalues $0, 1, 2, \dots$ while $\bar{N}$ is the average number in the system.

The average particle number grows with the chemical potential. We simply differentiate (5.68) with respect to μ. The result is

$$\frac{\partial \bar{N}}{\partial \mu} = \frac{\langle N^2 \rangle - \langle N \rangle^2}{k_B T}, \tag{5.69}$$

which is never negative.

The chemical potential is an equilibrium parameter. Consider two systems, as described by $H = H_1 + H_2 + H_{12}$ with a small interaction H_{12} and assume thermal equilibrium. This means that system 1 and system 2 have the same temperature T as the combined system. If they also may freely exchange particles, and are given sufficient time to settle, the system will be in the grand-canonical Gibbs state:

$$\begin{aligned} G^* &= \mathrm{e}^{(F^* + \mu N - H)/k_B T} \\ &\approx \mathrm{e}^{(F_1^* + \mu N_1 - H_1)/k_B T}\,\mathrm{e}^{(F_2^* + \mu N_2 - H_2)/k_B T}. \end{aligned} \tag{5.70}$$

We conclude that two loosely interacting systems are characterized by the same temperature and by equal chemical potentials. Whereas the temperature describes the eagerness to get rid of energy, the chemical potential is a measure for the zeal to dispose of particles. Equilibrium is attained if these desires are the same on both sides of the frontier.

5.2.6 Chemical Reactions

Particles are created in chemical reactions. As an example, consider a system with H_2O, H_2 and O_2 molecules and with H^+ and OH^- ions. We label them with $a = 1, 2, 3, 4, 5$, respectively. In a reaction of type r, ν^{ra} particles of species a are generated. The so called stoichiometric coefficients are integers without a common divisor. A negative sign indicates that the corresponding particle vanishes.

In our example we consider two reactions the stoichiometric coefficients of which are listed in Table 5.2.

Equation (5.67) says that the chemical potential is free energy decrease if one particle is added to the system, at constant temperature. We conclude that the free energy F^* increases by

$$A_r = -\sum_a \nu^{ra} \mu_a \tag{5.71}$$

if one reaction of type r takes place.

Chemical equilibrium is achieved if all so-called affinities A_r vanish. Only then the free energy is minimal. Put otherwise, non-vanishing affinities drive chemical reactions.

5.2.7 Particle Number as an External Parameter

In this subsection, and only here, we will write $\hat{N}$ for the particle number operator, $N = 0, 1, 2, \ldots$ for one of its eigenvalues, and $\bar{N} = \langle \hat{N} \rangle$ for the expectation value, or average.

The canonical way of thinking is in terms of quantum mechanics where systems are made up of a fixed number of particles. There is a Hamiltonian $H_N = H_N(\mathcal{V})$ for exactly N particles confined to a certain region $\mathcal{V}$. We assume a fluid medium, a liquid or gas, where the dependency on $\mathcal{V}$ is via its volume V only and write

$$F(T, V, N) = -k_B T \ln \mathrm{Tr}_N \, \mathrm{e}^{-H_N(\mathcal{V})/k_B T}. \tag{5.72}$$

The trace refers to the Hilbert space $\mathcal{H}_N$ of square normalizable wave functions with exactly N arguments $\boldsymbol{x}_1, \boldsymbol{x}_2, \ldots, \boldsymbol{x}_N$.

The grand-canonical approach refers to a Hilbert space $\mathcal{H}$ describing an arbitrary number of particles. It is realized as a Fock space. There are creation and annihilation operators, either for bosons or fermions, which are applied repeatedly to the unique vacuum state.

In this context one defines the grand-canonical free energy as

$$F^*(T, V, \mu) = -k_B T \ln \mathrm{Tr} \, \mathrm{e}^{(-H(\mathcal{V}) + \mu\hat{N})/k_B T}. \tag{5.73}$$

Now, the Fock space can be decomposed into the eigenspaces $\mathcal{H}_N$ of the particle number operator. Recall that we agreed to denote the eigenvalues of $\hat{N}$ by $N = 0, 1, \ldots$. The Fock space trace then becomes a sum over N and a trace over $\mathcal{H}_N$. Equation (5.73) can therefore be rewritten as

$$F^*(T, V, \mu) = -k_B T \ln \sum_{N=0}^{\infty} \mathrm{Tr}_N \, \mathrm{e}^{(-H_N(\mathcal{V}) + \mu N I)/k_B T}$$
$$= -k_B T \ln \sum_{N=0}^{\infty} \mathrm{e}^{(-F(T, V, N) + \mu N)/k_B T} \,. \tag{5.74}$$

One speaks of the thermodynamic limit if the particle number N is sent to infinity while all intensive quantities (temperature, chemical potential, particle density $n = N/V$ and so on) are kept constant. In this thermodynamic limit, the sum on the right hand side of (5.74) is dominated by its largest contribution. This follows from a theorem of functional analysis: the p-norm of a well-behaved function converges, with $p \to \infty$, towards the supremum norm.

For a large system we obtain

$$F^*(T, V, \mu) = \inf_N \{F(T, V, N) - \mu N\} \,. \tag{5.75}$$

Clearly, a Legendre transformation. The infimum is a minimum and will be attained at

$$\frac{\partial F(T, V, N)}{\partial N} = \mu, \tag{5.76}$$

so that

$$F^*(T, V, \mu) = F(T, V, \bar{N}) - \mu \bar{N} \tag{5.77}$$

holds true. Here $\bar{N} = \bar{N}(T, V, \mu) = \partial F^*/\partial \mu$ is the average particle number as a function of temperature, volume and chemical potential.

We have shown in this subsection that, in the thermodynamic limit, the grand-canonical free energy is the Legendre transform of the canonical free energy. In the thermodynamic limit the fixed number N of particles in a system and the mean value $\bar{N} = \langle \hat{N} \rangle$ is irrelevant since the relative fluctuations vanish with $N \to \infty$. This can be read off (5.69).

After this has been settled, we do not need to distinguish anymore between the particle number operator $N = \hat{N}$ and its eigenvalues $N \in \mathbb{N}$.

5.2.8 *Summary*

Phenomenological thermodynamics is an effort to describe the equilibrium of bulky matter. Bulk matter has internal properties described by temperature, pressure, chemical potentials, internal energy and so on, collectively called thermodynamic variables. If two systems with different temperature, or pressure, or chemical potential are brought into contact, they strive for equilibrium such that these thermodynamic variables become equal.

If the environment change slowly, the system will run through a sequence of equilibrium states. Such processes are called reversible.

The thermodynamic variables are related. Some are arguments of a thermodynamic potential, others are defined as first partial derivatives. Although the free energy is a preferred potential, because it can be calculated within a quantum theoretical framework, it is not unique. There are more thermodynamic potentials obtained by Legendre transformations. We present them all: free energy, internal energy, Gibbs potential and enthalpy, as long as only temperature and volume are involved.

The second derivatives describe material properties and relations between them. We studied the heat capacity, either for constant volume or constant pressure and likewise the compressibility, either for constant pressure or entropy. There are equalities and inequalities which hold true irrespective of the microscopic model that gave rise the free energy.

We also explain the meaning of chemical potentials and the notion of the affinity of a chemical reaction.

A final subsection discusses particle numbers as observables, their eigenvalues and averages, or expectation value. Strictly speaking, the results are valid only in the thermodynamic, or bulk limit.

5.3 Continuum Physics

This section explains how to describe continuously distributed matter. It is a condensed version of the theory part of the author's recent book.[5] We rewrite the laws, rules and definitions of traditional thermodynamics as a set of coupled differential equations for classical fields. Of central importance is the notion of a material point. We pick a tiny region which is so small that it has practically no extension—as perceived by an engineer. The same region, however, appears to be huge to a physicist. It contains a tremendous number of particles such that the thermodynamic limit is a good approximation. Fortunately, these two description are compatible.

Since the material point is so small it is always in thermodynamic equilibrium. The variables which describe the local equilibrium state, however, may change with location and time. Globally, the system need not be in equilibrium.

[5] *Continuum Physics* by Peter Hertel, Graduate Texts in Physics, Springer 2012.

There are two types of equations. Balance equations say that an increase in a quantity is caused by the inflow of this quantity or by production. The balance equation for internal energy is the first main law of thermodynamic in a very concrete form. Likewise, the balance equation for entropy allows a precise and concrete formulation of the famous second law. Balance equations have to be satisfied irrespective of the material.

Material equations provide additional information which is typical for the material under study. They are necessary because there are too few balance equations for too many fields.

The material equations of continuum physics cannot be established within the framework of thermodynamics; they require the models and methods of quantum physics. This will be the subject of the following subsections. Here we lay the ground for it.

5.3.1 Material Points

Let us consider water vapor. A tiny cube of 1.0 mm side length has a volume of $1.0 \times 10^{-9}\,\text{m}^3$. Since one mole of an ideal gas under standard conditions occupies a volume of $22.7 \times 10^{-3}\,\text{m}^3$, our cube contains 4.4×10^{-8} mole which amounts to $\bar{N} = 2.7 \times 10^{16}$ particles. Recall that $N_\text{A} = 6.022 \times 10^{23}$ particles make up a mole.

For an ideal gas, one finds $\langle N^2 \rangle = \langle N \rangle^2 + \langle N \rangle$. The root mean square fluctuation $\sqrt{\langle N^2 \rangle - \langle N \rangle^2}$ thus is given by $\sqrt{\langle N \rangle}$, the relative fluctuation by

$$\frac{\sqrt{\langle N^2 \rangle - \langle N \rangle^2}}{\langle N \rangle} = \frac{1}{\sqrt{\langle N \rangle}}. \tag{5.78}$$

In our case this works out to 6×10^{-9}. The number of H_2O molecules in the millimeter sized cube is well defined to at least eight digits. Sufficient for the engineer. On the other hand, 10^{16} or more particles are really very many and one may safely apply relations which are strictly valid only for $N \to \infty$. $\bar{N}$ is really close to infinity.

A material point, which has practically no extension, is situated at a position $\boldsymbol{x}$. It is associated with a number of properties, such as mass density, flow velocity, temperature and so on. The property f has a real value and may depend on time t and position $\boldsymbol{x}$. Therefore, $f = f(t, \boldsymbol{x})$ is a classical field. We assume that the reader knows how to differentiate fields and how to integrate them along one, two and three dimensional manifolds.

5.3.2 Balance Equations

Assume Y to denote an additive, transportable quantity. If system 1 has Y_1 of this quantity and a different system 2 has Y_2, then the combined system has $Y = Y_1 + Y_2$. Mass, linear and angular momentum and energy are examples of additive quantities. The quantity is transportable if it can be brought from one place to another. Mass and energy may flow just as linear or angular momentum. These properties are transportable.

Consider a sufficiently smooth fixed region $\mathcal{V}$. The quantity of Y in $\mathcal{V}$ at time t may be written as

$$Q(Y;t,\mathcal{V}) = \int_{x\in\mathcal{V}} \mathrm{d}V\,\varrho(Y;t,\boldsymbol{x}), \tag{5.79}$$

as a volume integral over the region $\mathcal{V}$. Here $\varrho(Y) = \varrho(Y;t,\boldsymbol{x})$ is the volume density[6] of this quantity, a field.

The quantity is transportable. It may flow from the back side of a surface $\mathcal{A}$ to its front side, the corresponding current is

$$I(Y;t,\mathcal{A}) = \int_{x\in\mathcal{A}} \mathrm{d}\boldsymbol{A}\cdot\boldsymbol{j}(Y;t,\boldsymbol{x}). \tag{5.80}$$

The scalar product $\mathrm{d}\boldsymbol{A}\cdot\boldsymbol{j}$ guarantees that only the current along the surface normal is taken into account. $\boldsymbol{j}$ is a current density.

Finally, the quantity under study may be produced within the region. The generation, or production rate[7] is given by

$$\Pi(Y;t,\mathcal{V}) = \int_{x\in\mathcal{V}} \mathrm{d}V\,\pi(Y;t,\boldsymbol{x}). \tag{5.81}$$

The prototype of all balance equations is

$$\frac{\mathrm{d}Q(Y;t,\mathcal{V})}{\mathrm{d}t} + I(Y;t,\partial\mathcal{V}) = \Pi(Y;t,\mathcal{V}). \tag{5.82}$$

The increase in content plus the outflow balance the production, per unit time. This balance equation must hold at any time and for an arbitrary region. Note that $\partial\mathcal{V}$ is the boundary of the region $\mathcal{V}$, a surface.

We now resort to Gauss' theorem:

$$\int_{\mathcal{V}} \mathrm{d}V\,\boldsymbol{\nabla}\cdots = \int_{\partial\mathcal{V}} \mathrm{d}\boldsymbol{A}\ldots, \tag{5.83}$$

[6] In this and following sections the symbol ϱ stands for a volume density. Not to be confused with a probability operator.

[7] Again, do not confuse with a projection operator.

to be applied to a scalar or vector field. With it one can rewrite (5.82) as

$$\nabla_t \varrho(Y,t,\boldsymbol{x}) + \sum_{i=1}^{3} \nabla_i j_i(Y,t,\boldsymbol{x}) = \pi(Y,t,\boldsymbol{x}). \tag{5.84}$$

The generic balance equation is a field equation. It holds at all locations $\boldsymbol{x}$ and all times t. These arguments are therefore simply dropped. Partial differentiation with respect to time is denoted by ∇_t. Likewise, ∇_i stands for partial differentiation with respect to the i'th space argument, here the component x_i of $\boldsymbol{x}$.

A further simplification is to sum automatically over doubly occurring indexes in the range from one to three. This useful convention has been introduced by Einstein.

The generic balance equation for an additive, transportable quantity Y in its short form is

$$\nabla_t \varrho(Y) + \nabla_i j_i(Y) = \pi(Y). \tag{5.85}$$

The second term is the Y-current divergence.

5.3.3 Particles, Mass and Electric Charge

N^a is the number of particles of species a. We will never enumerate them. Right from the beginning particles of the same species are correctly treated as identical.

We denote the particle densities at time t and location $\boldsymbol{x}$ by

$$n^a = n^a(t,\boldsymbol{x}) = \varrho(N^a; t,\boldsymbol{x}). \tag{5.86}$$

We may split off the particle density from the particle current density. If there are no particles, they cannot flow and the current density would vanish. If it does not vanish, there must be particles the density of which then is positive. We therefore define

$$\boldsymbol{j}^a = \boldsymbol{j}^a(t,\boldsymbol{x}) = n^a(t,\boldsymbol{x})\boldsymbol{v}^a(t,\boldsymbol{x}) = \boldsymbol{j}(N^a; t,\boldsymbol{x}). \tag{5.87}$$

We have just introduced the velocity field $\boldsymbol{v}^a = \boldsymbol{j}^a(t,\boldsymbol{x})$ for a-particles.

Particles may be created or may disappear in chemical reactions. In the previous Sect. 5.2 on Thermodynamics we have already introduced the notion of stoichiometric coefficients: ν^{ra} particles of species a will emerge in one reaction of type r or disappear, if ν^{ra} is negative.

Let $\Gamma^r = \Gamma^r(t,\boldsymbol{x})$ be the number of chemical reactions per unit time and unit volume, or volumetric reaction rate, as some authors call it. The volumetric production rate for a-particles thus should be written as

$$\pi^a = \pi(N^a) = \sum_r \Gamma^r \nu^{ra}. \tag{5.88}$$

The balance equation for particles of species a reads

$$\nabla_t n^a + \nabla_i n^a v_i^a = \sum_r \Gamma^r \nu^{ra}. \tag{5.89}$$

Let us now turn to the additive and transportable mass M. Its density occurs so frequently that we abbreviate it by

$$\varrho = \varrho(t, \boldsymbol{x}) = \varrho(M; t, \boldsymbol{x}). \tag{5.90}$$

A particle of species a has a mass m^a, so

$$\varrho = \sum_a m^a n^a. \tag{5.91}$$

The mass current density then is given by

$$\boldsymbol{j}(M) = \sum_a m^a n^a \boldsymbol{v}^a = \varrho \boldsymbol{v}, \tag{5.92}$$

where the mass stream velocity $\boldsymbol{v}$ reads

$$\boldsymbol{v} = \sum_a \frac{m^a n^a}{\varrho} \boldsymbol{v}^a. \tag{5.93}$$

It is an average over the average velocity of the particles. The mass production is

$$\pi(M) = \sum_a m^a \sum_r \Gamma^r \nu^{ra} = 0, \tag{5.94}$$

since

$$\sum_a m^a \nu^{ra} = 0. \tag{5.95}$$

Mass is conserved in each chemical reaction.

The mass balance equation reads

$$\nabla_t \varrho + \nabla_i \varrho v_i = 0. \tag{5.96}$$

Mass is conserved. It can be redistributed, but not created or destroyed.

Electric charge Q^{e} is another additive and transportable quantity. Its density is

$$\varrho^{\mathrm{e}} = \varrho(Q^{\mathrm{e}}) = \sum_a q^a n^a. \tag{5.97}$$

An a-particle carries a charge q^a which is transported together with the particle. The electric current density is written as

$$\boldsymbol{j}^{\mathrm{e}} = \boldsymbol{j}(Q^{\mathrm{e}}) = \sum_a q^a n^a \boldsymbol{v}^a. \tag{5.98}$$

We may not represent it as ϱ^{e} times a velocity. An electric current may flow although the charge density vanishes. Just think of an equal number of electrons and holes moving in opposite directions.

Since electric charge is preserved in chemical reactions, the volumetric charge production rate vanishes. We calculate with $\sum_a q^a \nu^{ra} = 0$ that

$$\pi(Q^{\mathrm{e}}) = \sum_a q^a \pi(N^a) = \sum_a q^a \sum_r \Gamma^r \nu^{ra} = 0. \tag{5.99}$$

The electric charge balance equation reads

$$\nabla_t \varrho^{\mathrm{e}} + \nabla_i j_i^{\mathrm{e}} = 0. \tag{5.100}$$

Recall Maxwell's equations (in usual notation)

$$\epsilon_0 \boldsymbol{\nabla} \cdot \boldsymbol{E} = \varrho^{\mathrm{e}} \tag{5.101}$$

and

$$\frac{1}{\mu_0} \boldsymbol{\nabla} \times \boldsymbol{B} - \epsilon_0 \nabla_t \boldsymbol{E} = \boldsymbol{j}^{\mathrm{e}}. \tag{5.102}$$

The electromagnetic field $\boldsymbol{E}, \boldsymbol{B}$ is generated by the electric charge density ϱ^{e} and current density $\boldsymbol{j}^{\mathrm{e}}$ which we have discussed just now. The remaining Maxwell equations

$$\boldsymbol{\nabla} \cdot \boldsymbol{B} = 0 \tag{5.103}$$

and

$$\boldsymbol{\nabla} \times \boldsymbol{E} + \nabla_t \boldsymbol{B} = 0 \tag{5.104}$$

allow to deduce the continuity equation (5.100) for electric charge. Our formulation of continuum physics and Maxwell's equations are compatible.

5.3.4 Conduction and Covariance

While matter flows it carries its properties with it. There is always a contribution $\varrho(Y)\boldsymbol{v}(Y)$ to the current density, the convection part. There may be more: the quantity Y can also be conducted, i.e. transported by short range interactions. We therefore split the current

$$j(Y) = \varrho(Y)\boldsymbol{v} + \boldsymbol{J}(Y), \tag{5.105}$$

into a convection and conduction current density, respectively.

Consider a Galilei transformations $t = t' + \tau$ and $x_j = a_j + R_{jk}x'_k + u_j t'$, where R is an orthogonal matrix. A scalar field S transforms according to $S(t, \boldsymbol{x}) = S'(t', \boldsymbol{x}')$, a vector field $\boldsymbol{V}$ as $V_j(t, \boldsymbol{x}) = R_{jk}V'_k(t', \boldsymbol{x}')$.

- The density $\varrho(Y)$ transforms as a scalar field.
- The volumetric production rate $\pi(Y)$ is a scalar field
- The combination $\nabla_t \varrho(Y) + \nabla_i \varrho(Y) v_i$ also transforms as a scalar.
- The conduction current $\boldsymbol{j}(Y) - \varrho(Y)\boldsymbol{v} = \boldsymbol{J}(Y)$ is a vector field.

These findings guarantee that the balance equation (5.85) or

$$\nabla_t \varrho(Y) + \nabla_i \varrho(Y) v_i = -\nabla_i J_i(Y) + \pi(Y) \tag{5.106}$$

survives a Galilei transformation. It is valid in all inertial frames.

Let us introduce the substantial time derivative by

$$\mathrm{D}_t f(t, \boldsymbol{x}) = \frac{f(t + \mathrm{d}t, \boldsymbol{x} + \boldsymbol{v}\mathrm{d}t) - f(t, \boldsymbol{x})}{\mathrm{d}t}, \tag{5.107}$$

or

$$\mathrm{D}_t = \nabla_t + \boldsymbol{v} \cdot \boldsymbol{\nabla}. \tag{5.108}$$

It describes the change with time as seen by a co-moving observer. It gives rise to an alternative formulation of balance equations which lends itself easily to a physical interpretation.

For this we introduce specific quantities σ by

$$\varrho(Y) = \varrho(M)\,\sigma(Y). \tag{5.109}$$

While densities are quantity per unit volume, specific quantities are given per unit mass. The generic balance equation (5.85) may be reformulated as

$$\varrho \mathrm{D}_t \sigma(Y) = -\boldsymbol{\nabla} \cdot \boldsymbol{J}(Y) + \pi(Y). \tag{5.110}$$

Equation (5.85) are partial differential equations which can be programmed in this form. Equation (5.106) should be viewed as a decomposition into properly transforming scalars. And finally, (5.110) is easy to interpret. On the left-hand side: the temporal change of a specific quantity, as noted by a co-moving observer and converted to a density. On the right of the equal sign: the inflow by conduction plus production.

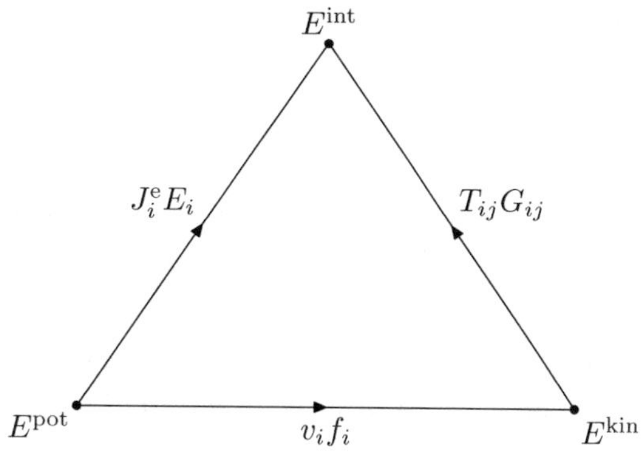

Fig. 5.2 Kinetic, potential and internal energy production. An *outgoing arrow* indicates energy export, an *incoming arrow* stands for import. Exports and imports sum up to zero; total energy is conserved

5.3.5 Momentum, Energy, and the First Main Law

Momentum and energy are additive, transportable properties. We split energy into kinetic and potential energy and a rest, namely internal energy. See Fig. 5.2 for a sketch of the energy balance.

Momentum

We shall study next the three components P_j of linear momentum. Their densities are mass times velocity per unit volume, i.e.

$$\varrho(P_j) = \varrho\, v_j. \tag{5.111}$$

Recall that the stream velocity was defined in (5.92). The momentum current density has two indexes, one for the flow direction, the other for the momentum component. It should be split into the convection contribution $\varrho v_j v_i$ and the conduction part:

$$j_i(P_j) = \varrho\, v_j v_i - T_{ji}. \tag{5.112}$$

T_{ji} is a second rank tensor, it describes stress.

The momentum production rate is the same as external force. One contribution couples to mass, the other to charge. We write

$$\pi(P_j) = f_j = \varrho\, g_j + \varrho^{\rm e}\, E_j. \tag{5.113}$$

g stands for the gravitational or any other pseudo force per unit mass, such as the Coriolis force, and $\boldsymbol{E}$ is the electric field strength.

The momentum balance equation may be written as

$$\varrho \mathrm{D}_t v_j = \nabla_i T_{ji} + f_j. \tag{5.114}$$

Evidently, the left hand side is mass times acceleration per unit volume, as felt by a co-moving observer. The right hand side is a decomposition into short and long range forces per volume. The divergence of the stress field indicates that more momentum enters a small volume than leaves at the opposite side, not by flow, but by short range interactions.

We skip the arguments for angular momentum. The balance equations are automatically satisfied provided

$$T_{ji} = T_{ij} \tag{5.115}$$

holds true. The stress tensor must be symmetric and can be diagonalized by an orthogonal coordinate transformation, at least locally.

Kinetic Energy

The kinetic energy E^{kin} has a density

$$\varrho(E^{\text{kin}}) = \frac{\varrho\, \boldsymbol{v}^2}{2}. \tag{5.116}$$

$\sigma(E^{\text{kin}}) = \boldsymbol{v}^2/2$ is the specific kinetic energy, its material time derivative reads as $v_j \mathrm{D}_t v_j$. Inserting the momentum balance equation (5.114) results in

$$\varrho \mathrm{D}_t \sigma(E^{\text{kin}}) = \nabla_i v_j T_{ji} - T_{ji} \nabla_i v_j + v_i f_i. \tag{5.117}$$

With the symmetric velocity gradient

$$G_{ij} = \frac{\nabla_i v_j + \nabla_j v_i}{2} \tag{5.118}$$

we arrive at

$$\pi(E^{\text{kin}}) = -G_{ij} T_{ij} + \boldsymbol{v} \cdot \boldsymbol{f} \tag{5.119}$$

as well as

$$j_i(E^{\text{kin}}) = -v_j T_{ji}. \tag{5.120}$$

Potential Energy

We put in that gravitational and electrostatic forces have a potential ϕ^g and ϕ^{e}, respectively:

$$g_j = -\nabla_j \phi^g \text{ and } E_j = -\nabla_j \phi^{\text{e}}. \tag{5.121}$$

Therewith the volumetric momentum production rate is given by

$$\pi(P_j) = -\varrho \nabla_j \phi^g - \varrho^{\text{e}} \nabla_j \phi^{\text{e}}. \tag{5.122}$$

One writes

$$\varrho(E^{\text{pot}}) = \varrho\,\phi^g + \varrho^{\text{e}}\phi^{\text{e}} \tag{5.123}$$

for the density of potential energy. Since both potentials are quasi-static, the following equality holds true:

$$\nabla_t \varrho(E^{\text{pot}}) = \phi^g \nabla_t \varrho + \phi^{\text{e}} \nabla_t \varrho^{\text{e}} = -\phi^g \nabla_i j_i(M) - \phi^{\text{e}} \nabla_i j_i(Q^{\text{e}}). \tag{5.124}$$

Rewriting this into a gradient and a rest finally results in

$$j_i(E^{\text{pot}}) = \phi^g j_i(M) + \phi^{\text{e}} j_i(Q^{\text{e}}) \tag{5.125}$$

and in

$$\pi(E^{\text{pot}}) = -\boldsymbol{v} \cdot \boldsymbol{f} - \boldsymbol{J}^{\text{e}} \cdot \boldsymbol{E}. \tag{5.126}$$

Internal Energy

There is more than kinetic and potential, namely internal energy E^{int}. The sum of them, the total energy $E^{\text{tot}} = E^{\text{kin}} + E^{\text{pot}} + E^{\text{int}}$ will be conserved.

Denote by u the specific internal energy. The internal energy current density has a conduction contribution which we call $\boldsymbol{J}^u$. This is the heat current density. Since total energy is conserved,

$$\pi(E^{\text{tot}}) = \pi(E^{\text{kin}}) + \pi(E^{\text{pot}}) + \pi(E^{\text{int}}) = 0, \tag{5.127}$$

we conclude

$$\pi(E^{\text{int}}) = T_{ij} G_{ij} + J_i^{\text{e}} E_i. \tag{5.128}$$

The balance equation for internal energy thus reads

$$\varrho \mathrm{D}_t u = -\nabla_i J_i^u + T_{ij} G_{ij} + J_i^{\text{e}} E_i. \tag{5.129}$$

There is one more refinement which we shall explain only superficially. Besides the Galilei transformations mentioned above there are discrete symmetries. Space inversion is one, time reversal $t \to -t$ another one. Velocity is odd, acceleration even with respect to time reversal, and so on. One may split the conduction contributions into a reversible part $\boldsymbol{J}'$ and an irreversible part $\boldsymbol{J}''$. The former transforms as the convection part, the irreversible part acquires an additional sign. Equation (5.129) then becomes

$$\varrho \mathrm{D}_t u = -\nabla_i J_i^u + T'_{ij} G_{ij} + T''_{ij} G_{ij} + J_i^{\text{e}\prime} E_i + J_i^{\text{e}\prime\prime} E_i. \tag{5.130}$$

This is the first main law of thermodynamics as a field equation. On the left—the change of internal energy as observed by a co-moving observer. On the right—five, and only five, contributions, namely

- heat conduction $-\nabla_i J^u_i$,
- compression $T'_{ij} G_{ij}$,
- internal friction $T''_{ij} G_{ij}$,
- electric polarization work $J^{e\prime}_i E_i$, and
- Joule's heat $J^{e\prime\prime}_i E_i$.

The second and forth term give rise to reversible changes, the remaining three terms cause irreversible behavior. Unfortunately we cannot go into details.

Multiplying (5.130) by $\mathrm{d}t$ and $\mathrm{d}V$ yields

$$\mathrm{d}U = \mathrm{d}W + \mathrm{d}Q \tag{5.131}$$

with

$$\begin{aligned} \mathrm{d}U &= \mathrm{d}t\mathrm{d}V\,\varrho \mathrm{D}_t u\;, \\ \mathrm{d}W &= \mathrm{d}t\mathrm{d}V\,\left(T'_{ij}G_{ij} + J^{e\prime}_i E_i\right)\;, \\ \mathrm{d}Q &= \mathrm{d}t\mathrm{d}V\,\left(\nabla_i J^u_i + T''_{ij}G_{ij} + J^{e\prime\prime}_i E_i\right). \end{aligned} \tag{5.132}$$

Admittedly, (5.131) is much more compact, but also less specific. From (5.132) we learn in addition what contributes to work $\mathrm{d}W$ and what to heat $\mathrm{d}Q$. We see that work is the reversible and heat the irreversible part of internal energy change. Last but not least, it is comforting to know that (5.132) is all[8] what is covered by the first main law.

5.3.6 *Entropy and the Second Main Law*

There is still another additive, transportable quantity, namely entropy S. We finally make use of the observation that each material points is always in thermal equilibrium.

Let us pick a material point with mass M. Its volume $V = V(t, \boldsymbol{x})$ will change in the course of time and consequently its mass density $\varrho = \varrho(t, \boldsymbol{x})$.

We shall denote by $s = \sigma(S)$ the specific entropy, the entropy per mass unit. One may write

$$\mathrm{d}S = \mathrm{d}t M \mathrm{D}_t s = \mathrm{d}t V \varrho \mathrm{D}_t s. \tag{5.133}$$

In the preceding section we have learned that the internal energy U is a thermodynamic potential such

$$\mathrm{d}U = T\mathrm{d}S + \mathrm{d}W + \sum_a \mu^a \mathrm{d}N^a \tag{5.134}$$

holds true. T is the temperature of the material point, the μ^a are chemical potentials.

[8] In the absence of external magnetic fields.

An explicit expression for $\mathrm{d}U = \mathrm{d}t\,V\,\mathrm{D}_t u$ has been derived before:

$$\mathrm{d}U = \mathrm{d}t\mathrm{d}V\left(-\nabla_i J_i^u + T'_{ij}G_{ij} + T''_{ij}G_{ij} + J_i^{\mathrm{e}\prime}E_i + J_i^{\mathrm{e}\prime\prime}E_i\right). \tag{5.135}$$

Recall that $\boldsymbol{J}^u$ is the heat current, G_{ij} the symmetrized velocity gradient, T_{ij} the stress tensor and $\boldsymbol{J}^{\mathrm{e}}$ the electric conduction current density. The latter two split into a reversible and irreversible, or elastic and dissipative part, as indicated by one or two primes.

The reversible, or elastic contributions are associated with work,

$$\mathrm{d}W = \mathrm{d}t\mathrm{d}V\left(T'_{ij}G_{ij} + J_i^{\mathrm{e}\prime}E_i\right). \tag{5.136}$$

Changes of chemical composition are expressed by

$$\mathrm{d}N^a = M\mathrm{d}t\,\mathrm{D}_t\frac{n^a}{\varrho} = \mathrm{d}t\mathrm{d}V\left(-\nabla_i J_i^a + \sum_r \Gamma^r \nu^{ra}\right). \tag{5.137}$$

Recall that $\boldsymbol{J}^a$ is the conduction part of the respective particle current density, or the diffusion current density. The ν^{ra} are stoichiometric coefficients and Γ^r the number of chemical reactions of type r per unit time and unit volume.

Now insert (5.133)–(5.137) into (5.131). One obtains

$$T\varrho\mathrm{D}_t s = -\nabla_i J_i^u + T''_{ij}G_{ij} + J_i^{\mathrm{e}\prime\prime}E_i + \sum_a \mu^a \nabla_i J_i^a + \sum_r \Gamma^r A^r. \tag{5.138}$$

Recall that

$$A_r = -\sum_a \nu^{ra}\mu_a \tag{5.139}$$

is the chemical affinity for a reaction of type r.

We just have to divide (5.138) by the temperature T and arrive at the balance equation $\nabla_t \varrho s + \nabla_i j_i(S) = \pi(S)$ for entropy. The entropy current is

$$\boldsymbol{j}(S) = \varrho s\boldsymbol{v} + \frac{1}{T}\boldsymbol{j}^u - \sum_a \frac{\mu^a}{T}\boldsymbol{J}^a. \tag{5.140}$$

The volumetric entropy production rate consist of five, and only five terms[9]:

$$\pi(S) = \pi^{\mathrm{hc}} + \pi^{\mathrm{df}} + \pi^{\mathrm{fr}} + \pi^{\mathrm{jh}} + \pi^{\mathrm{ch}}. \tag{5.141}$$

The first term

$$\pi^{\mathrm{hc}} = J_i^u \nabla_i \frac{1}{T} \tag{5.142}$$

[9] Again, if magnetic effects are neglected.

describes entropy production because of heat conduction. The next term

$$\pi^{\text{df}} = -\sum_a J_i^a \nabla_i \frac{\mu^a}{T} \tag{5.143}$$

takes diffusion into account. Internal friction contributes to entropy production by

$$\pi^{\text{fr}} = \frac{1}{T} T''_{ij} G_{ij}. \tag{5.144}$$

The forth term says that Joule's heat not only increases internal energy, but also entropy:

$$\pi^{\text{jh}} = \frac{1}{T} J_i^{\text{e}\,''} E_i, \tag{5.145}$$

with $\boldsymbol{E}$ as electric field strength. Entropy is also produced if chemical reactions take place, so the last term:

$$\pi^{\text{ch}} = \frac{1}{T} \sum_r \Gamma^r A^r. \tag{5.146}$$

And this is the second main law:

$$\pi(S) \geq 0. \tag{5.147}$$

Normally it is formulated as a general statement. But we have also nailed down the mechanisms of entropy production. Each single contribution to (5.141) must be positive. Take Ohm's law $\boldsymbol{j}^{\text{e}\,''} = \sigma \boldsymbol{E}$ for an example. The second main law requires that the conductivity σ cannot be negative.

It is not true that entropy must increase forever, as one can hear frequently. A system may loose entropy by outflow. However, if irreversible or dissipative processes occur, these are always associated with positive entropy production.

5.3.7 *Summary*

We presented a typical example for the concept of a material point. One cubic millimeter of water vapor still contains 10^{16} molecules leading to relative fluctuations of 10^{-8}. Quantities, currents and production rates are volume or surface integrals over appropriate densities.

The generic balance equation is derived and one for each additive and transportable quantity. These quantities are particle numbers, mass, electric charge, linear and angular momentum, kinetic, potential and internal energies as well as entropy. Currents are to be split into a convection and a conduction part. Only the latter transforms as a proper vector field, and we reformulate the balance equation with the aid of the substantial time derivative in a covariant manner.

A concrete definition of heat and work is derived, and we discuss the volumetric production rate of internal energy. This is the first main law. Likewise, the volumetric production rate of entropy is classified into effects. It cannot be negative: this is the message of the second main law. New to some readers might be the stress tensor, the diffusion and heat current densities as well as a description of chemical reactions.

5.4 Second Quantization

In this short section we will generalize a concept which already has been discussed in Sect. 2.5 on Small Oscillations. We define a particle number operator, generalize to many one-particle states and form the Fourier transform which turns out to be nothing else but a local quantum field. Boson fields obey commutation relations, fermions are describe by formally equal expression, with the commutator replace by an anti-commutator. As we shall see, a quantum field theory avoids the problem of wave functions with a fixed number of arguments and treats identical particles correctly, right from the start.

5.4.1 Number Operators

Assume a linear operator a which commutes with its adjoint $a^\star$ according to

$$[a, a^\star] = I. \tag{5.148}$$

Also assume a vector Ω such that

$$a\Omega = 0, \tag{5.149}$$

the zero vector. We shall now demonstrate why

$$N = a^\star a \tag{5.150}$$

is a number operator. Well,

$$N\Omega = 0 = 0\Omega \tag{5.151}$$

is obvious. The spectrum of N contains the eigenvalue 0. Let us define

$$\chi_n = \frac{1}{\sqrt{n!}}(a^\star)^n\Omega, \tag{5.152}$$

essentially applying n times the linear operator $a^\star$ to Ω, where $n \in \mathbb{N}$ is a whole number. $N\chi_0 = 0\chi_0$ has already be shown, see (5.151). Assume that the finding

$$N\chi_n = n\chi_n \tag{5.153}$$

were true up to the index n. Then assumption (5.152) is also true for $n + 1$ because of

$$N\chi_{n+1} = \frac{1}{\sqrt{n+1}} a^\star a a^\star \chi_n = \frac{1}{\sqrt{n+1}} a^\star (N+1)\chi_n = (n+1)\chi_{n+1}. \tag{5.154}$$

If Ω is normalized, $(\Omega, \Omega) = 1$, so is χ_n because of

$$(\chi_{n+1}, \chi_{n+1}) = \frac{1}{n+1}(a^\star\chi_n, a^\star\chi_n) = \frac{1}{n+1}(\chi_n, (N+1)\chi_n) = 1. \tag{5.155}$$

Again we have invoked the principle of complete induction.

N should be interpreted as the number of excitations, or particles, and Ω as the vacuum. $a^\star$ increases the number of particles by one, it is a creation operator. Likewise, the operator a decreases the number of particles by one, it is a annihilation or destruction operator. This can be inferred from

$$\chi_n = \frac{1}{\sqrt{n+1}}\, a\, \chi_{n+1}, \tag{5.156}$$

for $n \in \mathbb{N}$.

Assume a product of n' annihilation and n'' creation operators. The commutation rule (5.148) allows to simplify such an expression to a sum of products with $n' - 1$ annihilators and $n'' - 1$ creators. If n' differs from n'', either annihilators or creators survive. As a consequence, eigenvectors with different particle numbers are orthogonal,

$$(\chi_{n''}, \chi_{n'}) = \delta_{n''n'}. \tag{5.157}$$

The set of all linear combinations $\chi = \sum_n c_n\chi_n$ with complex numbers c_n is a Hilbert space, a Fock space.[10] It is generated by a normalized vacuum vector Ω, an annihilator a, a creator $a^\star$, its adjoint, and by the commutation relation $[a, a^\star] = I$. Within this space, the linear operator $N = a^\star a$ is self-adjoint and has non-degenerate eigenvalues which coincide with the set $\mathbb{N}$ of whole numbers.

[10] We omit technical details, e.g. completion by adding all convergent Cauchy sequences.

5.4.2 Plane Waves

In order to avoid non-essential complications we approximate the three-dimensional space by a finite cube with periodic boundary conditions.

The locations are in $\boldsymbol{x} \in [-L/2, L/2]^3$, and all functions f shall obey $f(x_1 + L, x_2, x_3) = f(x_1, x_2, x_3)$, and so on. The volume of this region is $V = L^3$.

Plane waves are described by

$$f_{\boldsymbol{k}}(\boldsymbol{x}) = \frac{1}{\sqrt{L^3}} \mathrm{e}^{\mathrm{i}\boldsymbol{k} \cdot \boldsymbol{x}}. \tag{5.158}$$

These functions obey the boundary conditions if $\boldsymbol{k} = 2\pi \boldsymbol{j}/L$ where the $\boldsymbol{j}$ are three integer numbers. The allowed wave vectors are evenly spaced in the entire three-dimension $\boldsymbol{k}$-space.

We calculate

$$\int \mathrm{d}^3x\, f^*_{\boldsymbol{k}''}(\boldsymbol{x}) f_{\boldsymbol{k}'}(\boldsymbol{x}) = \delta^3_{\boldsymbol{k}''\boldsymbol{k}'}. \tag{5.159}$$

Note that the integral extends over a cube of volume L^3. The δ^3 Kronecker symbol on the right hand side says that it evaluates to 1 if the two wave vectors are equal and to zero otherwise. This makes sense because the wave vectors in our model space are discrete, and not continuously distributed, as will be the case for the limit $L \to \infty$. The sum over all allowed wave vectors of plane waves at different positions can be worked out as

$$\sum_{\boldsymbol{k}} f^*_{\boldsymbol{k}}(\boldsymbol{x}'') f_{\boldsymbol{k}}(\boldsymbol{x}') = \delta^3(\boldsymbol{x}'' - \boldsymbol{x}'). \tag{5.160}$$

This time the symbol δ^3 stands for the well-known Dirac, or delta distribution. It is defined by

$$\int \mathrm{d}^3y\, \delta^3(\boldsymbol{x} - \boldsymbol{y})\, g(\boldsymbol{y}) = g(\boldsymbol{x}). \tag{5.161}$$

We say more about distributions in the chapter on Mathematical Aspects.

5.4.3 Local Quantum Field

For each allowed wave vector $\boldsymbol{k}$ there is an operator $a_{\boldsymbol{k}}$ and its adjoint $a^\star{}_{\boldsymbol{k}}$. They commute according to

$$[a_{\boldsymbol{k}''}, a_{\boldsymbol{k}'}] = 0 \text{ and } [a_{\boldsymbol{k}''}, a^\star{}_{\boldsymbol{k}'}] = \delta_{\boldsymbol{k}''\boldsymbol{k}'}. \tag{5.162}$$

Note that $[a^\star{}_{\boldsymbol{k}''}, a^\star{}_{\boldsymbol{k}'}] = 0$ is a consequence of (5.162). This equation expresses that the creation or annihilation of particles in different modes $\boldsymbol{k}''$ and $\boldsymbol{k}'$ are not correlated.

We define the quantum field $A = A(\boldsymbol{x})$ by

$$A(\boldsymbol{x}) = \sum_{\boldsymbol{k}} f_{\boldsymbol{k}}(\boldsymbol{x})\, a_{\boldsymbol{k}}. \tag{5.163}$$

Its commutation relations are

$$[A(\boldsymbol{x}''), A(\boldsymbol{x}')] = 0 \text{ and } [A^\star(\boldsymbol{x}''), A^\star(\boldsymbol{x}')] = 0 \tag{5.164}$$

as well as

$$[A(\boldsymbol{x}''), A^\star(\boldsymbol{x}')] = \delta^3(\boldsymbol{x}'' - \boldsymbol{x}'). \tag{5.165}$$

These relations follow from (5.162) and (5.160).

We speak of a field because $A = A(\boldsymbol{x})$ depends on location. It is a quantum field because its values are operators, or q-numbers in old terminology. It is called a local field because the field and its adjoint at different locations commute.

The vacuum state Ω is characterized by

$$A(\boldsymbol{x})\, \Omega = 0 \text{ and } (\Omega, \Omega) = 1. \tag{5.166}$$

Indeed, because of

$$a_{\boldsymbol{k}} = \int \mathrm{d}^3x\, f^*_{\boldsymbol{k}}(\boldsymbol{x}) A(\boldsymbol{x}), \tag{5.167}$$

Equation (5.166) guarantees that the vacuum Ω is void of particles, irrespective of their wave vector.

Note that the observable $N_{\boldsymbol{k}} = a_{\boldsymbol{k}}{}^\star a_{\boldsymbol{k}}$ is a number operator. It counts the number of particles of wave vector $\boldsymbol{k}$ or momentum $\boldsymbol{p} = \hbar\boldsymbol{k}$. The number of A-particles[11] is

$$N_A = \sum_{\boldsymbol{k}} N_{\boldsymbol{k}} = \int \mathrm{d}^3x\, A^\star(\boldsymbol{x}) A(\boldsymbol{x}), \tag{5.168}$$

as a consequence of (5.159). Note that the total number N_A of A-particles is an observable. If a process ϕ^{i} from an initial state to a final configuration ϕ^{f} is studied, the expectation values $(\phi^{\mathrm{i}}, N_A\phi^{\mathrm{i}})$ and $(\phi^{\mathrm{f}}, N_A\phi^{\mathrm{f}})$ may differ. A-particles may vanish or be created, N_A in general is not a conserved quantity.

5.4.4 Fermions

The discussion so far was about bosons. A plane wave state $f = f_{\boldsymbol{k}}$ can be occupied by $0, 1, 2, \ldots$ particles. As we know, any particle has a spin angular momentum

[11] Particles described by the quantum field A.

which may assume an integer multiple of $\hbar$ or a half-integer multiple $1/2, 3/2, \ldots$ Integer values characterize bosons, half-integer values describe fermions. The structure particles of matter are fermions: electrons, quarks, and neutrinos and their respective anti-particles. There are three, and only three families of such structure fermions. The interaction between them are transmitted by bosons, such as the mass-less photon, the three massive W^+, $W^0 = Z$, W^- bosons, gluons, possibly a graviton and also the Higgs boson.

The notion of being boson or fermion is not restricted to fundamental particles. Composite particles like the proton (made up of two up-quarks and a down-quark), the neutron (one up-quark, two down-quarks), the He^4 nucleus and so on are either fermions (proton, neutron) or bosons (He^4).

As contrasted with bosons, fermions may occupy the same one-particle state either not at all or at most once. Hence, the number operator N of (5.150) and its subsequent usage has to be changed: we simply have to replace the commutator $[X, Y] = XY - YX$ by the anti-commutator

$$\{X, Y\} = XY + YX. \tag{5.169}$$

Let us denote a fermion annihilator by b. It obeys the fermion counterpart of (5.149), namely

$$\{b, b\} = 0 \text{ and } \{b, b^\star\} = I. \tag{5.170}$$

Note that $\{b^\star, b^\star\} = 0$ is a consequence. Both expressions, bb as well as $b^\star b^\star$, vanish, and $bb^\star + b^\star b = I$.

As before, $N = b^\star b$ is the particle number observable. Because b annihilates the vacuum, we obtain $N\Omega = 0 = 0\Omega$. The value 0 is an eigenvalue of the particle number operator. For the state $b^\star\Omega$ we calculate

$$Nb^\star\Omega = b^\star bb^\star\Omega = b^\star(-b^\star b + I)\Omega = b^\star\Omega. \tag{5.171}$$

Therefore the value 1 is also an eigenvalue of the particle number operator. Larger values are not possible because $b^{\star n}$ vanishes for $n > 1$. Fermions can occupy a state either not at all or once. This finding is also known as Pauli's exclusion principle.

The above can easily be generalized to more than one single particle state. Such states are characterized by $(\sigma, \boldsymbol{k})$ where $\boldsymbol{k}$ is an allowed wave vector and σ denotes the spin projection onto a fixed axis. In the case of electrons or protons, σ has the values $+1/2$ and $-1/2$, or $\uparrow$ and $\downarrow$. There are fermion annihilators $b_{\sigma\boldsymbol{k}}$ which anti-commute according to

$$\{b_{\sigma''\boldsymbol{k}''}, b_{\sigma'\boldsymbol{k}'}\} = 0 \text{ and } \{b_{\sigma''\boldsymbol{k}''}, b^\star{}_{\sigma'\boldsymbol{k}'}\} = \delta_{\sigma''\sigma'}\delta^3_{\boldsymbol{k}''\boldsymbol{k}'}I. \tag{5.172}$$

Note that $\{b^\star{}_{\sigma''\boldsymbol{k}''}, b^\star{}_{\sigma'\boldsymbol{k}'}\} = 0$ is a consequence.

We define the corresponding fermion fields $B_\sigma = B_\sigma(t, \boldsymbol{x})$ by

$$B_\sigma(t, \boldsymbol{x}) = \sum_{\boldsymbol{k}} f_{\boldsymbol{k}}(t, \boldsymbol{x}) b_{\sigma \boldsymbol{k}}. \tag{5.173}$$

The commutation rules are similar to (5.164) and (5.165), with commutators replaced by anti-commutators, namely[12]

$$\{B_{\sigma''}(t, \boldsymbol{x}''), B_{\sigma'}(t, \boldsymbol{x}')\} = 0 \text{ and } \{B^\star{}_{\sigma''}(t, \boldsymbol{x}''), B^\star{}_{\sigma'}(t, \boldsymbol{x}')\} = 0 \tag{5.174}$$

as well as

$$\{B_{\sigma''}(t, \boldsymbol{x}''), B^\star{}_{\sigma'}(t, \boldsymbol{x}')\} = \delta_{\sigma''\sigma'}\, \delta^3(\boldsymbol{x}'' - \boldsymbol{x}'). \tag{5.175}$$

Again, the fermion annihilation operators can be constructed from the local quantum field by

$$b_{\sigma \boldsymbol{k}} = \int \mathrm{d}^3 x\, f^*_{\boldsymbol{k}}(t, \boldsymbol{x})\, B_\sigma(t, \boldsymbol{x}), \tag{5.176}$$

and the creation operators by forming the adjoint expression. With a normalized vacuum state Ω which is annihilated by the components of the fermion field,

$$B_\sigma(t, \boldsymbol{x})\Omega = 0, \tag{5.177}$$

the Fock space of fermions can be constructed as described above.

$$N(t) = \sum_\sigma \int \mathrm{d}^3 x\, B^\star{}_\sigma(t, \boldsymbol{x}) B_\sigma(t, \boldsymbol{x}) \tag{5.178}$$

is the total number of B-fermions, an observable.

5.4.5 Some Observables

Obviously, the particle density, an observable, is given by

$$n(\boldsymbol{x}) = A^\star(\boldsymbol{x}) A(\boldsymbol{x}). \tag{5.179}$$

This is compatible with (5.168). It follows that

$$\frac{\hbar}{\mathrm{i}} A^\star(\boldsymbol{x}) \nabla A(\boldsymbol{x}) \tag{5.180}$$

is the linear momentum density and

[12] The Poisson bracket and its quantum analog have been denoted by curly brackets as well.

$$-\frac{\hbar^2}{2m} A^\star(\boldsymbol{x}) \Delta A(\boldsymbol{x}) \tag{5.181}$$

the kinetic energy density. We conclude that

$$H = -\frac{\hbar^2}{2m} \int \mathrm{d}^3 x\, A^\star(\boldsymbol{x})\, \Delta\, A(\boldsymbol{x}) = \frac{\hbar^2}{2m} \int \mathrm{d}^3 x \parallel \nabla A(\boldsymbol{x}) \parallel^2 \tag{5.182}$$

represents the kinetic energy observable.

In general, the energy for particles in an external potential U and interacting by a potential V can be written as

$$\begin{aligned} H = &\frac{\hbar^2}{2m} \int \mathrm{d}^3 x \parallel \nabla A(\boldsymbol{x}) \parallel^2 \\ &+ \int \mathrm{d}^3 x\, A^\star(\boldsymbol{x})\, U(\boldsymbol{x})\, A(\boldsymbol{x}) \\ &+ \int \mathrm{d}^3 x' \int \mathrm{d}^3 x''\, A^\star(\boldsymbol{x}') A^\star(\boldsymbol{x}'')\, V(\boldsymbol{x}'' - \boldsymbol{x}')\, A(\boldsymbol{x}'') A(\boldsymbol{x}'). \end{aligned} \tag{5.183}$$

The corresponding expressions for fermions look similar, we do not write them here.

5.4.6 Time

We spared the time dimension so far and shall remedy this deficiency now.

The primary concept is a plane wave of a single particle. It propagates in space by a plain wave as described in (5.158). We will take time t into account by writing

$$f_{\boldsymbol{k}}(t, \boldsymbol{x}) = \mathrm{e}^{-\mathrm{i}\omega(\boldsymbol{k})\, t} f_{\boldsymbol{k}}(\boldsymbol{x}). \tag{5.184}$$

For each momentum $\hbar\boldsymbol{k}$ there is an energy $\hbar\omega = \hbar\omega(\boldsymbol{k})$. Slow free massive particles are characterized by $\omega(\boldsymbol{k}) = \hbar\boldsymbol{k}^2/2m$, where m is the particle mass. Free photons obey $\omega(\boldsymbol{k}) = c\,|\boldsymbol{k}|$ with c as the velocity of light in free space. Quasi-particles like electrons in the conduction band of a solid or phonons, the quanta of lattice vibrations, are governed by their own dispersion relations $\omega = \omega(\boldsymbol{k})$ which depend on the environment in which they propagate.

The time-dependent quantum field is defined as

$$A(t, \boldsymbol{x}) = \sum_{\boldsymbol{k}} f_{\boldsymbol{k}}(t, \boldsymbol{x})\, a_{\boldsymbol{k}}, \tag{5.185}$$

with the plane waves of (5.184). The commutation relations (5.164) and (5.165) remain true as long as they refer to the same time. This is an allowed statement in a Galilei world with its notion of a universal time. Relativistic field theories must do better.

5.4.7 *Summary*

Starting with creation and annihilation operators for a complete set of single particle states we introduced local fields. They are fields since they are functions of space and time. They are quantum fields because the field strength is a linear operator. The appropriate Hilbert space, a so-called Fock space, accommodates an arbitrary number of particles, its states automatically are symmetric or antisymmetric with respect to particle permutations. The fields obey local commutation relations, i.e. they commute or anti-commute at different locations. Bosons, or integer spin particles, respect ordinary commutation relations while fermions, particles with half-integer spin, are reigned by anti-commutators.

5.5 Gases

The molecules of a gas are far apart from each other at random locations and move with random velocities. Matter in its liquid state is made up of densely packed particles which may glide easily one over the other. Usually, there is also a solid phase where the particles are very close to each other and arranged regularly over large distances. Gases are distinguished by relatively low particle density and high compressibility. With liquids they have in common that there are no static shear forces. Gases and liquids are fluids: a body will not maintain its form.

Since normally the molecules of a gas are far apart, inter-molecular forces can be neglected, except for collisions, and we speak of an ideal gas. Although there is no direct interaction, it plays a role whether the identical particles are bosons or fermions, i.e. whether they have integer or half-integer spin.

We derive expressions for an ideal Fermi gas and an ideal Bose gas. At low values of the chemical potential both behave alike and we speak of an ideal Boltzmann gas.

Fermi gases exert a so-called degeneracy pressure even at zero temperature which may stabilize a star in its final stage as white dwarf or neutron star.

Bose gases are special: if the density surpasses a critical density, a growing fraction of particles condenses to the quantum-mechanical ground state in form of a superposition. Supra-fluidity and supra-conductivity may be observed, truly strange effects.

We also discuss the photon gas, or black-body radiation. It behaves as a boson gas at zero chemical potential since the number of photons cannot be controlled.

The heat capacity of the ideal gas is affected by internal degrees of freedom. We show how.

And finally, within the framework of classical statistical mechanics, we indicate how to correct for inter-particle interactions. The inter-atomic potential may be determined by fitting experimental thermodynamic data.

5.5.1 Fermi Gas

Denote by $b_{\sigma \boldsymbol{k}}$ a fermion annihilation operator for a single particle state (or mode)

$$f_{\boldsymbol{k}}(\boldsymbol{x}) = \frac{1}{\sqrt{L^3}} \mathrm{e}^{\mathrm{i}\boldsymbol{k} \cdot \boldsymbol{x}} . \tag{5.186}$$

The region within which the particles are confined is a cube of side length L with periodic boundary conditions, for simplicity. σ is a polarization index, for instance positive or negative helicity, if the particles have spin=1/2.

The adjoint $b^{\star}{}_{\sigma \boldsymbol{k}}$ is the associated creation operator. We study fermions and postulate

$$\{b_{\sigma' \boldsymbol{k}'}, b_{\sigma'' \boldsymbol{k}''}\} = 0 \tag{5.187}$$

for the annihilation operators; the same holds true for the creation operators. Creators and annihilators anti-commute as follows:

$$\{b_{\sigma' \boldsymbol{k}'}, b^{\star}{}_{\sigma'' \boldsymbol{k}''}\} = \delta_{\sigma' \sigma''} \delta^3_{\boldsymbol{k}' \boldsymbol{k}''}. \tag{5.188}$$

Recall that $\{X, Y\} = XY + YX$ is the anti-commutator, a symmetrized product.

The Hamiltonian of our ideal Fermi gas is

$$H = \sum_{\sigma \boldsymbol{k}} \epsilon_{\boldsymbol{k}} \, N_{\sigma \boldsymbol{k}}, \tag{5.189}$$

where the one-particle energies do not depend on polarization:

$$\epsilon_{\boldsymbol{k}} = \frac{\hbar^2}{2m} | \boldsymbol{k} |^2, \tag{5.190}$$

for slow massive particles. $N_{\sigma \boldsymbol{k}}$ is the occupation number of the one-particle state $\sigma \boldsymbol{k}$. Its eigenvalues are zero or one: void or occupied. This is a consequence of the fermion anti-commutation rules. Applying the creation operator twice[13] results in $(b^{\star}{}_{\sigma \boldsymbol{k}})^2 \Omega = 0$. No mode is occupied by more than one fermion of the same species.

The total number of particles in the system is the observable

[13] Recall that Ω is the vacuum which is void of all particles.

$$N = \sum_{\sigma \boldsymbol{k}} N_{\sigma \boldsymbol{k}}. \tag{5.191}$$

We have to calculate the trace of

$$\mathrm{e}^{(\mu N - H)/k_\mathrm{B}T} = \prod_{\sigma \boldsymbol{k}} \mathrm{e}^{(\mu - \epsilon_{\sigma \boldsymbol{k}})\, N_{\sigma \boldsymbol{k}}/k_\mathrm{B}T}. \tag{5.192}$$

The trace, for each mode, is a sum of the eigenvalues of the occupation number $N_{\sigma \boldsymbol{k}}$, i.e. 0 and 1. Hence

$$\mathrm{Tr}\,\mathrm{e}^{(\mu N - H)/k_\mathrm{B}T} = \prod_{\sigma \boldsymbol{k}} \left\{ 1 + \mathrm{e}^{(\mu - \epsilon_{\boldsymbol{k}})/k_\mathrm{B}T} \right\}. \tag{5.193}$$

Thus, the grand-canonical free energy of a gas of non-interacting fermions is

$$F^* = -k_\mathrm{B}T \sum_{\sigma \boldsymbol{k}} \ln \left\{ 1 + \mathrm{e}^{(\mu - \epsilon_{\boldsymbol{k}})/k_\mathrm{B}T} \right\}. \tag{5.194}$$

For large volumes the allowed wave vectors get so close that the sum over them may be replaced by an integral,

$$\sum_{\boldsymbol{k}} = V \int \frac{\mathrm{d}^3 k}{(2\pi)^3}. \tag{5.195}$$

$V = L^3$ is the volume of our region, a cube of sidelength L. Since the one-particle energy levels do not depend on polarization, the sum over σ simply gives a factor 2 for spin $= 1/2$ particles like electrons. The free energy then becomes

$$F^* = -2k_\mathrm{B}TV \int \frac{\mathrm{d}^3 k}{(2\pi)^3} \ln \left\{ 1 + \mathrm{e}^{(\mu - \epsilon_{\boldsymbol{k}})/k_\mathrm{B}T} \right\}. \tag{5.196}$$

By the way, in our case the grand-canonical free energy depends on the external parameter V and on the chemical potential μ as well as the temperature T. The latter two are intensive variables, F^* and V are extensive. It follows that F^* is proportional to the volume, at least in the thermodynamic limit. This is well born out by (5.196).

We should generalize the above expression to arbitrary dispersion relations $\epsilon = \epsilon_{\boldsymbol{k}}$. For this we define

$$z(\epsilon) = 2 \int \frac{\mathrm{d}^3 k}{(2\pi)^3}\, \delta(\epsilon_{\boldsymbol{k}} - \epsilon). \tag{5.197}$$

$z = z(\epsilon)$ is the specific density of states: there are $V \mathrm{d}z(\epsilon)$ one-particle states, or modes, in the energy interval $[\epsilon, \epsilon + \mathrm{d}\epsilon]$. An alternative formulation of (5.196) reads

$$F^* = -k_B T V \int d\epsilon\, z(\epsilon)\, \ln\left\{1 + e^{(\mu - \epsilon)/k_B T}\right\}. \tag{5.198}$$

The number $\bar{N}$ of particles in a region of volume V is the negative partial derivative of F^* with respect to the chemical potential μ. We find

$$n(T, \mu) = -\frac{1}{V}\frac{\partial F^*}{\partial \mu} = \int d\epsilon\, z(\epsilon) \frac{1}{e^{(\epsilon - \mu)/k_B T} + 1}. \tag{5.199}$$

The particle density n depends on temperature and chemical potential. It is never negative and grows with μ. Note that $n(T, -\infty) = 0$.

The pressure is given by $p = -\partial F^*/\partial V$, i.e.

$$p(T, \mu) = k_B T \int d\epsilon\, z(\epsilon)\, \ln\left\{1 + e^{(\mu - \epsilon)/k_B T}\right\}. \tag{5.200}$$

Since the expression in curly brackets is always larger than 1, the logarithm is positive, and with it the pressure. The pressure increases with chemical potential.

For non-relativistic massive spin=1/2 fermions with $\epsilon_{\boldsymbol{k}} = \hbar^2 \boldsymbol{k}^2/2m$ one finds

$$z_0(\epsilon) = \frac{1}{\pi^2}\left\{\frac{2m}{\hbar^2}\right\}^{3/2} \sqrt{\epsilon}. \tag{5.201}$$

The pressure of such a gas is given by

$$p(T, \mu) = \frac{2}{3}\int_0^\infty d\epsilon\, z_0(\epsilon) \frac{\epsilon}{e^{(\epsilon - \mu)/k_B T} + 1}. \tag{5.202}$$

In the limit $T \to 0$ we calculate

$$n(0, T) = \int_0^\mu d\epsilon\, z_0(\epsilon) = \frac{2}{3\pi^2}\left\{\frac{2m}{\hbar^2}\right\}^{3/2} \mu^{3/2} \tag{5.203}$$

and

$$p(0, \mu) = \frac{2}{3}\int_0^\mu d\epsilon\, z_0(\epsilon)\, \epsilon = \frac{4}{15\pi^2}\left\{\frac{2m}{\hbar^2}\right\}^{3/2} \mu^{5/2}. \tag{5.204}$$

By eliminating the chemical potential μ we arrive at the following relation between pressure and particle density at absolute zero:

$$p = a\,\frac{\hbar^2}{m} n^{5/3}, \tag{5.205}$$

where a is a numerical factor,

$$a = \frac{1}{5}\left(\frac{3\pi^2}{2}\right)^{2/3} = 1.2058. \tag{5.206}$$

If a relatively small star as our sun has burnt all its nuclear fuel, it will end up with carbon or oxygen since these nuclei are the most strongly bound. Both nuclei have as many protons as neutrons, so that each electron goes with two nucleons. Electrons and nuclei form a locally neutral plasma with a certain electron density n and a mass density $\varrho = 2m_\text{p}n$. Even at zero temperature there is a pressure

$$p = a\,\frac{\hbar^2}{m_\text{e}}\left\{\frac{\varrho}{2m_\text{p}}\right\}^{5/3}. \tag{5.207}$$

This material equation replaces the ideal gas pressure law $p \propto T\varrho$. The degeneracy pressure (5.205) may stabilize a star once it cools down. It cools down because its fuel for nuclear reactions is exhausted. There is no heat source for a sufficiently high temperature and the associated radiation and thermal pressure. Unfortunately there is no room here to discuss neutron stars and the mass limit for them. If the dying star is too massive it will end up as a neutron star or as a black hole.

5.5.2 Bose Gas

We now repeat the above arguments for bosons. The photon is a boson, but also the helium atom with its total angular momentum 0. The quasi-electrons of a metal may, via lattice deformations, bind to pairs which then behave as bosons. Boson fields are governed by commutation rules instead of anti-commutators.

The commutators of annihilators $a_{\sigma\boldsymbol{k}}$ and of associated creation operators $a^*_{\sigma\boldsymbol{k}}$ vanish,

$$[a_{\sigma'\boldsymbol{k}'}, a_{\sigma''\boldsymbol{k}''}] = [a^*_{\sigma'\boldsymbol{k}'}, a^*_{\sigma''\boldsymbol{k}''}] = 0, \tag{5.208}$$

while

$$[a_{\sigma'\boldsymbol{k}'}, a^*_{\sigma''\boldsymbol{k}''}] = \delta_{\sigma'\sigma''}\delta^3_{\boldsymbol{k}'\boldsymbol{k}''} \tag{5.209}$$

holds true. Nearly the same as for fermions, but with commutators $[X, Y] = XY - YX$ instead of anti-commutators $\{X, Y\} = XY + YX$.

As we have mentioned various times before, the one-particle occupation number $N_{\sigma\boldsymbol{k}}$ now has eigenvalues $n = 0, 1, 2, \dots$ It is a natural number operator.

The chain of arguments is as above, with one exception. Instead of (5.193) we work out

$$\text{Tr}\,\text{e}^{(\mu N - H)/k_\text{B}T} = \prod_{\sigma\boldsymbol{k}}\left\{\frac{1}{1 - \text{e}^{(\mu - \epsilon_{\boldsymbol{k}})/k_\text{B}T}}\right\}. \tag{5.210}$$

On the right hand side you find the result of $1 + x + x^2 + \cdots = 1/(1 - x)$, the sum of a geometric series. However: caution! The geometric series converges only for $|\,x\,| < 1$. In our case we must restrict the validity of (5.210) by

$$\mu < \epsilon_{\min} = \min_{\boldsymbol{k}} \epsilon_{\boldsymbol{k}}. \tag{5.211}$$

The grand-canonical free energy for bosons may be written as

$$F^* = k_B T V \int \mathrm{d}\epsilon\, z(\epsilon)\, \ln \left\{ 1 - \mathrm{e}^{(\mu - \epsilon)/k_B T} \right\}. \tag{5.212}$$

The particle density reads

$$n(T, \mu) = \int \mathrm{d}\epsilon\, z(\epsilon)\, \frac{1}{\mathrm{e}^{(\epsilon - \mu)/k_B T} - 1}. \tag{5.213}$$

The fraction is obviously the mean occupation number of a single particle state. It ranges from zero to infinity for bosons, while the corresponding expression in (5.199) for fermions is between zero and one.

Let us compute the maximal particle density for spin=0 non-relativistic particles. We obtain

$$\begin{aligned} n_{\max}(T) &= \frac{1}{2\pi^2}\left(\frac{2m}{\hbar^2}\right)^{3/2} \int_0^\infty \mathrm{d}\epsilon\, \sqrt{\epsilon}\, \frac{1}{\mathrm{e}^{\epsilon/k_B T} - 1} \\ &= \frac{1}{2\pi^2}\left(\frac{2m}{\hbar^2}\right)^{3/2} (k_B T)^{3/2} \int_0^\infty \frac{\mathrm{d}x\, \sqrt{x}}{\mathrm{e}^{x} - 1} \\ &= b \left(\frac{m k_B T}{\hbar^2}\right)^{3/2}, \end{aligned} \tag{5.214}$$

where $b = 0.33174$ is a numerical constant. Convince yourself that (5.214) describes in fact a particle density.

What happens if there are more particles to accommodate, if n is larger than $n_{\max}$? One may show that the excess $n - n_{\max}$ will coherently occupy the system's ground state while $n_{\max}$ is distributed according to (5.213). This mixed state—coherently occupied ground state and a Bose gas—is characterized by maximal entropy. The ground state portion grows if the temperature is lowered. At zero temperature all particles are in the one-particle state of lowest energy.

The phenomenon we just described was proposed by Bose and Einstein, and one speaks of Bose–Einstein condensation. It became clear much later that Bose–Einstein condensation may explain supra-fluidity and supra-conductivity.

5.5.3 Black-Body Radiation

The photon behaves as a boson. Although it has only two polarization states, right hand or left hand circular polarization, it is described by a vector-vector field A_μ with an auxiliary condition. For massive particles there is always a frame of reference where the particle is at rest. Galilei transformations which respect this are rotations. The unitary irreducible representations of this so-called little group SU_2 are characterized by a spin quantum number which is either $0, 1, \dots$ (bosons) or $1/2, 3/2, \dots$ (fermions). The photon has no mass and there is no frame of reference where the particle is at rest. It has a different little group, namely E_2, the unitary irreducible representations of which are either one- or infinite-dimensional. The photon transforms according to a one-dimensional representation which is characterize by the projection of angular momentum onto the direction of motion, namely $\sigma = 1$. Because space reflexion is a symmetry, there is an equivalent state with $\sigma = -1$.

$\sigma = 1/2$ would also be possible, as for the anti-neutrino, or $\sigma = -1/2$ for the neutrino. However, these particles are not the same. Neutrinos and anti-neutrinos are fermions. The photon, however, which coincides with its anti-particle, is a boson with two polarization states.

The Photon Gas

Recall how we deduced the grand-canonical Gibbs state. Entropy should be maximal with the constraints that close-by operators must be states and that the average energy and particle number are prescribed. This is why we encounter three Lagrange multiplier, namely free energy, temperature and chemical potential. For massless particles, however, their average number cannot be prescribed. We need no corresponding Lagrange multiplier. This boils down to choosing $\mu = 0$.

We may apply our result for bosons. The one-particle energy is

$$\epsilon_{\boldsymbol{k}} = \hbar c |\, \boldsymbol{k} \,| \tag{5.215}$$

where c is the speed of light. The corresponding density of states is easily worked out:

$$z(\epsilon) = \frac{1}{\pi^2} \frac{\epsilon^2}{(\hbar c)^3}. \tag{5.216}$$

Therefore, the free energy of a photon gas is

$$F(T,V) = \frac{1}{\pi^2} k_B T V \frac{1}{(\hbar c)^3} \int_0^\infty \mathrm{d}\epsilon\, \epsilon^2 \ln\left\{1 - e^{-\epsilon/k_B T}\right\}$$
$$= \frac{1}{\pi^2} V \frac{(k_B T)^4}{(\hbar c)^3} \int_0^\infty \mathrm{d}x\, x^2 \ln\left\{1 - e^{-x}\right\}$$
$$= -\frac{1}{3\pi^2} V \frac{(k_B T)^4}{(\hbar c)^3} \int_0^\infty \mathrm{d}x \frac{x^3}{e^x - 1}. \quad (5.217)$$

The last step was partial integration. The integral has a value $\pi^4/15$, and we finally arrive at

$$F(T,V) = -\frac{\sigma}{3} V T^4 \quad (5.218)$$

where the so-called Stefan–Boltzmann constant σ is

$$\sigma = \frac{\pi^2}{15} \frac{k_B^4}{\hbar^3 c^3} = 5.670 \times 10^{-8}\,\mathrm{W\,m^{-2}\,K^{-4}}. \quad (5.219)$$

The internal energy of the photon gas is calculated as

$$U = F + TS = \sigma V T^4. \quad (5.220)$$

This is the Stefan–Boltzmann law: the energy density of a cavity—the walls of which have a temperature T—grows with temperature as T^4, the proportionality factor being a constant of nature.

The spectral intensity of this cavity, or black body radiation can be read off from the integral in the last line of (5.217). It is given by

$$\mathrm{d}\omega\, S(\omega) = \frac{1}{\pi^2} \frac{\hbar}{c^3} \frac{\mathrm{d}\omega\, \omega^3}{e^{\hbar\omega/k_B T} - 1}, \quad (5.221)$$

normalized such that

$$\int_0^\infty \mathrm{d}\omega\, S(\omega) = \sigma T^4 \quad (5.222)$$

holds true. Equation (5.221) is Planck's black body radiation law for the spectral intensity, i.e. for the energy per unit volume in the angular frequency interval $\mathrm{d}\omega$.

5.5.4 Boltzmann Gas

For large negative chemical potential both expressions (5.198) and (5.212) become equal:

$$F^* = -k_B T\, V\, e^{\mu/k_B T} \int \mathrm{d}\epsilon\, z(\epsilon)\, e^{-\epsilon/k_B T}. \quad (5.223)$$

A Legendre transformation $F = F^* + \mu N$, where $N = -\partial F^*/\partial\mu$ results in

$$F(T, V, N) = -N\, k_B T \,\ln \frac{V}{N}\left(\frac{m k_B T}{2\pi\hbar^2}\right)^{3/2}. \tag{5.224}$$

The free energy is extensive just as the particle number. The rest is intensive, namely V/N and T. So this is all right. The expression in brackets is an inverse length squared, and raised to the power 3/2 makes it an inverse volume. This dimensionally compensates the V. Therefore, the free energy (5.224) is an energy.

In fact, (5.224) is the result of a classical calculation in the thermodynamic limit. In classical physics, observables are functions living on phase space

$$\Omega = \{\omega \in \mathbb{R}^{6N} \,\big|\, \omega = (\boldsymbol{x}_1, \boldsymbol{p}_1, \boldsymbol{x}_2, \boldsymbol{p}_2, \ldots, \boldsymbol{x}_N, \boldsymbol{p}_N)\}. \tag{5.225}$$

The volume element of this phase space is

$$\mathrm{d}\Gamma = \frac{\mathrm{d}^3x_1 \mathrm{d}^3p_1}{(2\pi\hbar)^3}\frac{\mathrm{d}^3x_2 \mathrm{d}^3p_2}{(2\pi\hbar)^3}\cdots\frac{\mathrm{d}^3x_N \mathrm{d}^3p_N}{(2\pi\hbar)^3}. \tag{5.226}$$

If the particles are identical, a factor $1/N!$ in front takes into account that permutations of configurations are not counted multiply. In this case, the observables must also be invariant under such permutations.

The Hamiltonian of a system of identical non-interacting particles is

$$H(\omega) = \sum_j \frac{\boldsymbol{p}_j^2}{2m} + \sum_j V(\boldsymbol{x}_j). \tag{5.227}$$

$V(\boldsymbol{x})$ vanishes for $\boldsymbol{x} \in \mathcal{V}$ and returns plus infinity if the particle should be outside the region $\mathcal{V}$. It has the effect of confining the particles.

The free energy in the framework of classical physics is

$$F = -k_B T \,\ln \frac{1}{N!}\int_{\Gamma\in\Omega} \mathrm{d}\Gamma \; \mathrm{e}^{-H(\omega)/k_B T}. \tag{5.228}$$

With

$$\int \frac{\mathrm{d}p}{2\pi\hbar}\,\mathrm{e}^{-p^2/2mk_B T} = \sqrt{\frac{mk_B T}{2\pi\hbar^2}} \tag{5.229}$$

we obtain

$$F(T, V, N) = -k_B T \,\ln \frac{1}{N!} V^N \left(\frac{mk_B T}{2\pi\hbar^2}\right)^{3N/2}. \tag{5.230}$$

With $\ln N! = N \ln N + \ldots$ for large N this simplifies to (5.224).

At low particle densities, the nature of identical particles, whether boson or fermion, plays no role, and a treatment by classical statistical mechanics is adequate. We have also seen that the shape of the confining region $\mathcal{V}$ plays no role, only its volume $V = \text{vol}\,\mathcal{V}$. Moreover, to get a perfect match, Planck's constant must be included in the phase space volume element and a factor $N!$ for identical particles. This behavior goes under the name of Boltzmann statistics. It is the common limit of Bose–Einstein and Fermi statistics.

Pressure and Heat Capacity

The free energy (5.224) of an ideal Boltzmann gas allows to calculate the pressure and other thermodynamic properties. By partially differentiation with respect to the volume we find

$$p(T, V, N) = p = N\frac{k_B T}{V}. \tag{5.231}$$

This ideal gas law is only valid for low particle density because then the intermolecular interaction may safely be neglected.

Recall that there are N_A particles in one mole (Avogadro's number). We introduce the universal gas constant R as

$$R = N_A k. \tag{5.232}$$

With it the pressure law reads

$$p = \nu \frac{RT}{V} \quad \text{where} \quad \nu = \frac{N}{N_A}. \tag{5.233}$$

ν stands for the amount of substance measured in mole.

The heat capacity at constant volume is calculated from the second partial derivative with respect to temperature. The heat capacity is

$$C_V = \frac{3}{2}\nu R. \tag{5.234}$$

It neither depends on volume nor on temperature. $3/2$ reflects that there are three translational degrees of freedom each contributing with $R/2$. This value changes if molecular rotations or vibrations are taken into account.

The ideal gas approximation (5.231) delivers

$$\kappa_T = \frac{1}{p} \tag{5.235}$$

for the isothermal compressibility, with p the ideal gas pressure.

5.5.5 Rotating and Vibrating Molecules

So far we discussed the translational degrees of freedom only. However, atoms and molecules have internal degrees of freedom as well.

First of all, atoms or molecules may become excited. However, with a typical excitation energy $\Delta\epsilon$ of one electron volt, the typical temperature Θ of $k_B\Theta = \Delta\epsilon$ is in the 10,000 Kelvin range. This plays no role for gases under normal conditions of 300 K and one bar.

Molecules may vibrate. For small oscillation an approximation by a parabolic potential is quite good, and we shall elaborate on this here. Typical temperatures are in the thousand Kelvin range, and we shall show how vibrations freeze at lower temperatures.

Molecules may also rotate. It turns out that an approximation as a rigid rotator is quite sufficient. Typical temperatures are in the ten to hundred Kelvin range, at normal temperature we encounter saturation.

Let us focus on the simplest of all molecules, that of hydrogen. We have studied it Sect. 3.2 on the Hydrogen Molecule. We found that, in Born–Oppenheimer approximation, the two protons have a common electron cloud which instantaneously adapts to the nuclei's configuration. In fact, the two H⁻ ions, or protons interact by a potential $V(R)$ which depends on the distance between the two ions. Around its minimum at R_0 it may be approximated by a parabola. See Fig. 3.2.

In the expression for the rotational degree of freedom we replace the inter-nuclear distance R by its equilibrium value R_0. This amounts to treating the molecule as a rigid rotator. In particular, the contributions H_{vib} for vibrations of the molecule and H_{rot} for rotations commute.

The eigenvalues of the former are

$$\epsilon_{\text{vib}}(n) = n\hbar\Omega \tag{5.236}$$

for $n = 0, 1, \ldots$ where $\hbar\Omega$ is the energy spacing. It is proportional to the curvature of the potential at its minimum. For the hydrogen molecule, a value $\hbar\Omega = 0.54$ eV has been measured and calculated.

The rotational spectrum is

$$\epsilon_{\text{rot}}(l) = l(l+1)\frac{\hbar^2}{2\mu R_0^2}. \tag{5.237}$$

Each level is $(2l + 1)$-fold degenerate. μ is the ion's reduced mass, and the typical temperature θ_{rot} is defined by

$$k_B\theta_{\text{rot}} = \frac{\hbar^2}{2\mu R_0^2}. \tag{5.238}$$

For a hydrogen molecule, the $l = 1$ rotational level is excited by $2k_B\theta_{rot} = 15.1$ meV, corresponding to a wave length of 82.1 μm, i.e. in the far infrared. The typical temperature is $\theta_{rot} = 88$ K.

The free energy associated with rotations of one molecule is

$$F_{rot} = -k_B T \ln \sum_{l=0}^{\infty} (2l+1)\, e^{-l(l+1)\theta_{rot}/T}, \tag{5.239}$$

a function of the temperature only, not of volume. The contribution to both heat capacities, C_V or C_p, is

$$C(T) = k_B \frac{\theta_{rot}^2}{T^2} \{ \langle L^4 \rangle - \langle L^2 \rangle^2 \} \tag{5.240}$$

where here

$$\langle \dots \rangle = \frac{\sum_{l=0}^{\infty} (2l+1) \dots e^{-l(l+1)\theta_{rot}/T}}{\sum_{l=0}^{\infty} (2l+1)\, e^{-l(l+1)\theta_{rot}/T}}. \tag{5.241}$$

For high temperatures one may approximate the sum over l by an integral. The result is

$$F_{rot} \approx -k_B T \ln \int_0^{\infty} dl\, 2l\, e^{-l^2\theta_{rot}/T} = -k_B T \ln \frac{T}{\theta_{rot}}. \tag{5.242}$$

The classical approximation gives a contribution

$$C \approx k \tag{5.243}$$

to the heat capacity, instead of the quantum theory result (5.240).

In fact, our result (5.240) is not yet the final one. The two nuclei were treated as distinct, but they are the same for a hydrogen molecule. A space reflexion should not result in a different state; the parity of the states must be even. Therefore, the sum in (5.239) must be restricted to even values of the angular momentum. The result is depicted in Fig. 5.3.

At very low temperatures the rotational degrees of freedom are frozen which is a typical quantum effect.

Air which is a mixture of mainly nitrogen and oxygen, has a much lower typical temperature than hydrogen. For air the classical approximation is good under normal conditions.

Molecular vibrations are characterized by much higher typical temperatures, such as $\theta_{vib} = 6300$ K for the hydrogen molecule H_2. At normal conditions, the vibrational

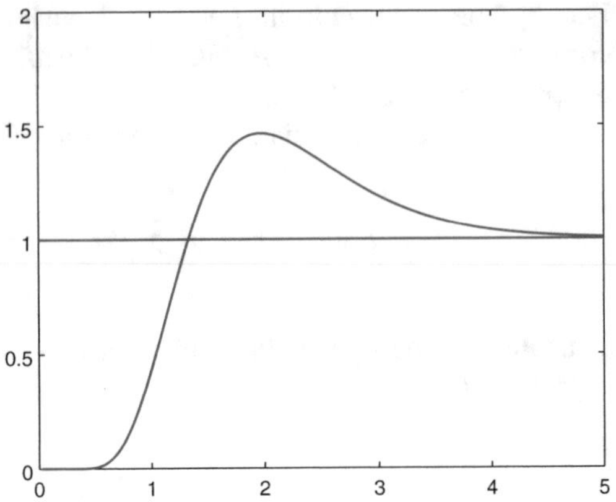

Fig. 5.3 Rotational heat capacity. The contribution to the heat capacity C (in units of k_B) by the rotational degree of freedom is plotted versus temperature in units of θ_{rot}. The *curve* is for homo-nuclear diatomic molecules like hydrogen. The *straight line* depicts the classical approximation

degrees of freedom are frozen and play no role. We here refrain from going into details.

5.5.6 Cluster Expansion

Up to now we have dealt with ideal gases: the interaction between molecules could be neglected. We shall now relax this restriction. To keep things simple we focus on a Boltzmann gas as described by the Hamiltonian function

$$H(\omega) = \sum_j \frac{\boldsymbol{p}_j^2}{2m} + \sum_{k>j} \phi(\boldsymbol{x}_k - \boldsymbol{x}_j). \tag{5.244}$$

There are N identical particles confined to a region $\mathcal{V}$ with volume V. $\phi = \phi(\boldsymbol{y})$ is the interaction potential, and the potential energy is a sum over pairs of particles.

The contribution of kinetic energy to the free energy is given by

$$\mathrm{e}^{-\beta F_{\text{kin}}} = \frac{V^N}{N!} \left(\frac{m}{2\pi^2\hbar^2\beta} \right)^{3N/2} \tag{5.245}$$

with $\beta = 1/k_B T$. For a large number N of particles at low particle density $n = N/V$ this amounts to

$$F_{\text{kin}} = -N\, k_B T \, \ln \frac{V}{N} \left(\frac{m k_B T}{2\pi\hbar^2} \right)^{3/2} . \tag{5.246}$$

Here we are interested in the addition F_{pot} to the free energy as described in

$$\mathrm{e}^{-\beta F_{\text{pot}}} = V^{-N} \int \mathrm{d}^3x_1 \mathrm{d}^3x_2 \ldots \mathrm{d}^3x_N \, \mathrm{e}^{-\beta \sum_{k>j} \phi(\boldsymbol{x}_k - \boldsymbol{x}_j)} . \tag{5.247}$$

Let us introduce the abbreviations

$$f(\boldsymbol{y}) = \mathrm{e}^{-\beta\phi(\boldsymbol{y})} - 1 \text{ and } f_{kj} = f(\boldsymbol{x}_k - \boldsymbol{x}_j). \tag{5.248}$$

With this (5.247) reads

$$F_{\text{pot}} = -k_{\text{B}}T \ln V^{-N} \int \mathrm{d}^{3N}x \prod_{k>j} (1 + f_{kj}). \tag{5.249}$$

The product is

$$\prod_{k>j} (1 + f_{kj}) = 1 + \sum f_{kj} + \sum\sum f_{lm} f_{kj} + \ldots \tag{5.250}$$

The single sum extends over pairs, the double sum over pairs of pairs, and so on.

We reorder the sum into clusters. The idea behind this is that f_{kj} vanishes unless the particles k and j are close-by. The potential ϕ and with it f vanish if $|\,\boldsymbol{x}_j - \boldsymbol{x}_k\,| > r_0$, i.e. if their distance exceeds the interaction range r_0.

If a term $f_{lj}\, f_{kj}$ does not vanish then a particle at $\boldsymbol{x}_k$ and another one at $\boldsymbol{x}_l$ are both close to a third particle at $\boldsymbol{x}_j$ and therefore close-by themselves. They form a 3-cluster. And so on.

Here we just work out the sum over all 2-clusters. These are products of f-factors without an index occurring twice or more. If such a contribution is made up of M factors we must calculate

$$V^{-N} \int \mathrm{d}^{3N}x \, \underbrace{f_{zy} \ldots f_{kj}}_{M} = V^{-2M} \left(\int \mathrm{d}^3x_1 \mathrm{d}^3x_2 \, f(\boldsymbol{x}_2 - \boldsymbol{x}_1) \right)^M . \tag{5.251}$$

We approximate the double integral by

$$V \int \mathrm{d}^3\xi \, f(\boldsymbol{y}) \tag{5.252}$$

with the following justification. Practically all spheres with interaction radius r_0 are in the region if its center is in the region. Exceptions are positions close to the boundary. Since for large volumes the surface layer of a finite small thickness r_0 becomes irrelevant the factor in front of (5.252) is simply V. Instead of (5.251) one may write

$$V^{-N} \int \mathrm{d}^{3N}x \, \underbrace{f_{zy} \ldots f_{kj}}_{M} = V^{-M} \left(\int \mathrm{d}^3\xi \, f(\boldsymbol{y}) \right)^M . \tag{5.253}$$

There are

$$\frac{1}{M!}\frac{(N-2)(N-3)}{2}\ \ldots,\ \frac{(N-2M+2)(N-2M+1)}{2} \tag{5.254}$$

such terms. In the thermodynamic limit $N \to \infty$, $V \to \infty$ but $N/V \to n$ we may simplify to

$$\frac{1}{M!}\left(\frac{N^2}{2}\right)^M . \tag{5.255}$$

The contributions of 2-clusters add up to

$$\sum_M \left(\frac{N^2}{2V}\right)^M \left(\int \mathrm{d}^3\xi\, f(\boldsymbol{x})\right)^M = \mathrm{e}^{(N^2/2V)\int \mathrm{d}^3\xi\, f(\boldsymbol{x})} . \tag{5.256}$$

The contribution of potential energy to the free energy thus expands as follows:

$$F_{\text{pot}} = k_{\text{B}}T\frac{N^2}{V}b_2(T) + \ldots \tag{5.257}$$

The so-called second virial coefficient $b_2 = b_2(T)$ is given by

$$b_2(T) = \frac{1}{2}\int \mathrm{d}^3\xi\, \left(1 - \mathrm{e}^{-\phi(\boldsymbol{y})/k_{\text{B}}T}\right). \tag{5.258}$$

The pressure law now read

$$p = k_{\text{B}}T\, \left\{n + b_2(T)n^2 + b_3(T)n^3 + \ldots\right\}. \tag{5.259}$$

The zeroth virial coefficient vanishes: no particles, no pressure. The first virial coefficient $b_1(T)$ has the value 1 for all gases. That justifies the name *ideal gas* because its behavior does not depend on the particle species.

Equation (5.259) is a virial expansion of the pressure into a power series in n, the particle density. It has a certain radius of convergence $\bar{n} = \bar{n}(T)$ which indicates a phase transition to the liquid state.

5.5.7 Joule–Thomson Effect

If a gas streams through a throttle, its velocity and pressure will change, but not its enthalpy. The gas enters the throttle of volume V with pressure p_1 and leaves it with pressure p_2. The passage is fast enough and there is no heat exchange. Before the throttle the internal energy is U_1, and after it $U_2 = U_1 + Vp_1 - Vp_2$. We conclude that the enthalpy $H = U + pV$ remains constant.

The pressure will change and with it the temperature. The Joule–Thomson coefficients $\mu = \mu(T, p)$ measures the temperature change $\mathrm{d}T$ brought about by a pressure increase $\mathrm{d}p$. It is defined by

$$\mu = \left.\frac{\mathrm{d}T}{\mathrm{d}p}\right|_{\mathrm{d}H=0} . \tag{5.260}$$

We require the enthalpy as a function of temperature and pressure which are not its natural variables. So we turn to the free energy of a real gas which we have just discussed. With entropy $S = -\partial F/\partial T$, pressure $p = -\partial F/\partial V$ and enthalpy $H = F + TS + pV$ one transforms

$$F(T, V, N) = -N k_B T \ln \frac{V}{N}\left(\frac{mk_B T}{2\pi\hbar^2}\right)^{3/2} - k_B T \frac{N^2}{V} b_2(T) \tag{5.261}$$

into

$$H(T, p, N) = \frac{5}{2} N k_B T + Np\left\{b_2(T) - T\, b_2'(T)\right\} . \tag{5.262}$$

$\mathrm{d}H = 0$ for small pressure p gives

$$\mu_{\mathrm{JT}}(T, 0) = \frac{2}{5k_B}\left\{T\, b_2'(T) - b_2(T)\right\} . \tag{5.263}$$

We restrict our result to small pressures since higher virial coefficients would be needed for higher pressures.

The Joule–Thomson effect—an expanding gas gets colder—is absent for the ideal gas! Historically, it served as an important hint that not all gases behave ideally. Technically, the effect is exploited in all sorts of cooling devices such as refrigerators, air conditioner and heat pumps.

Let us focus on the possibility to determine the inter-molecular potential.

To this purpose we assume a model potential with adjustable parameters with which we fit to experimental data. The Lenard-Jones 12-6 potential is often used:

$$\phi(r) = \phi_0 \left\{\left(\frac{r_0}{r}\right)^{12} - 2\left(\frac{r_0}{r}\right)^{6}\right\} . \tag{5.264}$$

The potential describes strong repulsion and short range attraction. It is minimal at $r = r_0$ with a minimum value $\phi(r_0) = -\phi_0$.

We have fitted experimental data for the second virial coefficient of argon by properly adjusting the potential parameters, see Fig. 5.4.

The best fitting potential parameters are $r_0 = 2.74$ Å and $\phi_0 = 18.8$ meV. You may easily convince yourself that such a potential cannot bind an Ar_2 molecule. Argon is a noble gas, after all.

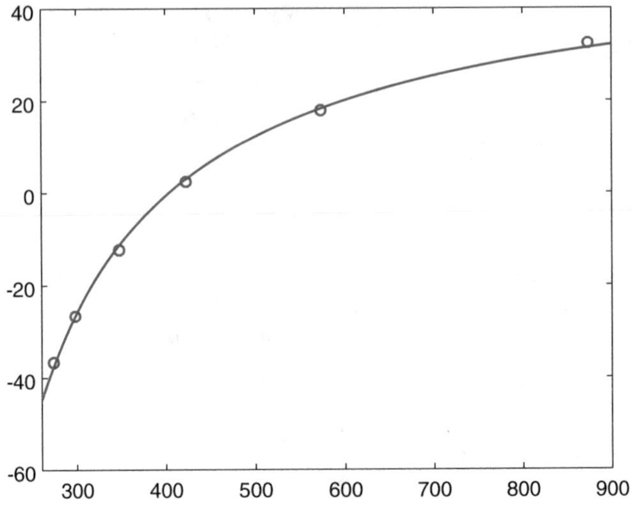

Fig. 5.4 Second virial coefficient of argon. Plotted over the temperature in Kelvin is the second virial coefficient of argon in Å^3. *Circles* represent experimental data. The parameters of a 12-6 Lennard-Jones potential were determined to fit the experimental data best. The resulting second virial coefficient is represented by a *full line*

With the potential we have just determined from fitting the second virial coefficient we now calculate the Joule–Thomson coefficient. It is depicted in Fig. 5.5.

5.5.8 Summary

In this section we have focused on gases. The free energy of all gases depends on the volume of the confining region only, not on its form; they are fluid media characterized by pressure.

A gas of fermions with negligible interaction, an ideal Fermi gas, will exert a pressure even at absolute zero. By this mechanism, white dwarfs or neutron stars are stabilized.

The ideal Bose gas, for instance of Helium atoms, has a critical, temperature dependent density. If there are too many particles, a fraction of them will condense in the one-particle ground state in a coherent manner. They cannot be scattered individually thereby exhibiting such strange phenomena as supra-fluidity or supra-conductivity.

Photons are bosons as well, but since they have no energy threshold, i.e. rest mass, their number cannot be controlled. They form a Bose gas at zero chemical potential. We have calculated the spectral intensity of such black body radiation.

At low densities, the difference between a Bose and a Fermi gas vanishes. The free energy of the so-called Boltzmann gas may be obtained either as a limit of a

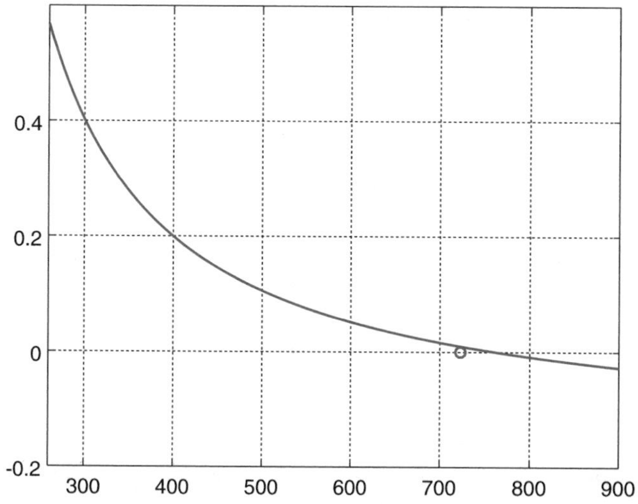

Fig. 5.5 Joule–Thomson coefficient. The low-pressure Joule–Thomson coefficient μ_{JT} (in Kelvin/bar) of argon is plotted versus temperature in Kelvin. The *solid line* represents values which have been calculated with an optimal 12-6 Lennard-Jones potential with just two adjustable parameters. The *circle* marks the measured inversion point. At lower temperatures, the gas becomes colder when expanding. Above the inversion point an expanding gas becomes hotter. The inversion point is predicted rather well

quantum gas for low chemical potential or within the framework of classical statistical mechanics. We found the ideal gas pressure law and discuss the additional degrees of freedom of molecular rotation and vibration. Rotational degrees of freedom freeze (become unnoticeable) at rather low temperatures and contribute with R to the molar heat capacity at high temperatures. In the intermediate range, subtle quantum effects are to be taken into account.

We finally dealt with a systematic procedure to incorporate inter-molecular forces. We expanded the potential energy contribution to the free energy into clusters. 2-clusters of close-by particles lead to a modification of the pressure law which is proportional to n^2, the particle density squared. 3-clusters modify it by an n^3 term, ans so on. The virial expansion of the pressure into powers of the particle density beaks down if a singularity shows up indicating a transition to the liquid phase.

The first correction to the ideal gas law, as described by the second virial coefficients, has been calculated and fitted to a molecular potential. With two adjustable parameters only, a wide range of experimental data can be described. The Joule–Thomson effect, which we discussed in some detail, is reproduced rather well. The best fitting potential for Argon cannot bind a molecule.

5.6 Crystal Lattice Vibrations

A crystal is nothing else but a huge molecule. In particular, the Born–Oppenheimer approximation is well justified. The nuclei are positioned at fixed places, and very rapidly the electrons attain there optimal distribution of lowest energy. The corresponding potential energy depends on the position of the nuclei, and the minimal total energy will describe the crystal's ground state.

Some of the electrons are strongly bound to their nucleus, others are only weakly localized or not at all; they form a cloud of shared electrons.

The crystal therefore may be thought of as a lattice of ions on the one hand and of a sea of quasi-free electrons. We here focus on the crystal lattice while the following section is about the quasi-free electrons.

We first present a phenomenological description of a solid medium. We than show that the excited states of the lattice are well described by independent oscillations, so called normal modes. Next, we discuss a simplified model, a periodic chain of ions. If the chain contains just one type of ion, a simple dispersion relation can be derived. If the chain is of the -A-B-A-B- type, the dispersion relation has two branches, namely an acoustical and an optical branch. For low wave numbers, the acoustical branch is the same as for sound waves. The optical branch is called so because it may cross the dispersion relation of infrared light.

Instead of saying that a certain oscillator is in its nth exited state—which are energetically equidistant—we prefer a parlance where the oscillator carries n phonons of the corresponding mode.

Phonons are Bose particles, and the vibrationally excited lattice is to be thought of as a phonon gas. In order to calculated its thermal properties one must know the density of phonon states. We present two standard approaches: the Debye model and its improvement, the Bose–Einstein model.

5.6.1 *Phenomenological Description*

A solid medium may support shear forces, in contrast to gases and liquids. Recall that the current density of linear momentum

$$j_i(P_j) = \varrho\, v_j v_i - T'_{ji} - T''_{ji} \tag{5.265}$$

has three contribution. The first in (5.265) is the convection current density: if matter flows, it carries its momentum with it. The rest, up to a conventional sign, is the stress tensor. It consists of a reversible contribution T'_{ij} and an irreversible part T''_{ij}.

Let us focus on a solid at rest. There is no momentum convection, no internal friction and no change of the momentum density. The balance equation for momentum reduces to

$$\partial_i T'_{ij} = f_j, \tag{5.266}$$

where f_j is the external force per unit mass.

Stress, Strain and Hooke's Law

In a solid, stress causes strain: the medium will be distorted. A material point at $\boldsymbol{x}$ in the undistorted state will be found at $\bar{\boldsymbol{x}} = \boldsymbol{x} + \boldsymbol{u}(\boldsymbol{x})$ after distortion. $\boldsymbol{u} = \boldsymbol{u}(\boldsymbol{x})$ is the displacement field. Two neighboring points $\boldsymbol{x}$ and $\boldsymbol{x} + \mathrm{d}\boldsymbol{x}$ before distortion have a distance ds as described by $ds^2 = |\,\mathrm{d}\boldsymbol{x}\,|^2 = \delta_{ij}\mathrm{d}x_j\mathrm{d}x_j$. After distortion their distance $\mathrm{d}\bar{s}$ squared will be $\mathrm{d}\bar{s}^2 = \mathrm{d}s^2 + 2S_{ij}\mathrm{d}x_i\mathrm{d}x_j$, with

$$2S_{ij} = \partial_i u_j + \partial_j u_i + (\partial_i u_k)(\partial_j u_k). \tag{5.267}$$

Because displacement gradients are usually small, the last term of (5.267) may safely be neglected, and we arrive at

$$\mathrm{d}\bar{s} = \mathrm{d}s + S_{ij}\mathrm{d}x_i\mathrm{d}x_j \tag{5.268}$$

with

$$S_{ji} = \frac{\partial_j u_i + \partial_i u_j}{2}, \tag{5.269}$$

the strain tensor. It is symmetric.

If the displacement field describes a rigid translation or a rigid rotation, the strain tensor vanishes. Hence, S_{ij} describes in fact the material's distortion.

No stress, by definition, produces no strain, and small stress will cause little strain. Strain and stress are proportional,

$$S_{kl} = \Lambda_{klij} T'_{ij}. \tag{5.270}$$

The forth rank tensor Λ is symmetric upon interchanging k with l, i with j, and the pairs $\{kl\}$ and $\{ij\}$. The latter holds true because Λ may be shown to be the second derivative of a free energy with respect to T_{ij} and T_{kl}. It follows that the fourth rank tensor Λ has at most 21 instead of 81 elements.

Elasticity Module and Poisson's Ratio

For an isotropic material there are just two independent contributions. They are usually introduced by the following relation:

$$S_{ij} = \frac{1+\nu}{E} T'_{ij} - \frac{\nu}{E}\delta_{ij} T'_{kk}. \tag{5.271}$$

E is called the elasticity constant or Young's modulus. It has the physical dimension of an energy per volume or a force per area, i.e. pressure. ν is called Poisson's ratio; it is dimensionless.

The mechanical properties of solids are characterized by two constants. For practical purposes, the limits of applicability of Hooke's law are more important. What is the maximal stress a material may support without irreversible changes?

Velocity of Sound

An oscillating solid medium is described by the following wave equation:

$$\varrho \ddot{u}_j = \frac{E}{2(1+\nu)} \left\{ \Delta u_j + \frac{1}{1-2\nu} \partial_j \partial_i u_i \right\}. \tag{5.272}$$

Friction has not been taken into account, we have neglected T''_{ij}. There are two types of acoustic waves

$$\boldsymbol{u}(t, \boldsymbol{x}) = \boldsymbol{a}\, \mathrm{e}^{-\mathrm{i}\omega t}\, \mathrm{e}^{\mathrm{i}\boldsymbol{k}\cdot\boldsymbol{x}}. \tag{5.273}$$

For $\boldsymbol{a} \parallel \boldsymbol{k}$ we find $\omega = c_{\mathrm{L}} |\, \boldsymbol{k}\, |$. The sound speed of such longitudinally polarized acoustic waves is

$$c_{\mathrm{L}} = \sqrt{\frac{1-\nu}{1+\nu)(1-2\nu)}} \sqrt{\frac{E}{\varrho}}. \tag{5.274}$$

For transversally polarized waves, i.e. for $\boldsymbol{a} \perp \boldsymbol{k}$, the sound speed is given by

$$c_{\mathrm{T}} = \sqrt{\frac{1}{2(1+\nu)}} \sqrt{\frac{E}{\varrho}}. \tag{5.275}$$

The two velocities of sound are easy to measure, so that the elasticity module E and Poisson's ratio ν may be determined. However, the latter values refer to adiabatic changes while the constants in Hooke's law are isothermal constant. An arbitrarily polarized sound wave is composed of a longitudinal mode and two degenerate transversal modes. They will split in T_1 and T_2 modes with different sound velocities if the solid is not isotropic.

5.6.2 *Phonons*

We will now discuss the propagation of vibrational excitations in the framework of quantum physics.

Born–Oppenheimer Approximation

Denote by $\bar{\boldsymbol{x}}_a$ the equilibrium position of ions, by $\boldsymbol{P}_a$ their linear momenta and by m_a and q_a the masses and charges, respectively. We introduce $\boldsymbol{X}_a$ as the deviation of the true ion position from its equilibrium value. The Hamiltonian of the solid made of N ions is

$$H = \sum_{a=1}^{N} \frac{\boldsymbol{P}_a^2}{2m_a} + V(\boldsymbol{X}_1, \boldsymbol{X}_2, \dots, \boldsymbol{X}_N). \tag{5.276}$$

The potential energy term contains the Coulomb repulsion between the ions and all terms referring to electrons. As said before, if the ion constellation changes, the potential energy will almost immediately assume its new minimum, so the Born–Oppenheimer approximation.

If the solid is slightly perturbed the ions will start to oscillated slightly about their equilibrium positions. The amplitudes of such oscillation will be small, and we may approximate the potential by

$$V(\boldsymbol{X}_1, \boldsymbol{X}_2, \dots, \boldsymbol{X}_N) = V_0 + \frac{1}{2}\sum_{ba} V''_{ba}\boldsymbol{X}_b\boldsymbol{X}_a + \dots. \tag{5.277}$$

Terms linear in deviations vanish because the potential energy in the equilibrium configuration is a minimum. The $3N \times 3N$ matrix V''_{ij} is obviously real and symmetric. Recall that there are $f = 3N$ degrees of freedom, namely three for each ion, f altogether.

Normal Modes

We shall now show that (5.277) may be rewritten as a sum of independent harmonic oscillators.

With this in mind we start from a rather general Hamiltonian

$$H = \frac{1}{2M}\sum_{ij} B_{ij} P_i P_j + \frac{m\Omega^2}{2}\sum_{ij} D_{ij} X_i X_j. \tag{5.278}$$

M is a typical mass, Ω a typical angular frequency.[14] The $f \times f$ matrices B and D are dimension-less. The first describes kinetic energy, it is therefore symmetric positive. The second is also symmetric and positive because the potential energy should have a stable minimum. X_i and P_i are canonical variables with[15]

$$\{X_i, X_j\} = \{P_i, P_j\} = 0 \;\text{ and }\; \{X_i, P_j\} = \delta_{ij} I. \tag{5.279}$$

In matrix notation—deviations and momenta are column vectors—the Hamiltonian reads

$$H = \frac{1}{2M}P^\dagger BP + \frac{M\Omega^2}{2}X^\dagger DX \tag{5.280}$$

where $A^\dagger_{ij} = A_{ji}$.

---

[14] Do not confuse with the vacuum state of a Fock space.

[15] $\{A, B\} = [A, B]/i\hbar$ here is the quantum Poisson bracket, not the anti-commutator.

Since B is positive, we may write $B = AA^\dagger$ for it, with a certain real and not-singular matrix A. Therefore, $P' = A^\dagger P$ is a natural choice for new momenta. In order to achieve a canonical transformation, the new deviations are $X' = A^{-1}X$. Indeed, it is easy to show that the canonical commutation relations (5.279) are valid for the new deviations and momenta as well. The Hamiltonian now is

$$H' = \frac{1}{2M} P'^\dagger P' + \frac{M\Omega^2}{2} X'^\dagger D' X'. \tag{5.281}$$

Just like D, the new matrix D' is real, symmetric and positive. It can be diagonalized orthogonally:

$$\Omega^2 D' = R(\omega^2) R^\dagger \text{ where } R^\dagger R = RR^\dagger = I \text{ and } (\omega^2)_{ij} = \omega_j^2 \delta_{ij}. \tag{5.282}$$

ω is a diagonal matrix with positive eigenvalues $\omega_1, \omega_2, \ldots, \omega_f$.

With $X'' = R^\dagger X'$ and $P'' = R^{-1}P' = R^\dagger P$ we define new canonical variables once more. Now the Hamiltonian is

$$H'' = \sum_{j=1}^{f} \left\{ \frac{P_j''^2}{2M} + \frac{M\omega_j^2 X_j''^2}{2} \right\}. \tag{5.283}$$

The canonical commutation relations hold true:

$$\{X_i'', X_j''\} = \{P_i'', P_j''\} = 0 \text{ and } \{X_i'', P_j''\} = \delta_{ij} I. \tag{5.284}$$

Thus we have reached our goal: to show that the general quadratic form (5.278) can be transformed into a sum of independent harmonic oscillators, or normal modes of vibrations.

Linear Chain

Let us study a simple model. There is a ring of ions of equal mass M with next-neighbor interaction. Its Hamiltonian reads

$$H = \frac{1}{2M} \sum_j P_j^2 + \frac{M\Omega^2}{2} \sum_j (X_{j+1} - X_j)^2. \tag{5.285}$$

The index j runs over integers in $[-m, m]$, and we identify $X_m = X_{-m}$ and $P_m = P_{-m}$. There are $f = 2m$ degrees of freedom. Matrix B in (5.278) is the unit matrix and D is given by

$$D = \begin{pmatrix} 2 & -1 & 0 & \dots & 0 & 0 & -1 \\ -1 & 2 & -1 & \dots & 0 & 0 & 0 \\ 0 & -1 & 2 & \dots & 0 & 0 & 0 \\ \vdots & & & & & & \\ 0 & 0 & 0 & \dots & 2 & -1 & 0 \\ 0 & 0 & 0 & \dots & -1 & 2 & -1 \\ -1 & 0 & 0 & \dots & 0 & -1 & 2 \end{pmatrix}. \quad (5.286)$$

Diagonalization is easy if one works with complex numbers:

$$X_j \propto \mathrm{e}^{\mathrm{i}kx_j - \mathrm{i}\omega t} \quad (5.287)$$

where $x_j = ja$. This *ansatz*, if inserted into (5.285), gives a solution if angular frequency and wave number are related by

$$\omega = 2\Omega | \sin \frac{ka}{2} |. \quad (5.288)$$

However, not every wave number k is allowed since the linear chain is a ring. X_{-m} and X_m are the same. The following wave numbers describe all distinct solutions:

$$k_r = \frac{r}{2m} \frac{2\pi}{a} \quad \text{for } r \in [-m, m]. \quad (5.289)$$

k_{-m} and k_m are to be identified as well. The wave numbers are equally distributed within the interval $[-\pi/a, \pi/a]$, the so called Brillouin zone.

One can also solve the linear chain model for two different ions with masses m_1 and m_2. This would be a one-dimensional model for the NaCl crystals. The mass ratio is $z = m_1/m_2 = 23.0/35.4$, and a the size of a NaCl unit. All what has been said before remains the same, but now there are two branches. One is called the acoustic, the other one the optical branch. The two dispersion branches of the dispersion relation are

$$\frac{\omega^2_{\text{O,A}}}{2\Omega^2} = 1 \pm \sqrt{1 - \gamma \sin^2(ka/2)}, \quad (5.290)$$

where

$$\gamma = \frac{4}{(\sqrt{z} + \sqrt{1/z})^2}. \quad (5.291)$$

We show the result in Fig. 5.6.

Phonon Dispersion Branches

The energy levels of a single harmonic oscillator with eigenfrequency ω are $n\hbar\omega$ for $n = 0, 1, \dots$ We drop an irrelevant additive term since energy is defined only up to

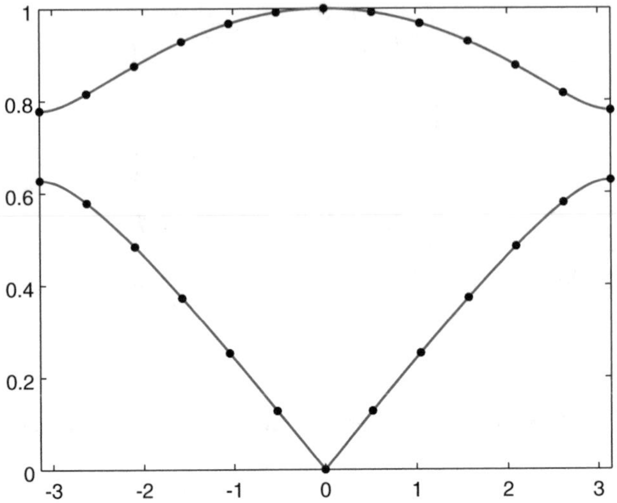

Fig. 5.6 Linear chain of alternating coupled oscillators. The dispersion relation has two branches, namely $\omega_{\rm LA}(k)$ and $\omega_{\rm LO}(k)$. They are plotted (normalized to the maximal value) over $ak \in [-\pi, \pi]$, the Brillouin zone. The upper branch refers to optical (O), the lower branch to acoustical phonons (A). The mass ratio is that for NaCl. The phonons are longitudinally (L) polarized. The chain is made up of twelve diatomic molecules, the allowed wave numbers have been marked by *full dots*. If the chain length approaches infinity, the Brillouin zone becomes densely populated

a constant. We have discussed this in Sect. 2.5 on Small Oscillations. One may also say that the oscillator carries n phonons of energy $\hbar\omega$ each. These phonons come in branches.

A crystal is composed of translated copies of a particular cell. We choose a primitive cell, the smallest possible master cell which allows to build the entire crystal. There are three linearly independent translation vectors $\boldsymbol{a}_1$, $\boldsymbol{a}_2$, and $\boldsymbol{a}_3$. An arbitrary crystal cell is a copy of the primitive cell translated by $g = i_1\boldsymbol{a}_1 + i_2\boldsymbol{a}_2 + i_3\boldsymbol{a}_3$ where the (i_1, i_2, i_3) are three integer numbers. The translation vectors $g = g(\boldsymbol{i})$ form the crystal lattice.

The reciprocal lattice consists of all wave vectors $\boldsymbol{k}$ for which $\boldsymbol{k} \cdot g(\boldsymbol{i}) = 2\pi n$ holds true, for all $\boldsymbol{i} \in \mathbb{Z}^3$, with n an integer. The Brioullin zone BZ is the set of wave vectors $\boldsymbol{k}$ which is closer to $\boldsymbol{k} = 0$ than to any other vector of the reciprocal lattice.

Without going into details, we report some facts on lattice vibrations. For each ion in the primitive cell, there are three branches of phonons, or dispersion relations. They are of the acoustic type or correspond to optical frequencies.

The former three branches describe vibrations of the primitive cell as a whole. They are reasonably well characterized by continuum physics in situations where the wave length is much larger than the dimension of a material point, i.e. the cell itself. For very low frequencies the dispersion relations are $\omega_{\rm LA}(\boldsymbol{k}) = c_{\rm L} |\, \boldsymbol{k} \,|$ and analogously $\omega_{\rm TA1,2}(\boldsymbol{k}) = c_{\rm T1,2} |\, \boldsymbol{k} \,|$. The A stands for acoustic, L and T indicate longitudinal and transversal polarization, respectively. The subscript 1 and 2 denote possibly different transversal sound velocities. They coincide for an isotropic medium.

If one considers the entire Brillouin zone, the group velocity $c = \partial\omega/\partial\boldsymbol{k}$ will not be constant as is the case close to $\boldsymbol{k} \approx 0$. There is true dispersion in the sense that phonons of different angular frequency travel with different speed. This is well born out by the linear chain model. See Fig. 5.6.

If the primitive cell contains r ions, there are $3(r-1)$ optical modes in addition. They come from oscillations within the primitive cell in opposite directions. Their frequencies are rather large, and they may be excited by infrared radiation. That explains why they are called optical. Frequencies in the optical branches do not vary much with the wave number.

Low Frequencies

The free energy of a boson gas is

$$F = k_{\mathrm{B}}TV \int \mathrm{d}\epsilon\, z(\epsilon)\, \ln\left\{1 - \mathrm{e}^{-\epsilon/k_{\mathrm{B}}T}\right\}, \tag{5.292}$$

as has been derived in the previous Sect. 5.5 on Gases. Since the number of phonons cannot be controlled, the chemical potential is absent, i.e. it must vanish. $z = z(\epsilon)$ is the density of states per volume with energy $\epsilon = \hbar\omega$.

For small values ϵ it can easily be calculated since the three acoustical dispersion branches are linear in $|\,\boldsymbol{k}\,|$.

We consider a crystal where the primitive unit cell is a cube of side length a. If there are N of them, there are also N uniformly distributed $\boldsymbol{k}$ vectors in the Brillouin zone. It follows that the number of phonon modes with an energy less than ϵ is

$$Z(\epsilon) = \frac{N}{(2\pi/a)^3}\,\frac{4\pi}{3}\left(\frac{\epsilon}{\hbar c}\right)^3 = \frac{V}{6\pi^2}\left(\frac{\epsilon}{\hbar c}\right)^3. \tag{5.293}$$

We have made use of the relation $\epsilon = \hbar\omega = \hbar c|\,\boldsymbol{k}\,|$ with c the velocity of sound for an acoustical branch. Since there are three such branches we must correct (5.293) into

$$Z(\epsilon) = \frac{V}{2\pi^2}\left(\frac{\epsilon}{\hbar c}\right)^3 \quad \text{where} \quad \frac{3}{c^3} = \frac{1}{c_1^3} + \frac{1}{c_2^3} + \frac{1}{c_3^3}. \tag{5.294}$$

c is a certain mean of the velocities of longitudinally polarized sound and the two transversally polarized modes. Dividing (5.294) by the volume and differentiating with respect to ϵ gives

$$z(\epsilon) = \frac{3}{2\pi^2}\,\frac{\epsilon^2}{(\hbar c)^2}. \tag{5.295}$$

This is the density of states per unit volume as required for (5.292).

Debye Model

Debye suggested to use the density (5.295) of states per unit volume also for higher frequencies. It has to be cut off at an energy $k_B\Theta$ such that all vibrational degrees of freedom are taken into account, $Z(k_B\Theta_D) = 3rN$, or

$$V \int_0^{k_B\Theta} d\epsilon\, z(\epsilon) = 3rN. \tag{5.296}$$

This results in the Debye energy Θ

$$k_B\Theta = \hbar c \left(\frac{6\pi^2 rN}{V}\right)^{1/3}. \tag{5.297}$$

With the Debye approximation we may calculate the free energy:

$$F = 9rNk_BT \int_0^1 dx\, x^2 \ln\left(1 - e^{-x\Theta/T}\right). \tag{5.298}$$

We obtain from it the following expression for the heat capacity

$$C = 3rNk_B \left(\frac{T}{\Theta}\right)^3 \int_0^{\Theta/T} ds\, 3s^2 \left(\frac{s/2}{\sinh s/2}\right)^2. \tag{5.299}$$

The limits are

$$C = \frac{12\pi^4}{5} rNk_B \left(\frac{T}{\Theta}\right)^3 \quad \text{if } T \ll \Theta \tag{5.300}$$

and

$$C = 3rNk_B \text{ if } T \gg \Theta. \tag{5.301}$$

Equation (5.300) is Debye's T^3 law for the heat capacity of a solid. Equation (5.301) is known as the rule of Dulong and Petit: at high temperatures each degree of freedom contributes with k_B, namely $k_B/2$ for the kinetic and $k_B/2$ for the potential energy.

Debye–Einstein Model

Einstein correctly recognized that the heat capacity of solids is a manifestation of lattice vibrations. A very crude model consisted in describing a fixed frequency to each mode of vibration. This resulted in the rule of Dulong and Petit.

Debye improved Einstein's approach. He took the dispersion relation of elasticity theory as a prototype and introduced a cut-off in order to accommodate all oscillators. Besides the rule of Dulong and Petit also the T^3 law could be explained.

Today the two models are combined. Debye's approach to acoustical phonons and Einstein's treatment of optical phonons. Besides the Debye constant—which can be determined from measurements of sound velocities—there is a typical frequency for each optical branch. With such a large number of adjustable parameters a good fit to experimental data is to be expected.

For NaCl the Debye temperature is 308 K if determined from the T^3 law and 320 K if calculated from sound velocities. For ice, the density of states (5.295) with $\Theta = 192$ K well describes its lattice vibrations.

5.6.3 Summary

We discussed the vibrational degrees of freedom of a regular lattice of ions. The Born–Oppenheimer approximation serves to define the potential energy as a function of the ions' positions. This potential energy is minimal at the equilibrium configuration and can therefore be expanded into a quadratic form of deviations. The quadratic form of momenta (kinetic energy) and the quadratic excitation potential may be reshaped into uncoupled harmonic oscillators by a sequence of transformations which preserve the canonical commutation relations. Each such oscillator describes a normal mode which can be excited to its nth energy level which is equal to $n\hbar\omega$. One may also say that each normal mode may carry n quanta of vibrational excitation, or phonons.

For a regular structure, a crystal lattice, the normal modes are characterized by a wave vector $\boldsymbol{k}$ and an angular frequency $\omega = \omega(\boldsymbol{k})$. Not all wave vectors are allowed, they must belong to the Brillouin zone. Such dispersion relations come in branches for acoustical and optical phonons which are either longitudinally or transversally polarized.

These findings are supported by sound waves and by a one-dimensional chain of equal or alternating ions.

Phonons are bosons since they may occupy the mode not at all, once, twice and so on, at least in the approximation of the potential at its minimum by a quadratic expression. In order to calculate the thermal properties of such a phonon gas we only need to know the density of vibrational modes per unit energy. We have discussed Einstein's crude approximation—constant frequencies for each branch—and Debye's refined approach to treat all modes as sound waves and cutting off at a particular energy $k_B\Theta$. The latter results in a T^3 law for the heat capacity, both may explain the rule of Dulong and Petit where each mode contributes with k_B to the heat capacity, at least for high enough temperatures.

5.7 Electronic Band Structure

The Born–Oppenheimer approximation describes a solid by two loosely coupled subsystems. One is the vibrating lattice, the other one a gas free quasi-electrons.

We shall recapitulate the one-dimensional hopping model of Chap. 2 with Simple Examples and describe the realistic three-dimensional crystal. The electrons come in energy bands which are usually separated by energy gaps.

Of particular interest is the density of one-particle states per unit volume. Since electrons are fermions, and quasi-electrons as well, the one-particle states can be occupied at most twice (if both spin polarization states are taken into account). A solid is electrically neutral, therefore, the electron density is fixed which in turn fixes the chemical potential. The chemical potential at zero temperature is the characteristic Fermi energy. If it lies within an energy band, the crystal is a good electric conductor. If the Fermi energy lies within a sufficiently broad gap, the solid is an electrical insulator. If, however, the band gap is narrow, we speak of a semi-conductor.

5.7.1 Hopping Model

Assume a one-dimensional chain of equally spaced identical ions and concentrate on a particular electron. If the ion were isolated, the electron would be bound to a certain level the energy of which is denoted by E. Since the ion is not isolated, the electron may hop easily to the same level of one of its neighbors. Jumping to an excited level is extremely improbable.

We enumerate the ions by $r = \dots, -1, 0, 1, \dots$. They are fixed at positions $x_r = ar$. The amplitude for the electron being located at ion r is u_r. R with $Ru_r = u_{r+1}$ describes hopping to the right neighbor, $Lu_r = u_{r-1}$ to the left neighbor.

The Hamiltonian for the model is

$$H = E - V\frac{R + L}{2}. \tag{5.302}$$

We have chosen the amplitude V the same for right and left hopping. It does not depend on the index r and thus the crystal's mirror and translational symmetry is realized.

The solutions of the Schrödinger equation are

$$w_r(k) \propto \mathrm{e}^{\,\mathrm{i}k\,ar} \tag{5.303}$$

such that

$$\{Hw(k)\}_r = \epsilon(k)\, w(k)_r \quad \text{with} \quad \epsilon(k) = E - V\cos ka \tag{5.304}$$

holds true. Wave numbers are restricted to the Brioullin zone $ak \in [-\pi, \pi]$. The dispersion relation $\epsilon = \epsilon(k)$ was shown in Fig. 2.4.

We have also studied the properties of the electron: it is a particle since it can be counted. It is a quasi-particle since the relation between energy and momentum is not $\epsilon = (\hbar k)^2/2m$, but more complicated. Its effective mass m_{eff} will depend on the wave number and can even become negative!

The number of states with energy below ϵ per unit cell is

$$Z(\epsilon) = 2 \int_0^\pi \frac{dk}{2\pi}\, \theta(\epsilon - \epsilon(k)) = \frac{1}{a\pi} \arccos \frac{E-\epsilon}{V}. \tag{5.305}$$

Its derivative is

$$z(\epsilon) = \frac{1}{a\pi} \frac{1}{\sqrt{V^2 - (\epsilon - E)^2}}. \tag{5.306}$$

The density $z = z(\epsilon)$ of one-particle states vanishes outside the energy band and is given by (5.306) within, for $\epsilon \in [E - V, E + V]$. See Fig. 5.7.

The discussion above may be generalized to more than one energy level. Assume there are E_1 and E_2 with hopping amplitudes V_1 and V_2. We mentioned already that hopping from level E_1 to E_2 at the same or to a neighboring location requires much energy and will therefore be excluded. There are now two dispersion bands $\epsilon_1(k)$ and $\epsilon_2(k)$ defined on the common Brillouin zone $ka \in [-\pi, \pi]$.

5.7.2 Fermi-Energy

The hopping model explains the basic features of electronic band structure remarkably well. To each energy level E of ions involved there is a band of electronic

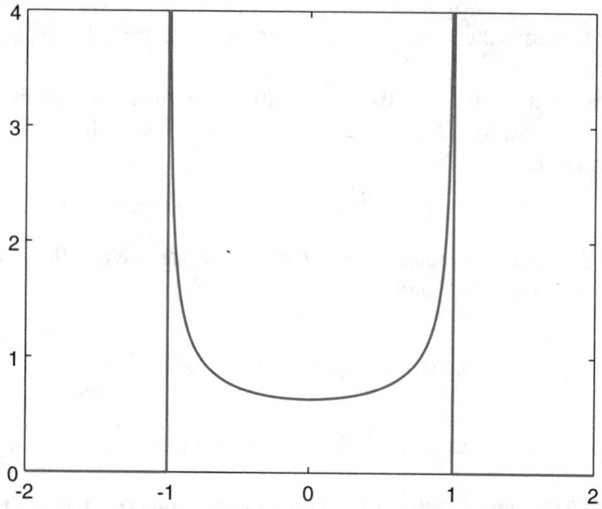

Fig. 5.7 Hopping model density of states. The number of states per unit energy per unit length is plotted versus energy. The band is arbitrarily centered at $E = 0$, energy is measured in units of V, and the density is normalized. The shape of $z = z(\epsilon)$ is specific for the one-dimensional hopping model with next neighbor interaction

one-particle states. Instead of being degenerate as often as there are ions, the levels split into values $\epsilon(k)$ for k in the Brillouin zone. If the electron is tightly bound, the band with is small; and it is large if the electrons are but loosely bound. This explains why the bands can be labeled by the quantum numbers of the bound states of the ions and why the bands are centered at different energies, and why their widths may differ substantially. Recall that the dispersion relations defining such bands allow to calculate the properties of the corresponding quasi-particles.

All this, for real-world three-dimensional solids, results in a structure which is difficult to calculate.

Summarily, the distribution of one-particle states is described by the density of states $z = z(\epsilon)$, i.e. their number per unit energy and per unit volume.

The distribution of electronic one-particle states determines the thermal properties of the electron cloud. Although the gas consists of quasi-electrons, they are still fermions. Quasi particles are linear combinations of free particles, and their fields anti-commute.

The free energy of the quasi-electrons is thus given by

$$F(T, V, \mu) = -V k_B T \int d\epsilon\, z(\epsilon)\, \ln \left\{ 1 + e^{(\mu - \epsilon)/k_B T} \right\}, \tag{5.307}$$

the particle density by

$$n(T, \mu) = \int d\epsilon\, z(\epsilon) \frac{1}{e^{(\epsilon - \mu)/k_B T} + 1}. \tag{5.308}$$

Note that the fraction, the average population of a one-particle level, is a number between zero and one.

The solid is electrically neutral, even locally. Therefore, the density $\bar{n}$ of quasi-electrons is a constant, the same for all temperatures. Hence, the chemical potential must be chosen to be

$$n(T, \mu) = \bar{n}. \tag{5.309}$$

In particular, the chemical potential for zero temperature will often show up. It is called Fermi's energy ϵ_F and defined by $n(0, \epsilon_F) = \bar{n}$ or

$$\int d\epsilon\, \theta(\epsilon_F - \epsilon)\, z(\epsilon) = \int_{-\infty}^{\epsilon_F} d\epsilon\, z(\epsilon) = \bar{n}. \tag{5.310}$$

This is so because the fraction in (5.308) becomes a jump function for $T \to 0$. See Fig. 5.8.

Equation (5.310) says that the ground state of the solid is such that all one-particle states are fully occupied up to the maximal Fermi energy. The same statement about atoms is called Hund's rule.

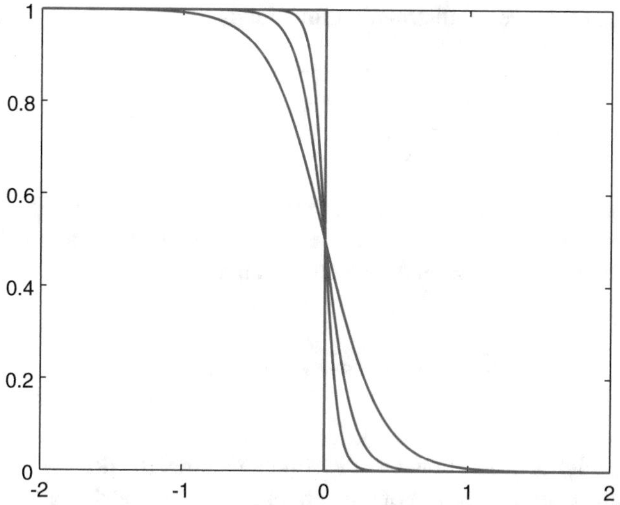

Fig. 5.8 Fermi Function. The Fermi function is plotted versus energy centered at the Fermi energy. The jump function corresponds to zero temperature, the smooth curves to $k_B T = 0.05, 0.1, 0.2$ in suitable units

5.7.3 The Cold Solid

Let us work out the heat capacity of a solid's electron. As we shall see, it is the dominant contribution for low temperatures.

Clearly, the internal energy of the electron gas is

$$U = V \int \mathrm{d}\epsilon\, z(\epsilon)\, \epsilon f(\epsilon) \;\text{ where }\; f(\epsilon) = \frac{1}{\mathrm{e}^{(\epsilon-\mu)/k_B T} + 1}. \tag{5.311}$$

The corresponding heat capacity is

$$C = \frac{\partial U}{\partial T} = \int \mathrm{d}\epsilon\, z(\epsilon)\, \epsilon\, \frac{\partial f}{\partial T}. \tag{5.312}$$

We have a look at (5.308) with (5.309) and conclude

$$0 = \frac{\partial \bar{n}}{\partial T} = \int \mathrm{d}\epsilon\, z(\epsilon)\, \frac{\partial f}{\partial T}. \tag{5.313}$$

It follows that (5.312) is the same as

$$\int \mathrm{d}\epsilon\, z(\epsilon)\, \epsilon\, \frac{\partial f}{\partial T} C = \int \mathrm{d}\epsilon\, z(\epsilon)\, (\epsilon - \epsilon_F)\, \frac{\partial f}{\partial T}. \tag{5.314}$$

The derivative of the Fermi distribution function is

$$\frac{\partial f}{\partial T} = \frac{s}{T} \frac{\mathrm{e}^{s}}{\left(\mathrm{e}^{s} + 1\right)^{2}} \quad \text{with} \quad s = \frac{\epsilon - \mu}{k_{\mathrm{B}} T}. \tag{5.315}$$

For a cold gas we may put $\mu \approx \epsilon_{\mathrm{F}}$. Moreover, since the factor of $z(\epsilon)$ in (5.312) is sharply peaked at $\epsilon \approx \epsilon_{\mathrm{F}}$, we may replace the density of state function by $z(\epsilon_{\mathrm{F}})$, its value at the Fermi energy. With these considerations we arrive at

$$C = V z(\epsilon_{\mathrm{F}})\, k_{\mathrm{B}}^{2} T \int \mathrm{d}s\, \frac{s^{2}\, \mathrm{e}^{s}}{\left(\mathrm{e}^{s} + 1\right)^{2}}. \tag{5.316}$$

The heat capacity is proportional to the solid's volume, the density of states at the Fermi level and to the first power of temperature. The physical dimensions are alright as well. The dimensionless integral in (5.316) evaluates to $\pi^2/3$.

5.7.4 *Metals, Dielectrics and Semiconductors*

The highest fully filled band is called the valence band, the one atop of it is the conduction band. There may be a band gap of forbidden energies. We encounter at least three different situations:

- The Fermi energy lies well inside the conduction band. At zero temperature, this is usually half-filled. Then $z(\epsilon_{\mathrm{F}})$ will not vanish, and there is a contribution to the heat capacity. The electrical conductivity is high and depends only weakly on temperature.
- The Fermi energy lies within the band gap. At zero temperature, the valence band is full, the conduction band is empty. At finite temperatures, some electrons are missing in the valance band which are thermally excited to the conduction band. This results in some electrical conductivity which, however, vanishes exponentially with growing band gap. Solids with this properties are electrical insulators with pronounced dielectric properties.
- At zero temperature, the conduction band is empty. However, the gap between valence and conduction band is rather small. At elevated temperatures, for instance at room temperature, the Fermi function allows for holes (missing electrons) in the valence band and for electrons in the conduction band. Therefore, there is an electric conductivity which strongly depends on temperature.

We shall not discuss the theory of conductivity here, in particular the role of impurities. Let us just mention one aspect.

Assume the solid to be exposed to an external electric field. We know that the energy levels of atoms and molecules are shifted, proportional to the field strength in cases of degeneracy, quadratic otherwise. This Stark effect is experienced by a solid as well. Now the bands get deformed. For a large band gap, an insulator remains to be an insulator and a electric conductor also remains to be a conductor. However, for small band gaps at zero field strength the conduction properties may change drastically. An insulator may become a conductor under the influence of an external electric field. This remark should open the way for understanding diodes and transistors and more complex devices. We stop here because a detailed discussion of the marvelous properties of semiconductors is outside the scope of this book, an overview of quantum theory.

5.7.5 *Summary*

We have recapitulated the one-dimensional hopping model. It well explains the splitting of an ion's energy level into a band of plane waves. Taking more levels into account, overlapping bands and the separation of bands by an energy gap of forbidden states can easily be demonstrated.

At zero temperature all states below a certain energy ϵ_F are fully occupied, and empty above this Fermi energy. With growing temperature, the average occupation of one-particle states will deviate more and more from the step function at zero temperature.

If the valence band is full and the conduction band empty, there is no electric conduction. The material is an insulator, or dielectric.

If the conduction band is neither full nor empty, there are many mobile electrons close to the Fermi level; the material is a good electric conductor. We have shown that there is a contribution to the heat capacity which grows linearly with temperature. At very low temperature, it surpasses the heat capacity of the vibrating lattice.

Most interesting are insulators at zero temperature with a small band gap, so called semiconductors. At elevated temperatures, the valence band—which was full—looses some electrons by thermal excitation to the conduction band. The missing electrons, so called holes, behave like quasi-electrons with a positive charge. The excited electrons move in the conduction band as normal quasi-electrons. Both contribute to the electrical conductivity which markedly depends on temperature. The charge transport by semiconductors may be manipulated by doping, thin layers of other constitution and by applying external electric fields. Semiconductors are the base materials for all kinds of applications. Just think about integrated circuits, processors and solid state memory, charge coupled devices, light emitting diodes, photovoltaic cells as well as power electronics.

5.8 External Fields

Matter can be exposed to an external field, either electric or magnetic. This matter then becomes polarized or magnetized, and we want to study the influence of temperature.

We first remark on the first law. The system we discuss should be split into matter proper and the electromagnetic field. As we shall see, the splitting is not unique, and therefore the first law may have various forms. All have their advantages and disadvantages.

We than describe the aligning of permanent electric dipoles. We shall see that the higher the temperature the less ordered are the individual dipole moments. The dielectric susceptibility is inversely proportional to the temperature.

Magnetic dipole moments—usually of electrons—in an external induction field exhibit similar behavior. With increasing field strength, they are aligned more and more until saturation is achieved. The temperature dependency of the magnetic susceptibility follow likewise Weiß's law. The functional form of the alignment degree, however, is markedly different for electric and magnetic dipoles. The former behave classically, the latter are quantized with two alternatives only.

If the magnetic dipoles do not only interact with the external induction field, but also among themselves, the dipole moments might stay aligned even after the external field is switched off. This phenomenon is rather pronounced with iron and one speaks of ferro-magnetism. As an example we discuss the Heisenberg ferromagnet. Its most significant feature, a phase transition from paramagnetic to ferromagnetic behavior, survives the mean field approximation.

5.8.1 Matter and Electromagnetic Fields

The electromagnetic field is defined by its action on matter, i.e. on charged particles. A point particle with charge q at location $\boldsymbol{x}$ and momentum $\boldsymbol{p}$ moves in a force field according to

$$\dot{\boldsymbol{p}} = q\,\{\boldsymbol{E} + \boldsymbol{v} \times \boldsymbol{B}\}\,. \tag{5.317}$$

Here $\boldsymbol{v} = \dot{\boldsymbol{x}}$ is the particle's velocity, $\boldsymbol{E}$ the electric field strength at the particle's location and $\boldsymbol{B}$ the magnetic induction. The energy $E = \sqrt{m^2c^4 + \boldsymbol{p}^2c^2}$ changes according to

$$\dot{E} = q\,\boldsymbol{v} \cdot \boldsymbol{E}. \tag{5.318}$$

The electromagnetic field obeys Maxwell's equations:

$$\begin{aligned}\epsilon_0 \nabla \cdot \boldsymbol{E} &= \varrho^{\mathrm{e}} \\ \nabla \cdot \boldsymbol{B} &= 0 \\ \frac{1}{\mu_0} \nabla \times \boldsymbol{B} &= \epsilon_0 \dot{\boldsymbol{E}} + \boldsymbol{j}^{\mathrm{e}} \\ \nabla \times \boldsymbol{E} &= -\dot{\boldsymbol{B}}.\end{aligned} \tag{5.319}$$

ϱ^{e} is the total electric charge density, $\boldsymbol{j}^{\mathrm{e}}$ the total current density.

Polarization and Magnetization

If the continuously distributed matter is exposed to an electric field it becomes polarized. There is a density of electric dipole moments which we denote by $\boldsymbol{P}$. It is associated with a polarization charge density

$$\varrho^{\mathrm{p}} = -\nabla \cdot \boldsymbol{P}. \tag{5.320}$$

Likewise, there is a polarization current density

$$\boldsymbol{j}^{\mathrm{p}} = \dot{\boldsymbol{P}} \tag{5.321}$$

if the polarization changes with time.

Another effect is that matter becomes magnetized. Denote by $\boldsymbol{M}$ the density of magnetic moments. Its curl is another contribution to the current density, namely

$$\boldsymbol{j}^{\mathrm{m}} = \nabla \times \boldsymbol{M}. \tag{5.322}$$

So, in general the charge density has two contributions,

$$\varrho^{\mathrm{e}} = -\nabla \cdot \boldsymbol{P} + \varrho^{\mathrm{f}}, \tag{5.323}$$

the current density consists of

$$\boldsymbol{j}^{\mathrm{e}} = \dot{\boldsymbol{P}} + \nabla \times \boldsymbol{M} + \boldsymbol{j}^{\mathrm{f}}. \tag{5.324}$$

The superscript f stands for 'free' or 'to be freely manipulated.'

Note that we have entered a vicious circle. The electric field strength must be known in order to find the polarization. This then produces a charge density which must be know to calculate the field strength according to (5.319). The same applies to the magnetic induction.

The standard way out is to introduce auxiliary fields, namely the displacement $\boldsymbol{D}$ and the magnetic field $\boldsymbol{H}$. They are defined as

$$\boldsymbol{D} = \epsilon_0 \boldsymbol{E} + \boldsymbol{P} \quad \text{and} \quad \boldsymbol{H} = \frac{1}{\mu_0} \boldsymbol{B} - \boldsymbol{M}. \tag{5.325}$$

Maxwell's equations may be reformulated as

$$\begin{aligned}\nabla \cdot \boldsymbol{D} &= \varrho^{\mathrm{f}} \\ \nabla \cdot \boldsymbol{B} &= 0 \\ \nabla \times \boldsymbol{H} &= \boldsymbol{j}^{\mathrm{f}} \\ \nabla \times \boldsymbol{E} &= -\dot{\boldsymbol{B}}.\end{aligned} \tag{5.326}$$

The good news is that this set of equations is driven by the charge density and current density of free charges only. The bad news is that we now have to deal with twelve fields instead of six. There must be further relations which relate $\boldsymbol{D}$ with $\boldsymbol{E}$ and $\boldsymbol{H}$ with $\boldsymbol{B}$.

As a consequence of (5.326) one might derive

$$\dot{\varrho}^{\mathrm{f}} + \nabla \cdot \boldsymbol{j}^{\mathrm{f}} = 0. \tag{5.327}$$

The free charge is conserved.

The First Law

Consider two parallel large metal plates of area A and separation s. In-between there is a homogeneous medium. On the upper plate a charge Q is deposited, and $-Q$ on the lower plate. The voltage U and the electric field strength E are related by $U = sE$, the displacement is $D = Q/A$. If a small charge $\mathrm{d}Q$ is moved from the lower to the upper plate, the work $U\mathrm{d}Q = sE\,A\,\mathrm{d}D$ is required, i.e. $E\,\mathrm{d}D$ per unit volume.

Here a similar argument for the magnetization. Consider a long coil with cross section A and length ℓ. Denote the number of turns by N. The coil is filled with a homogeneous magnetizable medium. If there flows a current I through the winding, the magnetic field strength in the core of the coil is $H = NI/\ell$. The induction flux is $\Phi = AB$. If the current changes, there will be a counter-voltage $U = N\dot{\Phi}$. Put otherwise, the power $UI = NA\dot{B}\ell H/N$ is fed to the system. Summed up, for changing the induction by $\mathrm{d}B$ the work $H\mathrm{d}B$ has to be spent, per unit volume.

In general, the fields $\boldsymbol{E}$ and $\boldsymbol{D}$ must not be parallel or spatially constant; and the same for $\boldsymbol{H}$ and $\boldsymbol{B}$. We therefore write

$$\mathrm{d}W = \int \mathrm{d}V\,\boldsymbol{E} \cdot \mathrm{d}\boldsymbol{D} + \int \mathrm{d}V\,\boldsymbol{H} \cdot \mathrm{d}\boldsymbol{B} \tag{5.328}$$

for the work to be spent. This is the first main law for matter in an electromagnetic field.

Field and Matter

Equation (5.328) is correct from a theoretical point of view, but does not correspond to the needs of experiment. It is the electric or magnetic field strength which usually can be controlled easily. There is another reason to modify (5.328). The expression contains the work to set up the vacuum field. The vacuum field energy does not depend on temperature and must not be part of a thermodynamic description of matter.

We therefore split the free energy into a field part F_{f} and into a matter part F_{m}. For the former we use the following expression:

$$F_{\mathrm{f}} = \int \mathrm{d}V \left(\frac{\boldsymbol{D}^2}{2\epsilon_0} - \frac{\boldsymbol{P}^2}{2\epsilon_0} \right) + \int \mathrm{d}V \left(\frac{\boldsymbol{B}^2}{2\mu_0} - \frac{\mu_0 \boldsymbol{M}^2}{2} \right) \tag{5.329}$$

for the electromagnetic field. The matter part of the free energy now is characterized by

$$\mathrm{d}F_{\mathrm{m}} = -S\mathrm{d}T - \int \mathrm{d}V\, \boldsymbol{P} \cdot \mathrm{d}\boldsymbol{E} - \mu_0 \int \mathrm{d}V\, \boldsymbol{M} \cdot \mathrm{d}\boldsymbol{H}. \tag{5.330}$$

This expression treats the field strengths $\boldsymbol{E}$ and $\boldsymbol{H}$ as thermodynamic variables, as desired. It must however be read together with $F = F_{\mathrm{f}} + F_{\mathrm{m}}$ and (5.329).

5.8.2 *Alignment of Electric Dipoles*

Think about a gas the molecules of which have a permanent dielectric moment. Water vapour is but one example. If H_0 is the Hamiltonian without field, it becomes

$$H(\boldsymbol{E}) = H_0 + \sum_a \boldsymbol{p}_a \cdot \boldsymbol{E}(\boldsymbol{X}_a) \tag{5.331}$$

in the presence of an electrostatic field $\boldsymbol{E} = \boldsymbol{E}(\boldsymbol{x})$. The subscript a labels the molecules, $\boldsymbol{p}_a$ are their dipole moments, $\boldsymbol{X}_a$ the location observables. We assume all dipole moments to be the same and the field to be constant.

The dipole moment is a sum of terms charge times location vector. Its components consists of commuting terms only and thus behave classically.

We best calculate the expectation value of a typical dipole:

$$\langle\, \boldsymbol{p}\,\rangle = \frac{\displaystyle\int \mathrm{d}\Omega\, \mathrm{e}^{\,\boldsymbol{p} \cdot \boldsymbol{\mathcal{E}}/k_{\mathrm{B}}T}\, \boldsymbol{p}}{\displaystyle\int \mathrm{d}\Omega\, \mathrm{e}^{\,\boldsymbol{p} \cdot \boldsymbol{\mathcal{E}}/k_{\mathrm{B}}T}}\,. \tag{5.332}$$

Intended is the average over all space directions. With $\boldsymbol{\mathcal{E}} = \mathcal{E}\boldsymbol{n}$ ($\boldsymbol{n}$ is a unit vector pointing in field direction) and $p = |\,\boldsymbol{p}\,|$ one obtains

$$\langle\,\boldsymbol{p}\,\rangle = \left\{\coth\left(\frac{pE}{k_{\mathrm{B}}T}\right) - \frac{k_{\mathrm{B}}T}{pE}\right\}\,p\boldsymbol{n}. \tag{5.333}$$

One sees that for low temperatures or a strong electric field, $k_{\mathrm{B}}T \ll p\,\mathcal{E}$, the factor in curly brackets becomes one which means complete alignment.

For small fields or high temperatures the polarization evaluates to

$$\boldsymbol{P} = \frac{N}{V}\langle\,\boldsymbol{p}\,\rangle = \frac{N}{3V}\frac{p^2\mathcal{E}}{k_{\mathrm{B}}T}\boldsymbol{n} + \dots, \tag{5.334}$$

where N is the number of molecules in a volume V. The dielectric susceptibility

$$\chi = \frac{1}{\epsilon_0}\frac{\partial P}{\partial\mathcal{E}} = \frac{N}{3V}\frac{1}{\epsilon_0}\frac{p^2}{k_{\mathrm{B}}T} \tag{5.335}$$

is proportional to the density and to the dipole moment squared; it falls off with temperature as $1/T$. The latter finding is Curie's law. Check that χ is in fact a dimensionless number.

5.8.3 Alignment of Magnetic Dipoles

The spin of a fermion is accompanied with a magnetic moment. The spin for the electron is $\boldsymbol{S} = \hbar\boldsymbol{\sigma}/2$ where the sigmas are the well-known Pauli matrices

$$\sigma_1 = \begin{pmatrix} 0 & 1 \\ 1 & 0 \end{pmatrix} \quad \sigma_2 = \begin{pmatrix} 0 & -\mathrm{i} \\ \mathrm{i} & 0 \end{pmatrix} \quad \sigma_3 = \begin{pmatrix} 1 & 0 \\ 0 & -1 \end{pmatrix}. \tag{5.336}$$

We have mentioned them previously when discussing the Zeeman effect of the hydrogen atom. The magnetic moment of spin one-half particles as a vector is $\boldsymbol{m} = m\boldsymbol{\sigma}$ with m the magnetic moment.

The system with the induction field $\boldsymbol{\mathcal{B}}$ is described by the Hamiltonian

$$H = H_0 + \int \mathrm{d}V\,\frac{\mathcal{B}^2}{2\mu_0} - \sum_a \boldsymbol{m}_a \cdot \boldsymbol{\mathcal{B}}(\boldsymbol{X}_a). \tag{5.337}$$

It consists of the part for matter without field, the field energy, and the interaction between matter and field. For N magnetic moments in a volume V and a homogeneous field pointing in direction $\boldsymbol{n}$ (a unit vector) this reads

$$H = H_0 + V\frac{\mathcal{B}^2}{2\mu_0} - m\mathcal{B}\sum_a \boldsymbol{\sigma}\cdot\boldsymbol{n}. \tag{5.338}$$

Now, the spin operators of different electrons commute with each other and with H_0. The eigenvalues for $\boldsymbol{\sigma}\cdot\boldsymbol{n}$ are plus and minus one. The free energy associated with (5.338) is

$$F = F_0 + V\frac{\mathcal{B}^2}{2\mu_0} - Nk_BT\ln 2\cosh\left(\frac{m\mathcal{B}}{k_BT}\right). \tag{5.339}$$

From

$$\frac{\partial F}{\partial\mathcal{B}} = VH = V\left(\frac{\mathcal{B}}{\mu_0} - M\right) \tag{5.340}$$

(see (5.328) and (5.325)) we calculate the following expression for the magnetization:

$$\boldsymbol{M} = \frac{N}{V}m\tanh\left(\frac{m\mathcal{B}}{k_BT}\right)\boldsymbol{n}. \tag{5.341}$$

Recall that the magnetization is the magnetic moment per unit volume. It is the product of the number of magnetic dipoles, their magnetic moment and a factor for the degree of alignment, here the hyperbolic tangent. It approaches one for low temperature or high induction.

The magnetic moment of an electron is approximately the negative of Bohr's magneton $\mu_B = 2\hbar/2m_B = 9.27\times10^{-24}$ J/T. Assume $T = 300$ K and an induction $\mathcal{B} = 1.0$ T. Then $x = mB/k_BT = 2.24\times10^{-3}$ is rather small—although the induction field strength is huge—and we may linearize (5.341):

$$\boldsymbol{M} = \frac{N}{V}\frac{m^2}{k_BT}\mathcal{B}\boldsymbol{n} + \dots \tag{5.342}$$

The magnetic susceptibility $\chi_m = \mu_0\partial M/\partial\mathcal{B}$ is always positive, the medium behaves para-magnetically.

Recall that we here studied independent freely rotatable magnetic dipoles. The electrons in a solid are not of this type.

In Fig. 5.9 we compare the alignment of dielectric and magnetic dipoles. The abscissa is $x = pE/k_BT$ and $x = mB/k_BT$, respectively. The ordinate ranges from zero to one, from no to total alignment, or saturation. The lower curve for the electric dipole moment markedly differs from the upper curve for the alignment of magnetic dipoles for freely rotatable electron spins.

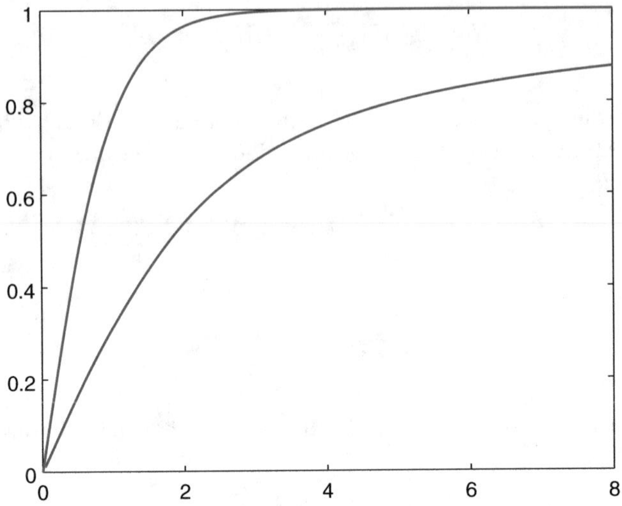

Fig. 5.9 Dielectric versus magnetic dipole alignment Plot of degree of alignment over $x = pE/k_BT$ or $x = mB/k_BT$, respectively. The *lower curve* refers to electric dipoles with permanent moment p in an electric field E. The *upper curve* is for freely rotatable electrons with magnetic moment m in an induction field B

5.8.4 Heisenberg Ferromagnet

There are materials with rotatable spins which, however, interact with each other. In particular, if parallel alignment of neighboring spins is favorable, the Gibbs state is spontaneously ordered such that a permanent magnetization results. The phenomenon is known as ferromagnetism.

Heisenberg has constructed a model which describes the essential features. The Hamiltonian is that of freely rotatable spins plus a spin interaction term:

$$H = H_0 + \int \mathrm{d}V \, \frac{\mathcal{B}^2}{2\mu_0} - \sum_a \boldsymbol{m}_a \cdot \boldsymbol{\mathcal{B}}(\boldsymbol{X}_a) - \frac{1}{2} \sum_{ba} g_{ba} \boldsymbol{\sigma}_b \cdot \boldsymbol{\sigma}_a . \tag{5.343}$$

The last term is a so-called exchange energy. It arises if a pair of spins are from indistinguishable particles, here: electrons. It depends on details of the electronic structure whether the exchange interaction is repulsive (normal case) or attractive. Here we assume that the matrix g_{ba} be positive.

A one-dimensional chain (Ising model) does not show ferromagnetic properties. A two-dimensional analytical solution was derived by Onsager in 1944. Although intensively investigated, no analytical solution could be found as of today for the three-dimensional Heisenberg ferromagnet. We here calculate an approximate solution by the mean field method.

We subtract $(\boldsymbol{\sigma}_b - \boldsymbol{s}_b) \cdot (\boldsymbol{\sigma}_a - \boldsymbol{s}_a)$ from $\boldsymbol{\sigma}_b \cdot \boldsymbol{\sigma}_a$. The additional terms contain a lot of so far undetermined parameters $\boldsymbol{s}_a$, and one hopes that they will affect the solution as little as possible.

Assume a homogeneous induction field $\boldsymbol{B} = \mathcal{B}\boldsymbol{n}$ and N spins in a volume V. The approximate Hamiltonian is

$$H_{\text{ap}} = H_0 + V\frac{\mathcal{B}^2}{2\mu_0} - \sum_a \left(m\mathcal{B}\boldsymbol{n} + \sum_b g_{ba}\boldsymbol{s}_b \right) \cdot \boldsymbol{\sigma}_a + \frac{1}{2}\sum_{ba} g_{ba}\boldsymbol{s}_a \cdot \boldsymbol{s}_a. \quad (5.344)$$

Since the matrix g_{ba} of coupling constants is real, symmetric and positive definite, the approximated Hamiltonian H_{ap} is larger than the original. It follows that the free energies obey the same relation. This may be inferred from the general definition

$$F = -k_B T \ln \operatorname{Tr} e^{-H/k_B T}. \quad (5.345)$$

If H increase, the exponential decreases, the trace and the logarithm as well, and the minus sign implies an increase of F.

The approximate free energy depends on the vectors $\boldsymbol{s}_1, \boldsymbol{s}_2, \ldots, \boldsymbol{s}_N$, and we want to determine the minimal value. One can show that it is attained for $\boldsymbol{s}_a = \boldsymbol{s}$ since the material under investigation is supposed to be translationally invariant. Moreover, $\boldsymbol{s}$ will point in direction of the only vector in the game, namely $\boldsymbol{n}$ in $\boldsymbol{B} = \mathcal{B}\boldsymbol{n}$. With this choice the approximate free energy is

$$F_{\text{ap}} = F(s) = F_0 + V\frac{\mathcal{B}^2}{2\mu_0} - Nk_B T \ln 2\cosh\left(\frac{m\mathcal{B} + gs}{k_B T}\right) + \frac{Ng}{2}s^2. \quad (5.346)$$

Here $g = \sum_b g_{ba}$ does not depend on the location index a, and we will express it as a temperature, $g = k_B \Theta_C$.

We want to minimize this expression with respect to s:

$$\tanh\left(\frac{m\mathcal{B}}{k_B T} + s\frac{\Theta_C}{T}\right) = s. \quad (5.347)$$

One calculates

$$\boldsymbol{M} = \frac{Nms}{V}\boldsymbol{n} \quad (5.348)$$

for the magnetization.

A plausible result. The magnetization is proportional to the density of freely rotatable electron spins. Its direction is that of the external magnetic induction field $\boldsymbol{B}$. s is the dimensionless alignment factor; it varies between minus and plus one.

Let us discuss first the behavior at large temperatures, i.e. $T > \Theta_C$. The equation $\tanh(s\Theta_C/T) = s$ then has one solution only, namely $s = 0$. An expansion of (5.347) for small induction gives

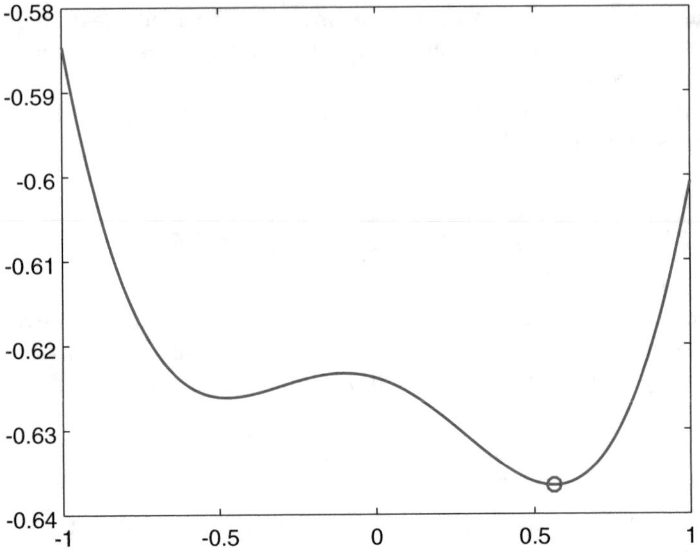

Fig. 5.10 Free energy of the Heisenberg model. The approximate free energy $F_{\text{ap}} = F(s)$ per spin (in units of $k_B\Theta_C$) is plotted versus the spin alignment parameter s. The temperature $T = 0.9\,\Theta_C$ is well below the Curie temperature Θ_C. A small induction field $\mathcal{B}$ of $m\mathcal{B}/k_B\Theta_C = 0.1$ has been applied for removing rotational symmetry. The free energy is minimal (mark) at $s^*(T)$

$$\boldsymbol{M} = \frac{N}{V}\frac{m^2\mathcal{B}}{k_B(T-\Theta_C)}\,\boldsymbol{n} + \dots \tag{5.349}$$

Above Curie's temperature the material is para-magnetic. The temperature dependence is described by the Curie–Weiss law. The magnetic susceptibility will become very large close to the Curie temperature. This points at a transition to another phase.

Below the Curie temperature, the condition (5.347) for a minimal free energy has three solutions with $\mathcal{B} = 0$. The equation $tanh(s\Theta_C/T) = s$ is solved by $s = 0$ and by $s = \pm s^*(T)$. One can easily convince oneself that the latter two solutions result in a smaller free energy, $F(\pm s^*) < F(0)$. See Fig. 5.10.

We obtain

$$\boldsymbol{M} = \frac{N}{V}\,s^*(T)\,m\,\boldsymbol{n} = M^*(T)\,\boldsymbol{n}. \tag{5.350}$$

There is a spontaneous magnetization. Its size—in the Heisenberg model with mean field approximation—can easily be calculated. If there was an induction field $\boldsymbol{\mathcal{B}} = \mathcal{B}\,\boldsymbol{n}$ which is gradually switched off, $\mathcal{B} \to 0$, the direction of the spontaneous magnetization is fixed by the formerly applied field. (In crystalline substances, however, there are intrinsic preferred directions.)

If the Curie temperature of the Curie–Weiss law is positive, the material behaves as a ferro-magnet.

The spontaneous magnetization of the Heisenberg model is shown in Fig. 5.11.

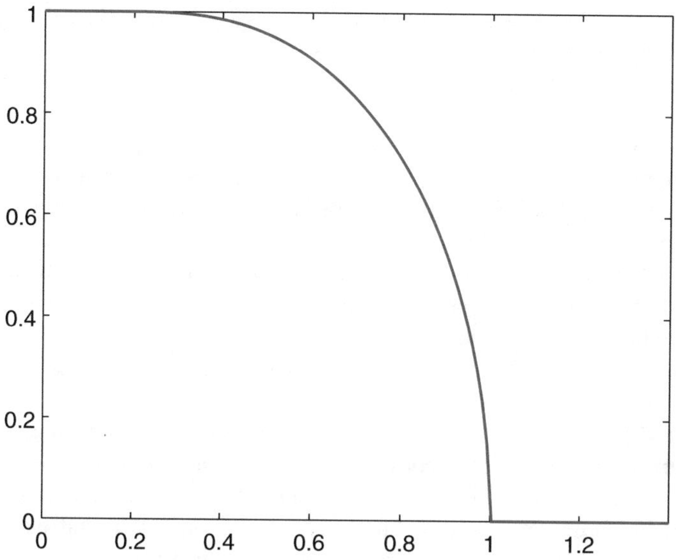

Fig. 5.11 Spontaneous magnetization. The Heisenberg model in mean field approximation provides the dependency of spontaneous magnetization on temperature. We have plotted the spontaneous magnetization $M^*(T)$ (in units of $M^*(0)$) versus the temperature T in units of the Curie temperature Θ_C

5.8.5 Summary

We set out to precisely describe the interaction of matter with a quasi-static electric or magnetic field. Matter in an external field becomes polarized such that there are polarization contributions to the electric charge and current densities. By introducing auxiliary fields, Maxwell's equations may be written in such a form that the fields are generated by free charges only.

However, since there are too many fields, relations are required between polarization $\boldsymbol{P}$ and magnetization $\boldsymbol{M}$ on the one hand and electric and induction field strengths $\boldsymbol{\mathcal{E}}$ and $\boldsymbol{\mathcal{B}}$ on the other hand. These relations are characteristic for the material under investigation and, their study belongs to the field of statistical thermodynamics proper.

The first law for the interaction of matter with an electric field reads $\text{d}W = \int \text{d}V\, \boldsymbol{E} \cdot \text{d}\boldsymbol{D}$. The dielectric displacement serves as independent variable, the electric field strength appears as a generalized force. The magnetic case is ruled by $\text{d}W = \int \text{d}V\, \boldsymbol{H} \cdot \text{d}\boldsymbol{B}$ with the magnetic induction as variable and the magnetic field strength as generalized force. Note tat the thermodynamic variables $\boldsymbol{D}$ and $\boldsymbol{B}$ are extensive while the fields $\boldsymbol{E}$ or $\boldsymbol{H}$ are intensive.

We have explained that there is an alternative form where the free energy is split into a term F_{m} for matter, a term F_{f} for the electromagnetic field, and a term F_{fm} describing the interaction. Choosing a particular expression for F_{m} one obtains

$\mathrm{d}F_{\mathrm{fm}} = -\int \mathrm{d}V\, \boldsymbol{P} \cdot \mathrm{d}\boldsymbol{E} - \mu_0 \int \mathrm{d}V\, \boldsymbol{M} \cdot \mathrm{d}\boldsymbol{H}$. Normally, the electric and the magnetic field strength are easy to control.

We have studied three examples for the interaction of matter with quasi-static fields.

The alignment of permanent electric dipoles shows the expected result for large field strength: complete alignment. Small fields cause a linear relationship between field strength and net polarization. The slope, the dielectric susceptibility, is inversely proportional to the temperature.

Freely rotatable electron spins are related with a magnetic moment which, like the spin, is quantized. We again encounter saturation and a linear relationship for small induction. The magnetic susceptibility likewise follows a $1/T$ law named after Curie. However, the functional forms of the dependency on field strength markedly differ for the electric and magnetic case.

The third model we have just studies is the Heisenberg ferromagnet. The interaction term is augmented by a correlation between pairs of spin which may raise or lower the energy. In the latter case the tendency to align in the same direction, as caused by an external field, is amplified to such an extent that there remains a magnetization even after the external field is switched off. The phenomenon of ferro-magnetism occurs for temperatures below a critical value, the Curie temperature. Above it, the material behaves para-magnetically. The susceptibility follows the Curie–Weiß law $\chi \propto 1/(T - \Theta_{\mathrm{C}})$.

Chapter 6
Fluctuations and Dissipation

In the previous chapter we have studied the Gibbs state. It describes the equilibrium between a system and its environment. Equilibrium with respect to energy exchange is characterized by equal temperatures, with respect to volume exchange by equal pressures, with respect to particle exchange by equal chemical potentials.

Thermodynamics proper studies relations between quantities or qualities which are described by real numbers. A quantum physical approach clearly distinguishes between thermodynamic parameters, such as temperature, volume or chemical potentials, and expectation values of observables, like energy, pressure or particle numbers. Thus, thermodynamics appears to be a classical theory although it is based on quantum physics.

Normally, it is not realized that thermodynamic variables are expectation values of observables. The corresponding probability distribution is sharply peaked at the expectation, or mean value. This is reflected in the concept of a material point: small enough to appear as a point to the engineer, large enough to be treated as a system with almost infinitely many particles.

There are however situations where fluctuations—deviations of an observable from its mean value—will play a role, and in this chapter we shall focus on them.

Expressions for fluctuations, in particular their variance, may be derived directly from thermodynamic potentials. The variance of energy is proportional to the heat capacity of a system. Likewise, in a system which is open with respect to particle exchange, the variance of the particle number is directly related to its compressibility. These findings are true for an arbitrary system. In the special case of an ideal Boltzmann gas we calculate the probability distribution directly for the particle number observable and derive the variance from it. Such density fluctuations are responsible for the clear sky to appear blue, and not black.

The forces on a conduction electron of an ohmic resistor are fluctuations; they produce a noisy voltage which is correctly described by the Nyquist formula. There is an intimate relation between the time correlation function of a fluctuation and its

P. Hertel, *Quantum Theory and Statistical Thermodynamics*,
Graduate Texts in Physics, DOI 10.1007/978-3-319-58595-6_6

power spectrum, as discovered by Wiener and Khinchin. We derive the theorem later once more in a quantum physical context.

The fluctuating force exerted on a small particle by a liquid causes it to move. We discuss this so called Brownian motion under various aspects in a section of its own. Although the particle moves, its motion is an equilibrium property! A path cannot be predicted, only that the particle is displaced from its original location by a distance growing with time according to $\sqrt{t}$.

The next section is dedicated to time-dependent perturbations of a Gibbs state. An external parameter, the electric field of incident light for instance, changes so rapidly that equilibrium is never reached.

We solve the equation of motion in lowest order of the interaction term. It is assumed that the system has been in equilibrium with its environment before the onset of the perturbation. The result is an expression where the response now is caused by all previous perturbations. The linear response in particular is mediated by a retarded Green's function which is proportional to a commutator of certain time-translated fluctuations. It satisfies so-called dispersion relations. We write down explicit expressions for the frequency dependent dielectric susceptibility of optical materials.

In the last chapter the relation between the response function and the time correlation function is closely studied. We derive the KMS formula which alternatively characterizes an equilibrium state. It allows to relate commutators with symmetrized products of fluctuations.

The final result of the discussion is the dissipation-fluctuation theorem as derived by Callen and Welton. It relates the absorptive part of the susceptibility with a spectral intensity function which is never negative. It follows that the absorptive part is never negative as well and causes absorption of electromagnetic field energy.

If a system in thermodynamic equilibrium is acted upon by an arbitrary perturbation of finite duration, work has to be spent and cannot be extracted, at least not in lowest order. This is the content of the second main law which we could prove finally. But there is a flaw: we have made use of it before, although in a much weaker form, when characterizing the Gibbs state.

6.1 Fluctuations

Fluctuations are non-regular deviations of an observable's value from its mean. Although one might think that in equilibrium all values are constant, such fluctuations are nevertheless an equilibrium property. In the following we will discuss some manifestations of fluctuations.

6.1.1 An Example

We speak about a system with Hamiltonian H which is in equilibrium with its environment. The system's state, the Gibbs state, is described by

$$G = \mathrm{e}^{(FI - H)/k_\mathrm{B}T}. \tag{6.1}$$

F is the free energy

$$F = -k_\mathrm{B}T \ln \mathrm{Tr}\, \mathrm{e}^{-H/k_\mathrm{B}T} \tag{6.2}$$

and serves as a thermodynamic potential. It depends on the temperature T of the environment and on all parameters $\lambda = \lambda_1, \lambda_2 \ldots$ of the Hamiltonian $H = H(\lambda)$. The temperature T, just like F, is a Lagrange multiplier. It has to be chosen such that the average, or internal energy U of the system has a prescribed value,

$$U = \frac{\mathrm{Tr}\, H\, \mathrm{e}^{-H/k_\mathrm{B}T}}{\mathrm{Tr}\, \mathrm{e}^{-H/k_\mathrm{B}T}} = \langle H \rangle. \tag{6.3}$$

One may differentiate this expression with respect to temperature and finds

$$\frac{\partial U}{\partial T} = \frac{\langle H^2 \rangle - \langle H \rangle^2}{k_\mathrm{B}T^2}. \tag{6.4}$$

The variance of a measurable quantity M is never negative:

$$\sigma^2(M) = \langle M^2 \rangle - \langle M \rangle^2 = \langle \left(M - \langle M \rangle\right)^2 \rangle \geq 0. \tag{6.5}$$

This explains why one writes $\sigma^2(M)$ for the variance. $\sigma(M)$ itself is the standard deviation. Since the variance of the Hamiltonian is positive, the internal energy grows with temperature, and $U = \langle H \rangle$ has a unique solution for the temperature. We have argued so before when discussing the significance of the Lagrangian multiplier T. Adding heat to a system always increases its temperature, as implied by the similarity of *heat* and *hot*, *Wärme* and *warm* or *calor* and *caliente*.

Think of some system in thermal contact with its environment. If one measures its energy content repeatedly, one obtains results $E_1, E_2, \ldots$. The measured values E_r are not all equal, but the mean converges towards a stable limit,

$$\lim_{N \to \infty} \frac{1}{N} \sum_{r=1}^{N} E_r = \bar{E} = \langle H \rangle. \tag{6.6}$$

This is a consequence of the law of large numbers. Likewise,[1]

$$\lim_{N\to\infty} \frac{1}{N-1} \sum_{r=1}^{N} \left(E_r - \bar{E}\right)^2 = \sigma^2(H) = \langle \left(H - \langle H \rangle\right)^2 \rangle. \tag{6.7}$$

Note that this is the heat capacity C_V because of

$$U = F + TS = F - T\,\frac{\partial F}{\partial T} \quad \text{and} \quad \frac{\partial U}{\partial T} = -T\,\frac{\partial^2 F}{\partial T^2} = C_V. \tag{6.8}$$

The heat capacity is an extensive property. If two systems are joined, their heat capacities add. It is remarkable that the expectation value of the fluctuation $\Delta H = H - \langle H \rangle$ squared is an extensive quantity although bilinear in H.

The system's energy just serves as an example. Naively, in the systems equilibrium state, everything is at rest. However, if one measures an observable, the measuring results scatter, or fluctuate, about their expectation value. This fluctuation is an equilibrium property! For any observable, it can be calculated if only the Gibbs state is known.

The name *fluctuation* is used in a double sense. 'An observable fluctuates about its mean' is a way of saying that the measured values of it scatter about the mean value. Secondly, the observable $\Delta M = M - \langle M \rangle I$ itself is the fluctuation of M.

In the following we will discuss some more manifestations of fluctuations.

6.1.2 Density Fluctuations

Think of a gas in a very large container. We concentrate on a certain region with volume V. The gas has a temperature T, and its particle density is determined by the constant chemical potential μ. The situation is described by the free energy

$$F^*(T, \mu, V) = -k_B T \ln \operatorname{Tr} \mathrm{e}^{(\mu N - H)/k_B T}. \tag{6.9}$$

For non-interacting particles we may rewrite (6.9) into

$$F^*(T, \mu, V) = \mp k_B T\, V \int \mathrm{d}\epsilon\, z(\epsilon)\, \ln \left\{1 \pm \mathrm{e}^{(\mu - \epsilon)/k_B T}\right\}. \tag{6.10}$$

This expression refers to Fermi and Bose statistics, plus and minus sign, respectively. $z = z(\epsilon)$ is the number of one-particles states, or modes, per unit energy and per unit volume.

For low particle densities (6.10) is to be approximated by

[1] Note the denominator $N - 1$ for the best estimation of the variance.

$$F^*(T, \mu, V) = -k_B T\, V\, e^{\mu/k_B T} \int d\epsilon\, z(\epsilon)\, e^{-\epsilon/k_B T}. \tag{6.11}$$

This is the classical limit (Boltzmann statistics). The integral is proportional to $T^{3/2}$, and we obtain

$$F^*(T, \mu, V) = -k_B T\, V \left\{ \frac{2mk_B T}{2\pi\hbar^2} \right\}^{3/2} e^{\mu/k_B T}. \tag{6.12}$$

Now, from (6.9) we deduce

$$\frac{\partial F^*}{\partial \mu} = -\langle N \rangle \text{ and } \frac{\partial^2 F^*}{\partial \mu^2} = -\frac{\langle N^2 \rangle - \langle N \rangle^2}{k_B T}. \tag{6.13}$$

This holds true in general.

For the low-density gas with negligible intermolecular interactions, as described by (6.12), one finds

$$\frac{\partial^2 F^*}{\partial \mu^2} = -\frac{\langle N \rangle}{k_B T}. \tag{6.14}$$

The particle number fluctuation, at least for an ideal gas, is given by

$$\langle N^2 \rangle - \langle N \rangle^2 = \langle N \rangle. \tag{6.15}$$

One may also calculate the probability p_n for n particles to be found in a volume V. The particles shall be randomly distributed such that there are $\bar{N}$ particles in V on the average. The situation is characterized by the Poisson distribution:

$$p_n \propto \frac{\bar{N}^n}{n!}. \tag{6.16}$$

One easily works out

$$\sum_{n=0}^{\infty} n\, p_n = \bar{N} \text{ and } \sum_{n=0}^{\infty} n^2\, p_n = \bar{N}^2 + \bar{N}. \tag{6.17}$$

This agrees with our previous result (6.15).

Again, the standard deviation $\sigma(N) = \sqrt{\langle N^2 \rangle - \langle N \rangle^2}$ grows as $\sqrt{\bar{N}}$ and not as $\bar{N}$.

As an example, consider light which has been scattered in the atmosphere. If the refractive index of the medium (air) would be truly constant, no scattering would occur. After all, electromagnetic plane waves propagate as plane waves in a homogeneous medium. Therefore, scattering occurs because the medium is inhomogeneous, because there are density fluctuations. An example is the color of the clear sky. An observer sees light coming from all sides because sunlight is scattered. The scattering cross section (in dipole approximation) is proportional to $|\ddot{\boldsymbol{d}}|^2 \propto \omega^4$, therefore more violet and blue light is scattered (Rayleigh scattering). However, the sun emits less violet than blue light and the human eye perceives blue better than violet. This explains why the sky appears to be blue and not violet.

By the way, the scattered light is partially polarized, the polarization depending on the line of view and on the sun's position. Insects use this effect for finding their way even when the sun is covered by clouds. Their eyes perceive violet and ultraviolet much better than humans.

The rising or setting sun—when rays pass a long distance through the atmosphere—appear reddish because the blue component has been scattered off the light path.

Incidentally, the particle number fluctuation may be calculated for an arbitrary thermodynamic system, not just for an ideal Boltzmann gas as above. For this purpose one resorts to the relation

$$F^*(T,\mu,V) = -Vp(T,\mu), \tag{6.18}$$

since F^*, an extensive quantity, must depend linearly on the only extensive variable V. Compare with (6.12). The isothermal compressibility κ_T was defined by (5.57), namely

$$\kappa_T = -\frac{1}{V}\left.\frac{\mathrm{d}V}{\mathrm{d}p}\right|_{\mathrm{d}T=0}. \tag{6.19}$$

By combining the latter two equations one will arrive at the following expression:

$$\langle N^2 \rangle - \langle N \rangle^2 = \bar{N}\, k_\mathrm{B} T\, \kappa_T. \tag{6.20}$$

If the substance is densely packed, as in a liquid, the particle fluctuations are small and with it the isothermal compressibility. The other extreme is an ideal gas. Its isothermal compressibility is given by $\kappa_T = 1/p$, i.e. large at low pressure.

6.1.3 *Correlations and Khinchin's Theorem*

Consider a process $t \to M_t = U_{-t} M U_t$. Here $M = M_0$ is an observable and its time evolution is called a process. The waiting operators form a one-parameter Abelian group

$$U_t = \mathrm{e}^{-\frac{\mathrm{i}}{\hbar} t H}. \tag{6.21}$$

This group is generated by the Hamiltonian H, the system's energy observable.

We concentrate on processes in thermal equilibrium. The latter is described by the Gibbs state

$$G = \mathrm{e}^{(FI - H)/k_{\mathrm{B}}T} \quad \text{with} \quad F = -k_{\mathrm{B}}T \ln \mathrm{Tr}\, \mathrm{e}^{-H/k_{\mathrm{B}}T}. \tag{6.22}$$

The free energy F serves to normalize the Gibbs state: $\mathrm{Tr}G = 1$. In the following text all expectation values refer to this Gibbs state.

A process in a thermal equilibrium state is stationary,

$$\langle M_t \rangle = \mathrm{Tr}GU_{-t}MU_t = \mathrm{Tr}\, U_{-t}GU_tM = \mathrm{Tr}GM. \tag{6.23}$$

We have performed a cyclic permutation within the trace and used the fact that both U_t and G are functions of one and the same H; they therefore commute. The expectation value $\langle M_t \rangle$ does not change with time. Therefore the fluctuation

$$\Delta M_t = M_t - \langle M \rangle \tag{6.24}$$

is of interest only. Clearly, its expectation value vanishes, $\langle \Delta M_t \rangle = 0$. The fluctuation as a process does not vanish, however. We set out to characterize it.

Let us define the temporal correlation function by

$$K(\tau) = \frac{\langle \Delta M_{t+\tau} \Delta M_t + \Delta M_t \Delta M_{t+\tau} \rangle}{2}. \tag{6.25}$$

Since it is symmetrized, the product is self-adjoint, and its expectation value will be real. It does not depend on the time instant t because we discuss a stationary process. τ is the time difference between $\Delta M_{t+\tau}$ and ΔM_t. Note that the instantaneous correlation $K(0) = \langle (\Delta M)^2 \rangle$ is always positive.

Let us Fourier transform the process:

$$\Delta M_t = \int \frac{\mathrm{d}\omega}{2\pi}\, \mathrm{e}^{-\mathrm{i}\omega t}\, \hat{M}_\omega \tag{6.26}$$

such that one may write

$$\Delta M_{t+\tau} = \int \frac{\mathrm{d}\omega'}{2\pi}\, \mathrm{e}^{\mathrm{i}\omega'(t-\tau)}\, \hat{M}_{\omega'}\star. \tag{6.27}$$

Insert this into (6.22). The result contains a factor

$$\int \frac{\mathrm{d}\omega}{2\pi} \int \frac{\mathrm{d}\omega'}{2\pi}\, \mathrm{e}^{\mathrm{i}(\omega'-\omega)t} \tag{6.28}$$

which should not depend on t. For this to be true the rest must contribute only if $\omega = \omega'$. In other words:

$$K(\tau) = \int \frac{\mathrm{d}\omega}{2\pi}\, \mathrm{e}^{\mathrm{i}\omega\tau}\, S(\omega) \tag{6.29}$$

where the spectral intensity $S = S(\omega)$ is defined by

$$\frac{\langle\, \hat{M} \star_{\omega'} \hat{M}_{\omega} + \hat{M} \star_{\omega} \hat{M}_{\omega'} \,\rangle}{2} = 2\pi\delta(\omega' - \omega) S(\omega). \tag{6.30}$$

Because the expectation value is of type $\langle\, A \star A \,\rangle \geq 0$ we conclude that the spectral intensity is never negative,

$$S(\omega) \geq 0. \tag{6.31}$$

Equation (6.29) with (6.31) is known as Khinchin's theorem or the Wiener–Khinchin theorem. The correlation function $K = K(\tau)$ for fluctuations in a stationary process is the Fourier transform of a non-negative spectral intensity $S = S(\omega)$. The Wiener–Khinchin theorem applies to stationary states in general and to the Gibbs state in particular.

The following two examples pertain to classical random variables instead of quantum observables. The former commute, the latter not. It follows that the Wiener–Khinchin theorem remains valid. Instead of a symmetrized product we may just write a product.

6.1.4 Thermal Noise of a Resistor

We present two examples of calculations which make use of the Wiener-Khintchin theorem. The discourse on Brownian movement is deferred to the next section because of its historical significance. Here we explain why ordinary ohmic resistors produce noise: random fluctuations of voltage.

Unless a system has been cooled down to zero temperature, its variables fluctuate about the equilibrium value. In particular, we discuss the simplest component of electric circuits, an ohmic resistor R. We shall see that the resistor produces a voltage although the circuit which it is part of is completely passive. Its frequency[2] distribution is described by the Nyquist formula. We comment on the role of ensemble averages as used in theoretical considerations and time averages which can be measured.

6.1.5 Langevin Equation

Think of a circuit made up of a resistor R and a parallel plate capacitor C. See Fig. 6.1 for a sketch. There is a charge Q_t on the upper plate and a charge $-Q_t$ on the lower. The electric current in the circuit is $I_t = \dot{Q}_t$. The voltage across the

---

[2] We follow electrical engineering tradition and talk of frequencies f instead of angular frequencies $\omega = 2\pi f$.

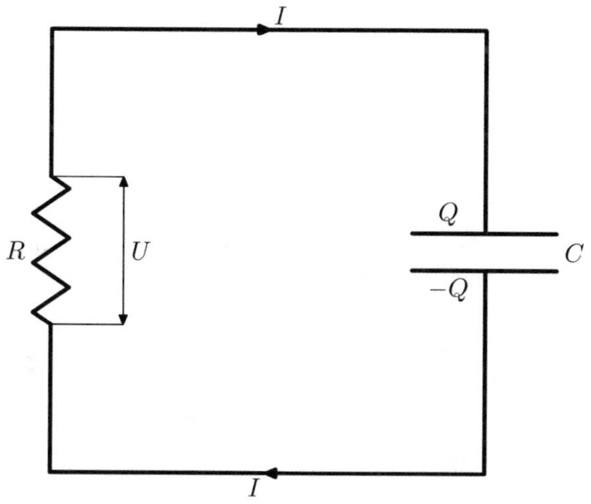

Fig. 6.1 RC circuit. A circuit made up of a resister R and a capacitor C. The charges on the parallel plates are $\pm Q$, respectively. $I = \dot{Q}$ is the current, the voltage across the resister is U

capacitor is $U_t = Q_t/C$. The same voltage is applied to the ohmic resistor such that current and voltage are related by $U_t = RI_t$. Charge Q_t, current I_t and voltage U_t are time-dependent variables while capacitance C and resistance R are constants. The differential equation

$$RC\dot{U}_t + U_t = 0 \tag{6.32}$$

has the obvious solution

$$U_t = U_0 \,\mathrm{e}^{-t/\tau} \tag{6.33}$$

where $\tau = RC$ is a time constant. But that is not all.

The free quasi-electrons in a conductor are not entirely free. Otherwise a voltage would accelerate them more and more and there would be an ever increasing current. Instead, the conduction electrons interact with lattice oscillations, or phonons, which brake, or decelerate them. In this way, a constant electric field will drive a stationary electric current.

Such interactions will also happen if there is no external electric field. Phonon-electron collisions will spontaneously accelerate conduction electrons, although randomly. There will be, per unit time, as many kicks to the left as to the right, and no net current may develop. However, although a particular conduction electron is not accelerated on average, there are random fluctuations. With a small probability many consecutive kicks into the same direction may occur which lead to observable currents and voltages.

We formulate this mathematically by a random electromotive voltage $V = V_t$, such that (6.30) now reads

$$\tau\dot{U}_t + U_t = V_t. \tag{6.34}$$

The V_t are random variables which are characterized by

$$\langle V_t \rangle = 0. \tag{6.35}$$

This says that the fanning electromotive voltages are fluctuations, they vanish on the average. Their correlation in time is described by

$$\langle V_{t'} V_{t''} \rangle = K_V(t' - t''). \tag{6.36}$$

With (6.36) we take into account that the circuit's parameters do not depend on time. Only time differences count. The Wiener–Khinchin theorem (6.26) with (6.28) says that the time correlation function

$$K_V(t) = \int \mathrm{d}f\, S_V(f)\, \mathrm{e}^{2\pi \mathrm{i} f t} \tag{6.37}$$

is the Fourier transform of a positive spectral intensity function $S_V = S_V(f)$. The latter is defined by

$$\langle \hat{V}^*(f'')\hat{V}(f') \rangle = \delta(f' - f'') S_V(f'). \tag{6.38}$$

Equation (6.34) is a Langevin equation, the random voltage being described by (6.35), (6.37) and (6.38). Its solution depends on initial conditions and the spectral intensity function $S_V = S_V(f) \geq 0$.

6.1.6 Nyquist Formula

Let us Fourier transform the Langevin equation:

$$-2\pi \mathrm{i} f \tau \hat{U}_f + \hat{U}_f = \hat{V}_f, \tag{6.39}$$

which results in

$$\hat{U}_f = \frac{\hat{V}_f}{1 - 2\pi \mathrm{i} f \tau}. \tag{6.40}$$

We conclude that the spectral intensity of the process $t \to U_t$ is

$$S_U(f) = \frac{S_V(f)}{1 + 4\pi^2 f^2 \tau^2}, \tag{6.41}$$

see (6.40).

We now assume that random electromotive voltages are correlated only for an extremely short time[3] as compared with the time constant τ of the circuit:

$$K_V(t) = \kappa\,\delta(t) \text{ or } S_V(f) = \kappa \text{ with } \kappa > 0. \tag{6.42}$$

This behavior is called *white noise*. Noise, because the signal $t \to \langle V_t \rangle$ vanishes. White, because the spectral intensity in the frequency region of interest is constant. This terminology comes from optics where white light is characterized by a constant spectral intensity within the frequency range of visible light.

For white noise the time correlation function $K_U(t)$ is

$$K_U(t) = \int \mathrm{d}f\, \frac{\kappa}{1 + 4\pi^2\tau^2 f^2}\, \mathrm{e}^{2\pi\mathrm{i}ft} = \frac{\kappa}{2\tau}\,\mathrm{e}^{-|t|/\tau}\,. \tag{6.43}$$

As is well-known, the field energy stored in a capacitor is $W = UQ/2 = CU^2/2$ which is equal to $k_BT/2$ in thermal equilibrium. Because of $K_U(0) = \langle U^2 \rangle$ we calculate

$$k_BT = C\langle U^2 \rangle = CK_U(0) = C\frac{\kappa}{2\tau} = \frac{\kappa}{2R} \tag{6.44}$$

or

$$S_U(f) = 2Rk_BT. \tag{6.45}$$

$S_U(f)$, as it appears in the Wiener–Khinchin theorem (6.38), is the power per unit frequency interval. Because the interval $[f, f + \mathrm{d}f]$ and $[-f - \mathrm{d}f, -f]$ are equivalent (the frequency f here is positive), we must write

$$\bar{S}_U(f) = S_U(f) + S_U(-f) = 4Rk_BT. \tag{6.46}$$

This is the famous Nyquist formula for Johnson noise. J.B. Johnson discovered this form of noise in 1927 and published rather accurate measurements in 1928. H. Nyquist, an electrical engineer of Swedish roots who emigrated to the USA, explained it shortly afterwards. Johnson noise is caused by the interactions between conduction electrons and thermally induced photons, or lattice vibrations. Its spectral intensity $\bar{S}_U(f)$ does not depend on frequency, at least not up to many GHz. Its dependency on T is obvious: higher temperature, more phonons, more interactions. The dependency on R is as well simple to explain: the noise of resistors in series adds up. And: fluctuations in the equilibrium state are always proportional to the Boltzmann constant k_B.

[3]Electronics is limited by micrometer dimensions, therefore times below $\approx 10^{-14}$ s cannot be resolved. The natural time for phonon-electron interactions is the natural atomic time unit of $\approx 10^{-17}$ s.

6.1.7 Remarks

Note that the capacitance C does not show up in Nyquist's formula. We could have analyzed a circuit made up of a resistor and an inductor as well. Then the inductivity L would have dropped out. In fact, any circuit of arbitrary impedance Z will serve. The decisive ingredient is the equipartition theorem. If the Hamiltonian of a system is a sum of independent observables squared, then each degree of freedom has an energy $k_B T/2$ in thermal equilibrium. This result of classical statistical mechanics, however, is true only for large enough temperatures. In our case, hf should be small if compared with $k_B T$. For $T = 300$ K this amounts to $f \ll 6$ THz. For higher frequency Nyquist's formula must be corrected by a factor which depends on $x = hf/k_B T$, the limit $x \to 0$ of which is 1. For details see the dissipation-fluctuation theorem to be discussed shortly.

Electronic circuits suffer from many sources of noise. Johnson noise produced by ohmic resistors is just one of them. Any electric current, which is a stream of charged particles, causes fluctuations because charge is quantized. Such shot noise was discovered by W. Schottky in 1918. It is particular important if only few (n) electrons pass a barrier in a short time. Then the current density fluctuation, which is proportional to $\sqrt{n}$, cannot be ignored; it delimits the performance of many electronic devices.

6.1.8 Summary

So far we have studied the Gibbs state and its expectation values. Such values appear to have a fixed value which is not the case. They fluctuate about the mean, or expectation value with a certain variance. This was demonstrated for the internal energy. Its fluctuation is even observable, namely proportional to the specific heat. For a large system, the internal energy grows with the number of particles N. It turns out that the variance $\sigma = \sqrt{\langle \Delta H^2 \rangle}$ grows like $\sqrt{N}$ only.

We also have discussed particle density fluctuations. They may be calculated from thermodynamics or from the assumption of independent particles with a specified mean (Poisson distribution). Density fluctuations are responsible for the scattering of light by the earth's atmosphere and explain why the clear sky looks blue.

Another aspect of fluctuations is noise in a very broad sense. We have presented a description of the noisy voltage produced by an ohmic resistor and derived Nyquist's formula for the power per unit frequency interval.

In order to understand the nature of noise we discussed the Wiener–Khinchin theorem. It states that the time correlation function of an observable quantity is the Fourier transform of a positive spectral intensity. This holds true for an arbitrary stationary process. For explaining the noise of a resistor a classical description for random variables is sufficient.

6.2 Brownian Motion

In 1905 Albert Einstein published three seminal papers in *Annalen der Physik*. One, *On the Electrodynamics of moving bodies*, revolutionizing the concept of space and time. Another one, *On a Heuristic Point of View Concerning the Production and Transformation of Light*, suggested that light is made up of particles which we call photons today. The third important paper[4] showed that the molecular theory of heat predicts stochastic motion of particles suspended in a resting liquid.

Small particles which are suspended in water in fact move in a chaotic way. The Scottish botanist Robert Brown had discovered this in 1827 when observing pollen grains, although he was not the first. This motion[5] was interpreted as a sign of life, but it soon turned out that powders from ancient rocks showed the same phenomenon. It was Albert Einstein who demonstrated that the small particles did not move on their own, but were pushed by the moving molecules of the environment. He predicted that the displacement grows proportional to the square root of time and calculated an expression for the proportionality constant.

His predictions were later experimentally verified and allowed to determine the value of the Boltzmann constant k_B. Since Avogadro's number N_A, the number of particles for one mole, is related with the universal gas constant[6] by $R = N_A k_B$, the mass of the proton could be calculated. Einstein's investigation of Brownian motion was the first concrete step to relate continuum physics with particle physics. Diffusion is the Brownian motion of a cloud of particles. Brownian motion is diffusion of a single particle. Fig. 6.2 shows such a (simulated) random walk.

6.2.1 *Einstein's Explanation*

So let us reconstruct Einstein's arguments. In order to keep the discussion as simple as possible, we restrict ourselves to the motion in one dimension.

At time $t = 0$ the Brownian particle is at rest and situated at $x = 0$. After a time interval τ_1 it suffers a collision with a molecule of the environment which translates it by s_1, until, after a time interval τ_2, it suffers another collision which translates it by s_2, and so on.

After the nth collision the Brownian particle has been moved by the environment to

$$x_n = \sum_{j=1}^{n} s_j, \tag{6.47}$$

[4]*Über die von der molekularkinetischen Theorie der Wärme geforderte Bewegung von in ruhenden Flüssigkeiten suspendierten Teilchen* a rough translation of which is *On the movement of particles suspended in a resting liquid as required by the molecular-kinetic theory of heat.*

[5]The term *Brownian movement* is used in the literature as well.

[6]A diluted gas of ν moles at temperature T within a vessel of volume V exerts a pressure $p = \nu RTV$.

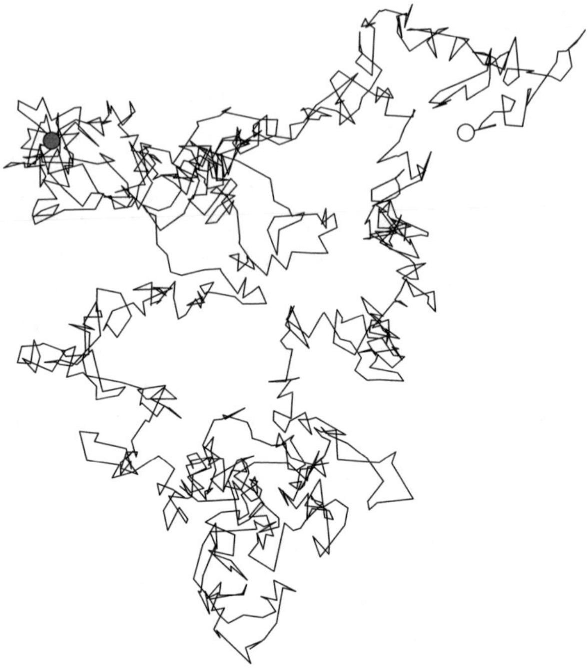

Fig. 6.2 Random walk of a Brownian particle. The starting point $(x, y) = (0, 0)$ is marked by a *filled circle*, the end point after 1000 observations by a *white circle*

and this location corresponds to time

$$t_n = \sum_{j=1}^{n} \tau_j . \tag{6.48}$$

We assume that the τ_j are independent and equally distributed random variables with expectation value $\langle \tau_j \rangle = \tau$. We likewise assume that the displacements s_j are independent and equally distributed random variables with expectation values $\langle s_j \rangle = 0$ and variance $\langle s_j^2 \rangle = s^2$.

The law of large numbers states

$$\lim_{n\to\infty} \frac{t_n}{n} = \tau . \tag{6.49}$$

The central limit theorem states that the probability distribution $p_n(z)$ for $x_n/\sqrt{n}$ converges towards the normal distribution $p(z)$ with variance s^2 which is

$$p(z) = \frac{1}{\sqrt{2\pi s^2}}\, \mathrm{e}^{-z^2/2s^2} . \tag{6.50}$$

The probability distribution for the location x_n therefore is

$$G_n(x) \approx \frac{1}{\sqrt{2\pi n s^2}}\, \mathrm{e}^{-x^2/2ns^2}. \tag{6.51}$$

We replace n by t/τ, according to (6.49), and obtain for the limit of G_n the following expression:

$$G(t,x) = \frac{1}{\sqrt{4\pi D t}}\, \mathrm{e}^{-x^2/4Dt}. \tag{6.52}$$

We have set

$$D = \frac{s^2}{2\tau}. \tag{6.53}$$

Recall that $\tau = \langle\, \tau_j \,\rangle$ is the average time between collisions of the environment's molecules with the Brownian particle. The displacement of the Brownian particle between two subsequent collision $\langle\, s_j \,\rangle$ vanishes on the average. However, its variance $\langle\, s_j^2 \,\rangle = s^2$ is positive.

The Brownian particle, which at time $t = 0$ was located at $x = 0$, will be found later within the interval $[x, x + \mathrm{d}x]$ with probability $\mathrm{d}x\, G(t, x)$.

In three dimensions (6.52) reads

$$G(t,\boldsymbol{x}) = G(t,x_1)\, G(t,x_2)\, G(t,x_3) = (4\pi D t)^{-3/2}\, \mathrm{e}^{-\boldsymbol{x}^2/4Dt}. \tag{6.54}$$

This function obeys the diffusion equation

$$\dot{G} = D\, \Delta G \tag{6.55}$$

with the initial condition

$$G(0,\boldsymbol{x}) = \delta^3(\boldsymbol{x}). \tag{6.56}$$

With $n_0 = n(0,\boldsymbol{x})$ as the initial distribution of particles we easily arrive at the following expression for the particle distribution at time t:

$$n(t,x) = \int \mathrm{d}^3 y\, G(t,\boldsymbol{x} - \boldsymbol{y})\, n_0(\boldsymbol{y}). \tag{6.57}$$

Indeed, (6.55) warrants that (6.57) satisfies the diffusion equation, and (6.56) guarantees $n(0,\boldsymbol{x}) = n_0(\boldsymbol{x})$.

We have plotted in Fig. 6.3 a simulated cloud of identical Brownian particles which all started at the same point and randomly walked for a certain time in the x, y-plane.

Back to (6.52). We work out

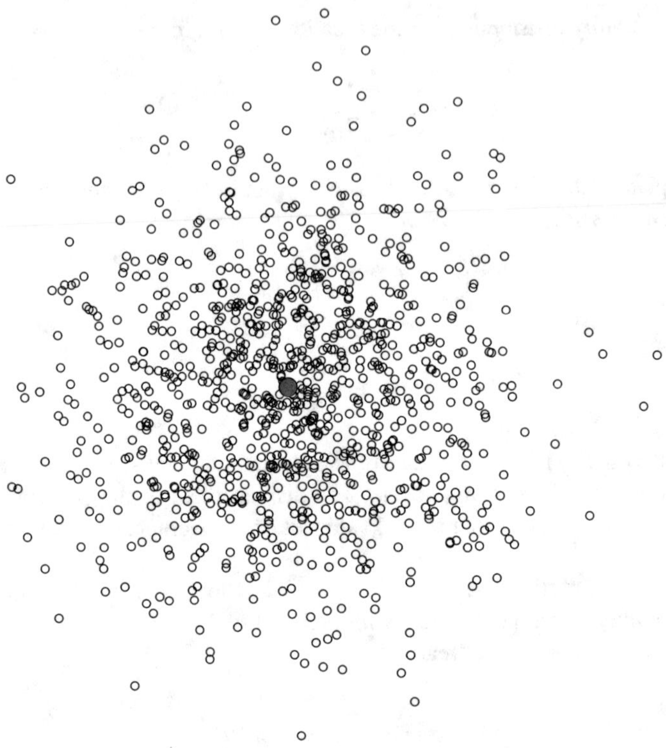

Fig. 6.3 Cloud of Brownian particles. 1000 Brownian particles all started at the same location (*filled circle*). Their positions in the (x, y)-plane after a certain time are represented by *small white circles*. Diffusion is Brownian motion en masse

$$\langle x(t)^2 \rangle = \int dx\, G(t, x)\, x^2 = 2Dt. \tag{6.58}$$

Hence the Brownian particle, as observed in the x, y plane of the microscope, drifts away from its origin according to

$$r(t) = \sqrt{\langle x(t)^2 \rangle + \langle y(t)^2 \rangle} = 2\sqrt{Dt}. \tag{6.59}$$

This formula was Einstein's first contribution to the theory of Brownian motion: the square root time dependence, and not to underestimate, the interpretation in terms of expectation values. A measurement of Brownian motion must be repeated very often such that the averaged results become reliable.

6.2.2 The Diffusion Coefficient

We now recapitulate Einstein's arguments for the interpretation of the diffusion coefficient D.

The Brownian particles are spheres of radius R and effective mass[7] m. They float in a medium with viscosity η and temperature T.

According to the rules of thermodynamics the density of Brownian particles at height z is

$$n(z) = n(0)\,e^{-mgz/k_BT}. \tag{6.60}$$

This will produce a diffusion current density

$$j^d(z) = -Dn'(z) = \frac{Dmg}{k_BT}n(z). \tag{6.61}$$

On the other hand, Brownian particles fall down at a velocity v which is governed by $mg = 6\pi\eta Rv$ (Stokes' law). η is the viscosity of the medium. The corresponding sedimentation current density is

$$j^s(z) = -\frac{mg}{6\pi\eta R}n(z). \tag{6.62}$$

In equilibrium the two current densities must add up to zero. This is in fact so if

$$D = \frac{k_BT}{6\pi\eta R} \tag{6.63}$$

holds true.

This result is rather plausible. The higher the temperature, the more the environment will push the Brownian particle. If its radius (and also its mass) becomes larger, the less it will respond to such kicks. And likewise, the larger the viscosity of the environment, the less the particle will move when hit.

Since the diffusion constant D, temperature T, viscosity η and the radius R for spherical test particles can be measured, (6.64) in fact allows to determine the value of the Boltzmann constant k_B and therewith Avogrado's number N_A. Perrin's experiments won him the 1926 physics Nobel prize 'for his work on the discontinuous structure of matter, and especially for his discovery of sedimentation equilibrium'.

[7] True mass minus mass of displaced liquid.

6.2.3 Langevin's Approach

Paul Langevin, a French physicist, has discussed Brownian motion as a stochastic process. The velocity $v = v(t)$ of the Brownian particle is a random variable depending on time. It obeys the following stochastic differential equation:

$$m\dot{v} + \Gamma v = f(t), \tag{6.64}$$

where m is the particle's mass, $\Gamma = 6\pi\eta R$ the friction constant, and $f = f(t)$ is the force on the particle exerted by the medium. It is a stochastic process as well.

Since we have to do with correlation functions of more than one random process $t \to A(t)$ we will write

$$K_A(\tau) = \langle \Delta A(t+\tau)\Delta A(t) \rangle. \tag{6.65}$$

Note that classical random variables commute; there is no need to write a symmetrized product.

Expectation values do not depend on time t, but only on the time difference τ. The liquid within which the Brownian particle moves is assumed to be in thermal equilibrium. Its state is stationary.

With

$$\Delta A(t) = \int \frac{\mathrm{d}\omega}{2\pi}\,\mathrm{e}^{-\mathrm{i}\omega t}\,\hat{A}(\omega) \tag{6.66}$$

we find, by the same reasoning as before,

$$K_A(\tau) = \int \frac{\mathrm{d}\omega}{2\pi}\,\mathrm{e}^{-\mathrm{i}\omega \tau}\,S_A(\omega) \tag{6.67}$$

with

$$S_A(\omega) \ge 0. \tag{6.68}$$

The correlation function is the Fourier transform of a positive spectral intensity, so the Wiener–Khinchin theorem.

Let us now return to our discussion of Langevin's approach to Brownian motion. We assume[8]

$$K_f(\tau) = \langle f(t+\tau)\, f(t) \rangle = \kappa\,\delta(\tau). \tag{6.69}$$

There are two time scales. The interaction of a colliding water molecule with the Brownian particle lasts only an extremely short time, and succeeding collisions are statistically independent. On the other hand, the position of the Brownian particle can be observed only in time intervals of milliseconds, or so. Therefore, from the observers point of view, the correlation function of the force can be well approximated

[8]Note that $t \to f(t)$ describes a fluctuation since $\langle f(t) \rangle = 0$.

by a δ-function. By the way, the Fourier transform of (6.71) is κ, a positive constant. Because it does not depend on ω we call $f = f(t)$ a white noise[9] process.

We Fourier transform Langevin's equation (6.65) and obtain

$$(-\mathrm{i}m\omega + \Gamma)\,\hat{V}(\omega) = \hat{f}(\omega), \tag{6.70}$$

hence the spectral intensities are related by

$$S_v(\omega) = \frac{\kappa}{m^2\omega^2 + \Gamma^2}. \tag{6.71}$$

The Forier transform of S_v is

$$K_v(\tau) = \frac{\kappa}{m\Gamma}\,\mathrm{e}^{-\Gamma|\tau|/m}. \tag{6.72}$$

As is well known, the translational degree of freedom of a particle is associated with an energy $k_B T/2$, therefore

$$K_v(0) = \langle\, v(t)^2 \,\rangle = mk_B T \tag{6.73}$$

holds true, i.e.,

$$K_v(\tau) = \frac{k_B T}{m}\,\mathrm{e}^{-\Gamma|\tau|/m}. \tag{6.74}$$

Now, we are primarily interested in the displacement of the Brownian particle which is

$$x(t) = \int_0^t \mathrm{d}s\; v(s). \tag{6.75}$$

Clearly, its expectation value $\langle\, x(t) \,\rangle$ vanishes because of $\langle\, v(s) \,\rangle = 0$. So let us calculate[10]

$$\langle\, x(t)^2 \,\rangle = \int_0^t \mathrm{d}s \int_0^t \mathrm{d}s'\, \langle\, v(s)\, v(s') \,\rangle, \tag{6.76}$$

the integrand being the velocity correlation function $K_v(s - s')$. Because it is symmetric, we rewrite the right hand side as

[9]The electromagnetic field of white light has a spectral intensity which, at least in the visible region, does not depend on ω, or color.

[10]We cannot rely on the statement that expectation values do not depend on time. In fact, demanding $x(0) = 0$ breaks time translation symmetry.

$$2\int_0^t \mathrm{d}s \int_0^s \mathrm{d}s'\, K_v(s-s'), \tag{6.77}$$

i.e.

$$\frac{2k_\mathrm{B}T}{m} \int_0^t \mathrm{d}s\ \mathrm{e}^{-\Gamma s/m} \int_0^s \mathrm{d}s'\ \mathrm{e}^{\Gamma s'/m}. \tag{6.78}$$

The result is

$$\langle\, x(t)^2 \,\rangle = \frac{2k_\mathrm{B}T}{\Gamma} \left\{ \frac{m}{\Gamma} - \frac{m}{\Gamma} \mathrm{e}^{-\Gamma t/m} + t \right\}. \tag{6.79}$$

Indeed, $\langle\, x(0)^2 \,\rangle = 0$. However, after a short start-up phase, $\langle\, x(t)^2 \,\rangle$ increases linearly with time with a proportionality constant which coincides with Einstein's finding.

6.2.4 Summary

We introduced the subject by a brief history of Brownian motion. Initially the random walk of suspended pollen grains was seen as a manifestation of life. However, the phenomenon also happened for ground wood, remnants of a mummy and for glass and stone powder. Einstein supposed that the suspended particles do not move by themselves but were driven by random forces of the suspending liquid. They do not swim, they float. He applied the law of large numbers and the central limit theorem and could show that the average distance squared grows linearly with time, the coefficient being the diffusion constant of the suspended particles. He also related the diffusion constant to the size of the particles, the viscosity of the liquid and its temperature.

Langevin has solved the problem by a different approach. Position, velocity and the force exerted on the particles are random variables. The differential equation is linear and can be solved by Fourier-transforming it so that it becomes an algebraic equation; an equation for spectral intensities may be derived. Since the correlation between forces is of very short duration on a macroscopic scale it should be approximated by a delta function which amounts to white noise (constant spectral intensity). The result describes the particle's location not only in the long run, but also in the initial phase.

6.3 Linear Response Theory

If a system is left alone it will strive for its equilibrium state. Well isolated from its environment, but not totally, its state becomes mixed more and more until the entropy has reached its maximum. 'Well isolated' means that the internal energy remains the same, and entropy, a measure of a state's degree of mixing, will grow.

A process, a sequence of states, is called reversible if it is in equilibrium all the time, namely in a Gibbs state. How fast may the variables of a reversible process change? There is no general answer because the relaxation time of a non-equilibrium state depends on many things. The larger the system, the slower equilibrium is achieved. Pressure differences tend to vanish fast, but heat conduction is a slow process.

As we have explained earlier, there is an entire branch of physics, namely Continuum Physics, the central issue of which is the material point: a point, as an engineer would view it, a huge region with many atoms or molecules as seen by the physicist. Each material point is in an equilibrium state, but different points are not. The material point must be small enough to rapidly attain its equilibrium state, but it must also be large enough so that quantities like particle density, pressure and so on make sense. The relative variations of such quantities, caused by unavoidable fluctuations, should be negligible.

Even if one were content with local equilibrium there is a maximal speed by which external parameters may change for the material point to be quasi instantaneously in equilibrium. If external parameters vary too rapidly, the system cannot follow and will never reach equilibrium because the parameters have changed in the meantime. And this will be the topic of this section: matter under the influence of rapidly changing external parameters.

In order to be specific we shall discuss the interaction of light with matter. The typical time for oscillations of the electric field is 10^{-17} s which is certainly much faster than any relaxation of a region of micrometer dimensions.

We shall carefully formulate the problem. The equations of motion are rewritten (interaction picture) to form a fixed point problem which invites a power series expansion. The first order, or linear response is then studied in closer detail.

6.3.1 Perturbations

The Hamiltonian may depend on external parameters, $H = H(\lambda)$. Such parameters describe the action of the environment onto the system. However, the influence of the system on the environment can safely be neglected. If external parameters, like an electric field strength, change slowly, the system will always be in an equilibrium state corresponding to the current values of the external parameters, and we speak of a reversible process. We shall now find out how a system reacts if the external parameters change rapidly.

Interaction Picture

Let us recapitulate Heisenberg's and Schrödinger's view on dynamics. Heisenberg argued: waiting a time t before measuring an observable M is the same as measuring another observable, namely

$$M_t = U_t \star M U_t \tag{6.80}$$

where the waiting operator

$$U_t = \mathrm{e}^{-\frac{\mathrm{i}}{\hbar} t H} \tag{6.81}$$

is unitary,

$$U_t \star U_t = U_t U_t \star = I, \tag{6.82}$$

since the energy H is self-adjoint. We may therefore write (6.80) as

$$M_t = U_{-t} M U_t. \tag{6.83}$$

Schrödinger thought differently. Waiting a time span t before preparing a state ϱ means preparing a state ϱ_t. Because of

$$\mathrm{Tr}\, \varrho\, M_t = \mathrm{Tr}\, \varrho\, U_{-t} M U_t = \mathrm{Tr}\, U_t \varrho\, U_{-t}\, M \tag{6.84}$$

one concludes

$$\varrho_t = U_t M U_{-t}. \tag{6.85}$$

Expectation values are the same in the Heisenberg and the Schrödinger picture. The respective equations of motion are the Schrödinger equation

$$\dot{\varrho}_t = \frac{\mathrm{i}}{\hbar} [\varrho_t \,, H] \tag{6.86}$$

or the Heisenberg equation

$$\dot{M}_t = \frac{\mathrm{i}}{\hbar} [H, M_t], \tag{6.87}$$

respectively.

There is a third view on dynamics. Often the energy can be split into a managable part H plus a perturbation V which may possibly depend on time. Primarily, we think as Schrödinger of states ϱ_t. However, we transform to the Heisenberg picture with the unperturbed Hamiltonian H only. If there were no perturbation, the transformed states would not depend on time. With perturbation they depend only weekly on time.

Define

$$A_t(t) = U_{-t} A_t U_t \tag{6.88}$$

for an arbitrary operator A_t where $U = U_t$ is given by (6.81), i.e. without the perturbation term. Equation (6.88) obeys the following equation of motion:

$$\frac{\mathrm{d}}{\mathrm{d}t}\, \varrho_t(t) = -\frac{\mathrm{i}}{\hbar} [V_t(t), \varrho_t(t)]. \tag{6.89}$$

Indeed, a small perturbation will result in a small rate of change. Since V_t describes the interaction of the unperturbed system with the perturbation, one speaks of the interaction picture. We shall make use of it in the next subsection.

Linear Response

We assume that our system has been in a Gibbs state before the perturbation was switched on:

$$V_t \to 0 \text{ and } \varrho_t \to G \text{ for } t \to -\infty. \tag{6.90}$$

With this initial condition (6.89) can be integrated to

$$\varrho_t(t) = G - \int_{-\infty}^{t} \mathrm{d}s \, \frac{\mathrm{i}}{\hbar} [V_s(s), \varrho_s(s)]. \tag{6.91}$$

We have rewritten the equation of motion plus initial condition in an integral equation for $t \to \varrho_t(t)$, a fixed point problem, because the unknown appears on the left and on the right hand side. It easily lends itself to an iteration procedure.

Without any perturbation the solution would be $\varrho_t(t) = G$. We calculate the next best approximation by inserting this into the right hand side:

$$\varrho_t(t) = G - \int_{-\infty}^{t} \mathrm{d}s \, \frac{\mathrm{i}}{\hbar} [V_s(s), G]. \tag{6.92}$$

Equation (6.92) describes the linear response to the perturbation $t \to V_t$. We shall process further the result.

Let us first undo the transformation to the interaction picture. The result is

$$\varrho_t = G - \int_{-\infty}^{t} \mathrm{d}s \, \frac{\mathrm{i}}{\hbar} [V_s(s-t), G]. \tag{6.93}$$

We now think in terms of Schrödinger, i.e. the states and not the observables change with time. The expectation value of an observable M is therefore

$$\mathrm{Tr}\, \varrho_t M = \mathrm{Tr} G M - \int_{-\infty}^{t} \mathrm{d}s \, \mathrm{Tr} G \frac{\mathrm{i}}{\hbar} [M, V_s(s-t)], \tag{6.94}$$

or

$$\langle M \rangle_t = \langle M \rangle - \int_{-\infty}^{t} \mathrm{d}s \, \langle \frac{\mathrm{i}}{\hbar} [M, V_s(s-t)] \rangle. \tag{6.95}$$

Expectation values without a subscript always refer to the Gibbs state.

Let us introduce the age $\tau = t - s$ of a perturbation. Then (6.95) reads

$$\langle M \rangle_t = \langle M \rangle - \int_0^\infty \mathrm{d}\tau \, \langle \frac{\mathrm{i}}{\hbar} [M(\tau), V_{t-\tau}] \rangle. \tag{6.96}$$

Note that time shifting the terms of the expectation value by a certain time has no effect.

Let us now specialize to a time-dependent perturbation of the form

$$H_t = H - \sum_k \lambda_k(t) \Lambda_k \tag{6.97}$$

where the $\lambda_k = \lambda_k(t)$ are real valued functions while the Λ_k are observables, i.e. self-adjoint linear operators. The linear response, as felt by an observable M, to such a perturbation is given by

$$\langle M \rangle_t = \langle M \rangle + \int_0^\infty \mathrm{d}\tau \sum_k \langle \frac{\mathrm{i}}{\hbar} [M(\tau), \Lambda_k] \rangle \, \lambda_k(t - \tau). \tag{6.98}$$

The commutator describes the influence of a perturbation Λ_k on the observable $M(\tau)$ at a later time, on the average. If both operators commute they would not influence each other. The influence factor is to be multiplied by the perturbation strength λ_k at the time of the perturbation which was τ seconds before now. This effect has to be summed over the components k of the perturbation and integrated over all ages τ. This then gives the deviation of the current value of M from the mean value. Note that (6.98) expresses a causal relation. Only past perturbations can be felt now.

Perturbation by Light

Normal matter consists of nuclei and electrons which we will enumerate be $a = 1, 2, \ldots$ Particle a is located at $\boldsymbol{X}_a$, has momentum $\boldsymbol{P}_a$, mass m_a and electric charge q_a. Its Hamiltonian is

$$H = \sum_a \frac{\boldsymbol{P}_a^2}{2m_a} + \frac{1}{4\pi\epsilon_0} \sum_{b>a} \frac{q_b q_a}{|\, \boldsymbol{X}_b - \boldsymbol{X}_a \,|}. \tag{6.99}$$

In dipole approximation the interaction with an external electric field $\boldsymbol{E}$ is expressed as

$$H_t = H - \int \mathrm{d}^3x \, \boldsymbol{P}(\boldsymbol{x}) \cdot \boldsymbol{E}(t, \boldsymbol{x}), \tag{6.100}$$

where

$$\boldsymbol{P}(\boldsymbol{x}) = \sum_a q_q \boldsymbol{X}_a \delta^3(\boldsymbol{x} - \boldsymbol{X}_a) \tag{6.101}$$

is the polarization, a field of observables.

Indeed, the perturbation is of the form (6.97), the 'sum' being a sum over the vector components and a spatial integration.

In optics, the polarization caused by a light wave is of particular interest. Hence we have to discuss

$$\langle P_j(\boldsymbol{x}) \rangle_t = \int_0^\infty \mathrm{d}\tau \sum_k \int \mathrm{d}^3 y \, \Gamma_{jk}(\tau, \boldsymbol{x}, \boldsymbol{y}) \, E_k(t - \tau, \boldsymbol{y}), \tag{6.102}$$

where

$$\Gamma_{jk}(\tau, \boldsymbol{x}, \boldsymbol{y}) = \langle \frac{\mathrm{i}}{\hbar} [P_j(\tau, \boldsymbol{x}), P_k(0, \boldsymbol{y})] \rangle. \tag{6.103}$$

Note that $P_j(\tau, \boldsymbol{x}) = U_{-\tau} P_j(\boldsymbol{x}) U_\tau$ is the time shifted polarization operator.

We add one more assumption, namely that the Gibbs state is not only invariant with respect to temporal, but also spatial translations. In this case the influence functions (6.103) depend on $\boldsymbol{\xi} = \boldsymbol{x} - \boldsymbol{y}$ only. Equation (6.103) simplifies to

$$\langle P_j(\boldsymbol{x}) \rangle_t = \int_0^\infty \mathrm{d}\tau \sum_k \int \mathrm{d}^3 \xi \, \Gamma_{jk}(\tau, \boldsymbol{\xi}) \, E_k(t - \tau, \boldsymbol{x} - \boldsymbol{\xi}), \tag{6.104}$$

and (6.103) becomes

$$\Gamma_{jk}(\tau, \boldsymbol{\xi}) = \langle \frac{\mathrm{i}}{\hbar} [P_j(\tau, \boldsymbol{\xi}), P_k(0, 0)] \rangle. \tag{6.105}$$

We have dropped a possible static polarization contribution $\langle \boldsymbol{P}(\boldsymbol{x}) \rangle$ since we are interested in optics. We shall now simplify the linear response formulas even further by decomposing into harmonic contributions.

Susceptibility Tensor

Let us Fourier transform the polarization[11] $P_j(t, \boldsymbol{x}) = \langle P_j(\boldsymbol{x}) \rangle_t$. It is decomposed as follows:

$$P_j(t, \boldsymbol{x}) = \int \frac{\mathrm{d}\omega}{2\pi} \int \frac{\mathrm{d}^3 q}{(2\pi)^3} \, \mathrm{e}^{-\mathrm{i}\omega t} \, \mathrm{e}^{\mathrm{i}\boldsymbol{q} \cdot \boldsymbol{x}} \, \hat{P}_j(\omega, \boldsymbol{q}). \tag{6.106}$$

The electric field is transformed in the same way.

Since (6.104) is a convolution we easily establish

[11]It should be clear from the context whether we speak of an observable or its expectation value in the perturbed state ϱ_t.

$$\hat{P}_j(\omega, \boldsymbol{q}) = \epsilon_0 \sum_k \chi_{jk}(\omega, \boldsymbol{q})\, \hat{E}_k(\omega, \boldsymbol{q}). \tag{6.107}$$

$\chi_{jk} = \chi_{jk}(\omega, \boldsymbol{q})$ is the dielectric susceptibility tensor. It is dimensionless because the factor ϵ_0 has been split off. Equation (6.105) becomes

$$\chi_{jk}(\omega, \boldsymbol{q}) = \frac{1}{\epsilon_0} \int_0^\infty \mathrm{d}\tau \int \mathrm{d}^3\xi\; \mathrm{e}^{\mathrm{i}\omega\tau}\, \mathrm{e}^{-\mathrm{i}\boldsymbol{q}\cdot\boldsymbol{\xi}}\, \Gamma_{jk}(\tau, \boldsymbol{\xi}) \tag{6.108}$$

with the Γ tensor of (6.103).

6.3.2 Dispersion Relations

The response of a system to perturbations is causal. Only perturbations from the past contribute. Let us therefore recapitulate a remarkable property of causal functions: the real and imaginary parts of the Fourier transformed are intimately related.

Causal Functions

A real valued causal function f may be written as

$$f(\tau) = \theta(\tau) f(\tau), \tag{6.109}$$

where θ is the Heaviside step function. The latter vanishes for $\tau < 0$ and has the constant value 1 for $\tau > 0$. Since it shows up in integrals only, its value for $\tau = 0$ is irrelevant.

We shall show in Sect. A.11 on Fourier Transforms that a causal function has a Fourier transform $g = \hat{f}$ the real part of which can be calculated from the imaginary part, and vice versa.

From the characterization (6.109) of a causal function one may derive the following representation of its Fourier transform:

$$g(\omega) = -\mathrm{i}\,\mathrm{Pr} \int \frac{\mathrm{d}u}{\pi} \frac{g(u)}{u - \omega} = \left(\int_{-\infty}^{\omega-\epsilon} + \int_{\omega+\epsilon}^{\infty} \right) \frac{\mathrm{d}u}{\pi} \frac{g(u)}{u - \omega} \tag{6.110}$$

with $\epsilon > 0$, $\epsilon \to 0$. The right hand side defines the principal integral $\mathrm{Pr} \int \dots$ For a real causal function we have

$$g^*(\omega) = g(-\omega) \tag{6.111}$$

which allows to write

$$g'(\omega) = \mathrm{Pr} \int_0^\infty \frac{\mathrm{d}u}{\pi} \frac{2ug''(u)}{u^2 - \omega^2} \tag{6.112}$$

and

$$g''(\omega) = 2\omega\, \mathrm{Pr} \int_0^\infty \frac{\mathrm{d}u}{\pi} \frac{g'(u)}{\omega^2 - u^2}. \tag{6.113}$$

g' and g'' are the real and the imaginary parts of the Fourier transformed causal function.

Kramers–Kronig Relations

Each component $\chi_{jk}(\omega, \boldsymbol{q})$ of (6.108) is the Fourier transform (with respect to ω) of a causal function. We may therefore write

$$\chi_{jk}(\omega, \boldsymbol{q}) = -\mathrm{i}\ \mathrm{Pr} \int \frac{\mathrm{d}u}{\pi} \frac{\chi_{jk}(u, \boldsymbol{q})}{u - \omega}. \tag{6.114}$$

Let us decompose the susceptibility tensor into a Hermitian (prime) and an anti-Hermitian part (double prime):

$$\chi'_{jk} = \frac{\chi_{jk} + \chi^*_{kj}}{2} \quad \text{and} \quad \chi''_{jk} = \frac{\chi_{jk} - \chi^*_{kj}}{2\mathrm{i}} \tag{6.115}$$

such that $\chi_{jk} = \chi'_{jk} + \mathrm{i}\chi''_{jk}$, always at $(\omega, \boldsymbol{q})$.

The Hermitian contribution χ'_{jk} causes light refraction, it is therefore called the refractive part. χ''_{jk} is called the absorptive or dissipative part of the susceptibility. It describes light absorption by the medium under study.

Apart from the phenomenon of optical activity, only the susceptibilities at $\boldsymbol{q} = 0$ are of interest. We shall therefore drop the argument altogether from now on. Instead of $\chi_{jk}(\omega, 0)$ we shall write $\chi_{jk}(\omega)$.

Because of

$$\Gamma^*_{kj}(\tau, \boldsymbol{\xi}) = \Gamma_{jk}(-\tau, -\boldsymbol{\xi}) \tag{6.116}$$

we conclude

$$\chi^*_{jk}(\omega) = \chi_{kj}(\omega). \tag{6.117}$$

This is the same as (6.111) except that now the tensor indexes must be flipped. We may therefore write the following expressions for the Kramers–Kronig rules:

$$\chi'_{jk}(\omega) = \mathrm{Pr} \int_0^\infty \frac{\mathrm{d}u}{\pi} \frac{2u\chi''_{jk}(u)}{u^2 - \omega^2} \tag{6.118}$$

and

$$\chi''_{jk}(\omega) = 2\omega \, \Pr \int_0^\infty \frac{du}{\pi} \frac{\chi'_{jk}(u)}{\omega^2 - u^2}. \tag{6.119}$$

Equations (6.118) and (6.119) are called dispersion relations because the refractive part necessarily depends on ω. Recall that a wave-length dependent refractive index $n = \sqrt{1 + \chi}$ causes dispersion, the splitting of a beam of white light into colors.

The Kramers–Kronig relations also say that there is no refraction without absorption. Although there may be frequency regions with only negligible absorption, there must be regions at different frequencies with strong absorption.

6.3.3 Summary

Fluctuations of a system in the Gibbs state are nevertheless equilibrium properties. If the Hamiltonian of a system depends on external parameters which change rapidly, the system will not reach its equilibrium state. We have studied this scenario for small perturbations V_t.

We assumed that the system was in a Gibbs state with temperature T before the perturbation was switched on. The equation of motion can be rewritten into an integral equation combining the differential equation and the initial condition. As a fixed-point problem it is suited for an expansion into powers of the perturbation. We have studied the first non-trivial approximation which describes the linear response to the perturbation.

We then specialized to a perturbation $V_t = -\lambda(t)\, V$. The expectation value of an arbitrary observable in the perturbed state is a retarded integral over $\lambda(t - \tau)$ multiplied by an influence, or Green's function. The latter is an expectation value of a commutator of the observable and V, time-translated by the age τ of the perturbation. We expressed our findings in terms of Fourier transforms and introduced the notion of (generalized) susceptibility.

Being the Fourier transform of a causal function, relations between the real and imaginary parts can be established. We have learned that there is no real part without an imaginary part and that both necessarily depend on the angular frequency. This, in optics, is called dispersion.

Of particular interest is the perturbation of matter by the electric field of light. We have derived an explicit expression for the dielectric susceptibility tensor. The associated dispersion relations are named after Kramers and Kronig.

6.4 Dissipation

In the previous section we have studied the linear response of matter to a perturbation by rapidly changing external fields. The system can no more be in an equilibriums state because the thermodynamic variables have already changed before the system could reach its equilibrium state. In particular, we have studied how an oscillating electric field affects the polarization. The relation is linear and retarded. The factor of proportionality between the Fourier transformed electric field and the Fourier transformed polarization is the system's susceptibility. An explicit expression for it has been derived.

In this section we shall demonstrate that the dissipative part of the susceptibility is tightly linked with the spectral intensity of a fluctuation. We shall show that work is required to perturb an initial equilibrium state which will relax to the same final state. Thus we shall prove the second law of thermodynamics, although for a special case.

6.4.1 Wiener–Khinchin Theorem

The susceptibility is a certain integral over the expectation value

$$\Gamma_{jk}(\tau, \boldsymbol{\xi}) = \langle \frac{\mathrm{i}}{\hbar}[P_j(\tau, \boldsymbol{\xi}), P_k(0, 0)] \rangle. \tag{6.120}$$

The time shift is accomplished by the waiting operator

$$U_\tau = \mathrm{e}^{-\frac{\mathrm{i}}{\hbar}\tau H} \tag{6.121}$$

according to

$$M(\tau) = U_{-\tau} M U_\tau. \tag{6.122}$$

H is the unperturbed Hamiltonian, the system's energy without electric field. Space translations are defined in the same way. Here we focus on the commutator expectation value

$$\Gamma(\tau) = \langle \frac{\mathrm{i}}{\hbar}[M(\tau), M(0)] \rangle. \tag{6.123}$$

As we have indicated before, the correlation function

$$K(\tau) = \langle \frac{M(\tau)M(0) + M(0)M(\tau)}{2} \rangle \tag{6.124}$$

is a similar construct. We have written the symmetrized product which is a self-adjoint operator, its expectation value therefore is a real number. Also note that the observables $M(t)$ are supposed to be fluctuations which can always be achieved by subtracting the constant expectation value. Equation (6.123) will not change.

In the following subsections we show that $\Gamma = \Gamma(\tau)$ and $K = K(\tau)$ are linearly related. Here we recapitulate an important property of the correlation function.

One Fourier-transforms the process $t \to M(t)$,

$$M(t) = \int \frac{d\omega}{2\pi} e^{-i\omega t} \hat{M}(\omega) \tag{6.125}$$

or

$$M(t+\tau) = \int \frac{d\omega'}{2\pi} \hat{M}\star(\omega). \tag{6.126}$$

Since the Gibbs state is stationary, the expectation values $\langle M(t+\tau)M(t)\rangle$ and $\langle M(t)M(t+\tau)\rangle$ do not depend on t, but merely on the time difference τ. This is the case if and only if there are contributions for $\omega = \omega'$ only. We conclude

$$\langle \frac{\hat{M}\star(\omega')\hat{M}(\omega) + \hat{M}(\omega)\hat{M}\star(\omega')}{2} \rangle = 2\pi(\omega - \omega')S(\omega) \tag{6.127}$$

and

$$K(\tau) = \int \frac{d\omega}{2} e^{i\omega\tau} S(\omega) \quad \text{where} \quad S(\omega) \geq 0. \tag{6.128}$$

The correlation function $K = K(\tau)$ is the Fourier transform of a non-negative spectral intensity $S = S(\omega)$. This finding plays an important role in proving that a passive medium will absorb electromagnetic energy, but cannot generate it at the expense of cooling a heat reservoir. Note however that the Wiener–Khinchin theorem holds true for any stationary state.

6.4.2 *Kubo–Martin–Schwinger Formula*

There is a formal similarity between the waiting operator and the Gibbs state. Both are exponentials of the Hamiltonian, one with a factor $-it$ in front, the other with $-\beta$. Somewhat metaphorically, time is an imaginary inverse temperature, and the inverse temperature is an imaginary time. In what follows we shall make use of this. We will set aside mathematical rigor because things would quickly become too complicated.

Let us define a complex shift by

$$A(z) = e^{-izH} A\, e^{izH} \tag{6.129}$$

for a complex number z. We easily may show

$$A(z)\,\mathrm{e}^{-\beta H} = \mathrm{e}^{-\beta H}\,\mathrm{e}^{\beta H}\,A(z)\,\mathrm{e}^{-\beta H} \tag{6.130}$$

or

$$A(z)\,G = G A(z - \mathrm{i}\hbar\beta). \tag{6.131}$$

G is a Gibbs state with $k_B T = 1/\beta$. We multiply from the right with an observable B and form the trace. The result is

$$\langle\, B A(z)\,\rangle = \langle\, A(z - \mathrm{i}\hbar\beta) B\,\rangle. \tag{6.132}$$

This is the famous KMS formula discovered independently by Kubo, Martin and Schwinger. It allows to relate a product of two observables with a similar expression where the operators are in reverse order. For us, it is an intermediary step for relating a commutator with an anti-commutator.

6.4.3 Callen–Welton Theorem

Assume the time-dependent Hamiltonian $H_t = H - \lambda(t)M$ and a Gibbs state at $t \to -\infty$. The linear response to this perturbation, as felt by the observable M, is given by

$$M_t = \langle\, M\,\rangle + \int_0^\infty \mathrm{d}\tau\,\Gamma(\tau)\,\lambda(t-\tau), \tag{6.133}$$

where

$$\Gamma(\tau) = \frac{\mathrm{i}}{\hbar}\langle\, M(\tau)M - MM(\tau)\,\rangle \tag{6.134}$$

is a Green's function. Note that only fluctuations contribute. Compare this with the correlation function

$$K(\tau) = \frac{1}{2}\langle\, M(\tau)M + MM(\tau)\,\rangle \tag{6.135}$$

where $M(t)$ again is a fluctuation. The former expression is the expectation value of a commutator, the latter of an anti-commutator. There should be a relation between (6.134) and (6.135).

Define

$$\phi(\tau) = \langle\, M(\tau)\,M\,\rangle, \tag{6.136}$$

the Fourier transform of which is

$$\hat{\phi}(\omega) = \int \mathrm{d}\tau\;\mathrm{e}^{\mathrm{i}\omega\tau}\,\phi(\tau). \tag{6.137}$$

By declaring

$$f(z) = \int \frac{d\omega}{2\pi}\, e^{-i\omega z}\, \hat{\phi}(\omega) \tag{6.138}$$

we analytically continue (6.136) to complex arguments z.

On the other hand, $A(z)$ as defined by (6.129), gives rise to another function

$$g(z) = \langle\, M(z)\, M\, \rangle \tag{6.139}$$

which can be shown to be analytic within a sufficiently broad stripe around the real axis. Since f and g coincide on the real axis, they are equal in the complex plane as well. We may therefore write

$$\langle\, M(\tau)\, M\, \rangle = f(\tau) \text{ and } \langle\, M\, M(\tau)\, \rangle = f(\tau - i\hbar\beta). \tag{6.140}$$

After these preparations we may represent the influence function and the correlation function as

$$\Gamma(\tau) = \frac{i}{\hbar}\{f(\tau) - f(\tau - i\hbar\beta)\} \tag{6.141}$$

and

$$K(\tau) = \frac{1}{2}\{f(\tau) + f(\tau - i\hbar\beta)\}, \tag{6.142}$$

respectively. Inserting (6.138) yields

$$\Gamma(\tau) = \frac{i}{\hbar}\int \frac{d\omega}{2\pi}\, e^{-i\omega\tau}\left\{1 - e^{-\beta\hbar\omega}\right\}\hat{\phi}(\omega) \tag{6.143}$$

and

$$K(\tau) = \frac{1}{2}\int \frac{d\omega}{2\pi}\, e^{-i\omega\tau}\left\{1 + e^{-\beta\hbar\omega}\right\}\hat{\phi}(\omega). \tag{6.144}$$

We can eliminate the unknown function $\hat{\phi}$ by equating

$$\hat{\phi}(\omega) = \frac{\hbar}{i}\,\frac{\hat{\Gamma}(\omega)}{1 - e^{-\beta\hbar\omega}} = 2\,\frac{\hat{K}(\omega)}{1 + e^{-\beta\hbar\omega}}. \tag{6.145}$$

Note that the Fourier transform of the correlation function is the spectral intensity $S = S(\omega)$, see (6.128). Therefore

$$\hat{\Gamma}(\omega) = \frac{2i}{\hbar}\, S(\omega)\, \tanh\frac{\beta\hbar\omega}{2} \tag{6.146}$$

holds true. We have solved our problem.

Not quite. We are not primarily interested in the Fourier transform of the influence function $\Gamma = \Gamma(\tau)$, but in the (generalized) susceptibility

$$\chi(\omega) = \int \mathrm{d}\tau\, \theta(\tau)\, \mathrm{e}^{\mathrm{i}\omega\tau}\, \Gamma(\tau) = \frac{1}{2\pi\mathrm{i}} \int \mathrm{d}u\, \frac{\hat{\Gamma}(u)}{u - \omega - \mathrm{i}\epsilon}. \tag{6.147}$$

ϵ in this formula denotes an infinitely small positive number. We easily deduce from this

$$2\mathrm{i}\chi''(\omega) = \hat{\Gamma}(\omega) \tag{6.148}$$

where $\chi''(\omega)$ is the imaginary, or dissipative contribution to the susceptibility. Equations (6.146) and (6.148) amount to

$$\chi''(\omega) = \frac{1}{\hbar} \tanh(\frac{\beta\hbar\omega}{2})\, S(\omega). \tag{6.149}$$

This is the famous dissipation-fluctuation theorem as derived by Callen and Welton. $\chi''(\omega)$ describes the dissipation of energy at angular frequency ω, and $S(\omega)$ is the fluctuation spectral intensity at the same angular frequency. Recall $1/\beta = k_{\mathrm{B}} T$.

Note that (6.149) reads

$$\chi''(\omega) = \frac{\omega}{2k_{\mathrm{B}} T}\, S(\omega) \tag{6.150}$$

for hight temperatures, or small β. Now $\hbar$ has vanished, we deal with the limiting case of classical statistical mechanics. The limiting case for small temperatures reads

$$\chi''(\omega) = \frac{\mathrm{sgn}\,(\omega)}{\hbar}\, S(\omega). \tag{6.151}$$

Bear in mind that the spectral intensity and the dissipative part of the susceptibility depend on temperature because the expectation values in a Gibbs state do so.

6.4.4 Interaction with an Electromagnetic Field

If the Gibbs state is perturbed by more than one external parameter, according to

$$H_t = H - \int \mathrm{d}^3x\, \boldsymbol{P}(\boldsymbol{x}) \cdot \boldsymbol{E}(t, \boldsymbol{x}), \tag{6.152}$$

say, the preceding arguments may easily be adapted.

We not only assume the Gibbs state to be invariant under time translations, but also under space translations. Therefore the expectation value $\langle\, P_j(t, \boldsymbol{x})\,\rangle$, where $P_j(t, \boldsymbol{x}) = U_{-t} P_j(\boldsymbol{x}) U_t$, does not depend on the space and time coordinates. $P_j(t, \boldsymbol{x}) = P_j(t, \boldsymbol{x}) - \langle\, P_j\,\rangle$ is a fluctuation because its expectation value vanishes.

Its space-time correlation functions are

$$K_{jk}(\tau, \boldsymbol{\xi}) = \frac{\langle\, P_j(t+\tau, \boldsymbol{x}+\boldsymbol{\xi})\, P_k(t, \boldsymbol{x}) + \dots \,\rangle}{2}, \tag{6.153}$$

where the dots stand for the preceding factors in reverse order. The Wiener–Khinchin theorem now reads

$$K_{jk}(\tau, \boldsymbol{\xi}) = \int \frac{\mathrm{d}\omega}{2\pi} \frac{\mathrm{d}^3 q}{(2\pi)^3}\, \mathrm{e}^{\mathrm{i}(\omega\tau - \boldsymbol{q}\cdot\boldsymbol{\xi})}\, S_{jk}(\omega, \boldsymbol{q}) \tag{6.154}$$

where S_{jk} is a non-negative matrix, $S \geq 0$.

The subsequent arguments can be translated one-by-one as well.[12] The fluctuation-dissipation theorem now reads

$$\chi''_{jk}(\omega, \boldsymbol{q}) = \tanh\left(\frac{\beta\hbar\omega}{2}\right) \frac{S_{jk}(\omega, \boldsymbol{q})}{\hbar\epsilon_0}. \tag{6.155}$$

For each argument $(\omega, \boldsymbol{q})$ the dissipative part χ''_{jk} of the dielectric susceptibility is a non-negative matrix. The dissipative part χ''_{jk} is responsible for the attenuation of electromagnetic excitations in passive matter.

By the way, (6.155) also passes the dimensional check. The spectral intensity S is a Fourier transform of the polarization correlation function K which has the dimension of polarization squared. Hence, S/ϵ_0 is an energy times seconds. Divided by $\hbar$, we obtain a dimensionless quantity, namely a susceptibility.

Energy Dissipation

The Hamiltonian may depend on external parameters $\lambda_1, \lambda_2, \dots$ which can oscillate rapidly thus preventing the system from being in equilibrium. To keep the notation simple we assume just one $\lambda = \lambda(t)$ which shall be always small.

Let us calculate the total work for a perturbation $H_t = H - \lambda(t)\Lambda$. We assume $\lambda(-\infty) = 0$ (far past) and $\lambda(+\infty) = 0$ (far future). In other words, the perturbation shall last a finite time only. The work to be spent is

$$W = \int \mathrm{d}\lambda(t) \langle\, \Lambda \,\rangle_t = \int \mathrm{d}t\, \dot{\lambda}(t) \int_0^\infty \mathrm{d}\tau\, \Gamma(\tau)\lambda(t-\tau). \tag{6.156}$$

Without loss of generality we may assume that Λ is a fluctuation; a constant expectation value would not contribute.

Let us insert the Fourier transforms:

$$W = -\int \frac{\mathrm{d}\omega}{2\pi}\, \mathrm{i}\omega\lambda(\omega)\, \chi(\omega)\, \lambda^*(\omega). \tag{6.157}$$

[12] Note however that the dielectric susceptibility $\chi_{jk}(\omega, \boldsymbol{q})$ is defined by splitting off a factor $1/\epsilon_0$ because of $\boldsymbol{D} = \boldsymbol{P} + \epsilon_0 \boldsymbol{E}$, in usual notation.

This expression is real, so one may add the complex conjugate and divide by two. This leads to

$$W = \int \frac{d\omega}{2\pi}\,\omega|\,\lambda(\omega)\,|^2\,\frac{\chi(\omega)-\chi^*(\omega)}{2i}. \tag{6.158}$$

By making use of the dissipation-fluctuation theorem (6.149) one arrives at

$$W = \frac{\beta}{2\hbar}\int_0^\infty \frac{d\omega}{2\pi}\,\omega\,\tanh\beta\hbar\omega\,|\,\lambda(\omega)\,|^2\,S(\omega). \tag{6.159}$$

$\beta = 1/k_B T$ is positive, ω as well, the hyperbolic tangent is positive just as $|\,\lambda(\omega)\,|^2$, and by the Wiener–Khinchin theorem the spectral intensity never becomes negative as well. Thus we have proven that the total work W associated with the perturbation is never negative! The dissipative part $\chi''(\omega)$ of the susceptibility causes in fact energy dissipation.

Needless to say that the perturbation brought about be an oscillating electric field fits into this scheme. The interaction of the field with matter causes field energy to vanishes. In an environment of constant temperature it is impossible to transform heat to work.

6.4.5 Summary

The susceptibility is proportional to the commutator of $V(\tau)$ with V, where V causes the perturbation and also feels it. τ is the time span between cause and effect. The correlation function is nearly the same with the commutator replaced by the anti-commutator.

We have derived and discussed the so called Kubo-Martin-Schwinger (KMS) formula. Since the Gibbs state and the waiting operator are both exponential functions of the Hamiltonian, a translation by a complex time-argument is possible, and one obtains $\langle\, BA(z)\,\rangle = \langle\, A(z - i\hbar\beta)B\,\rangle$ for two observables A and B, with β as inverse temperature.

A lengthy argument based on the KMS formula lead to the Callen–Welton theorem: the dissipative part of the susceptibility is an integral—with positive weight—of the spectral intensity. Since the latter is never negative, the dissipative part is also never negative: for all Hamiltonians H, for all tempertures T, for all perturbations V.

As an example we discussed the interaction of an oscillating electric field with matter. The susceptibility tensor can be split in an Hermition, refractive part and an anti-Hermitian dissipative part. The work to be spent for a perturbation is always positive. This is very close to a proof of the second main law of thermodynamics. Very close because our argument is valid only for the linear response to a perturbation by rapidly changing external parameters, not for heat conduction or diffusion.

Chapter 7
Mathematical Aspects

The essence of quantum physics is that observables and states are represented by linear operators, not by real valued functions on a phase space. Historically, the necessity for a new framework arose from the growing evidence for inherent contradictions. The objects of the new theory may or may not commute upon multiplication, in contrast to the old theory. Hence the mathematical behavior of linear operators is of particular interest, and this chapter collects, on a few pages, the essentials.

We do not strive for mathematical rigor. For this the reader must consult the huge body of good mathematical textbooks ore even the original literature. Instead we pay attention to clear definitions, coherence, relevance—and readability. Many topics have already been dealt with in the text, sometimes even in detail. Here we put them into a mathematical context.

The main subject—linear operators—is supported by an discourse on topological spaces, by an article on the Lebesgue integral and by a treatise on generalized functions.

The subsequent topics bring an overview over linear spaces, Hilbert spaces, projectors, normal operators, translations, the Fourier transform, position and momentum, to ladder operators and to representations of the rotation group.

7.1 Topological Spaces

Topology is a cross-sectional branch of mathematics. It deals with sets of points and relations between them like neighborhood. By adding more structure, more relations show up.

P. Hertel, *Quantum Theory and Statistical Thermodynamics*,
Graduate Texts in Physics, DOI 10.1007/978-3-319-58595-6_7

7.1.1 Abstract Topological Space

We start from some set Ω of points. Points in this context are entities without internal structure. There is a system $\mathfrak{T}$ of subsets[1] with the following properties:

- The empty set $\emptyset$ and Ω itself belong to $\mathfrak{T}$.
- Unions $A_1 \cup A_2 \cup \dots$ of arbitrarily many subset $A_j \subset \Omega$ belong to $\mathfrak{T}$.
- Intersections $A_1 \cap A_2 \cap \dots \cap A_n$ of finitely many subsets belong to $\mathfrak{T}$.

The set Ω and such a system $\mathfrak{T}$ is a topological space $(\Omega, \mathfrak{T})$. The subsets $A \in \mathfrak{T}$ are called open.

A set B is closed if it is the complement of an open set: $B = A^{\mathsf{C}} = \Omega \setminus A$ with $A \in \mathfrak{T}$. It follows that the empty set and Ω itself are open and closed at the same time. There are, however, subsets $A \subset \Omega$ which are neither open nor closed.

A_x is a neighborhood of $x \in \Omega$ if x is a member of an open set B contained in A_x, i.e. $x \in B \subset A_x$ with $B \in \mathfrak{T}$. In particular, an open set A is a neighborhood of all its member points.

Let us now discuss functions. Consider two topological spaces $(\Omega, \mathfrak{T})$ and $(\Omega', \mathfrak{T}')$. A mapping $f : \Omega \to \Omega'$ is continuous if the inverse image

$$f^{-1}(B) = \{x \in \Omega \mid f(x) \in B\} \tag{7.1}$$

of any open image $B \in \mathfrak{T}'$ is open, i.e. belongs to $\mathfrak{T}$. In short; the inverse image of any open set is open.

For example consider the constant mapping. For a $c \in \Omega'$ we set $f(x) = c$ for all $x \in \Omega$. This mapping is continuous. Take an arbitrary open set $B' \in \mathfrak{T}'$. If c is not in B', the inverse image is the empty set. If c is in B, the inverse image is Ω. Either set is open, therefore the constant function is continuous.

Another example. Let Ω be a set with topology $\mathfrak{T}$. The identity mapping $I : \Omega \to \Omega$, defined by $I(x) = x$, is continuous. Take an arbitrary open set $B \in \mathfrak{T}$. Its inverse image is $I^{-1}(B) = B$ is open, and I is continuous.

A continuous function can also be characterized by the property that the inverse image of a closed set is closed. Here the proof. Select a closed set $\bar{B} \subset \Omega'$. Its inverse image is $\bar{A} = f^{-1}(\bar{B})$. Its complement $A = \Omega \setminus \bar{A}$ is the set of all points which are <u>not</u> mapped into $\bar{B}$. In other words, $A = f^{-1}(B)$ where $B = \Omega' \setminus \bar{B}$. B is open, and the same holds true for A, since f is continuous. Therefore the set $\bar{A}$ is closed. For a continuous function, the converse image of an arbitrary closed set is closed. The reverse is also true: if the inverse image of an arbitrary closed set is closed, the function is continuous.

Let us come back to the topological space $(\Omega, \mathfrak{T})$. Let $M \subset \Omega$ be an arbitrary subset. The boundary ∂M is the set of points in the neighborhood of which are points in M and outside M. Put otherwise, for all neighborhoods U_x the condition $U_x \cup M \neq \emptyset$ is fulfilled as well as $U_x \cup (\Omega \setminus M) \neq \emptyset$.

---

[1] $A \subset B$ includes the case $A = B$.

It is easy to show that $A = M \setminus \partial M$ is open and that $\bar{A} = M \cap \partial M$ is closed. The set without its boundary is open, the set with its boundary is closed. Sets with an empty boundary are both open and closed. Check this for $M = \emptyset$ and $M = \Omega$.

The converse is also true: a set with an empty boundary is open. A set which includes its boundary—i.e. $\partial M \subset M$—is closed.

The topology $\mathfrak{T}$ as a system of subsets of Ω is almost the simplest structure one may imagine. Nevertheless, important concepts such as open and closed, continuity, neighborhood, boundary may be formulated in an abstract way. And many more, like accumulation points, covering and compactness.

7.1.2 Metric Space

Let us now become more concrete. We again start with a basic set Ω of points. For each pair $x, y \in \Omega$ a distance $d = d(x, y)$ is defined, a real, non-negative number. The distance function has to obey quite natural rules:

- The (shortest) way from x to y and the way back are equally long.
- The way to a different point always has a positive length.
- Detours do not pay.

Put otherwise,

$$\begin{aligned} d(x, y) &= d(y, x) \\ d(x, y) &\geq 0 \\ d(x, y) &= 0 \text{ implies } x = y \\ d(x, z) &\leq d(x, y) + d(y, z). \end{aligned} \tag{7.2}$$

With this, one may define open balls around x with radius R by

$$K_R(x) = \{y \in \Omega \mid d(x, y) < R\}. \tag{7.3}$$

The topology $\mathfrak{T}$ of open subsets consists of arbitrary units of such open balls. The properties of the distance function guarantee that $\mathfrak{T}$ defines in fact a topology which we will not prove here.

For $\Omega = \mathbb{R}$ as the basic set one may introduce a distance function by $d(x, y) = |\, y - x \,|$. You should convince yourself that $|\, y - x \,|$ in fact properly defines a distance. With this one declares the standard topology on $\mathbb{R}$. When saying that a function $f : \mathbb{R} \rightarrow \mathbb{R}$ be continuous, one always has the standard topology in mind where open intervals and unions thereof are the open subsets.

The distance function $d : \Omega \rightarrow \Omega$ generates the natural topology for Ω. The corresponding topological space is denoted by (Ω, d) or simply Ω if it is clear from the context how to measure distances.

7.1.3 Linear Space with Norm

Now we become even more concrete: the basic set Ω shall be a linear space $\mathcal{L}$. As usual, its points are called vectors. There is a field of scalars, either real or complex numbers. For any two scalars λ_1, λ_2 and for any pair x_1, x_2 of vectors, the linear combination $\lambda_1 x_1 + \lambda_2 x_2$ is again a vector in $\mathcal{L}$. For this the usual rules for multiplication with scalars and the addition of vectors apply. In particular, addition of vectors is commutative.

There is a unique zero vector 0 such that $x + 0 = 0 + x = x$.

The set $\mathcal{C}$ of complex valued continuous functions $f : [0, 1] \to \mathbb{C}$ living on the interval [0, 1] may serve as an example. Such functions are added pointwise and multiplied pointwise with complex number and are still continuous.

A norm associates a non-negative number $\| x \|$ with each vector x. The following rules shall apply:

$$\begin{aligned} \| x \| &= 0 \text{ implies } x = 0 \\ \| \lambda x \| &= | \lambda | \| x \| \\ \| x + y \| &\le \| x \| + \| y \|. \end{aligned} \tag{7.4}$$

It is not difficult to show that $d(x, y) = \| y - x \|$ fulfills the requirements of a distance function. Therefore, a linear space with norm is a metric space which, in turn, is a topological space.

For the aforementioned set $\mathcal{C}$ of continuous complex-valued functions one may define the norm

$$\| f \|_\infty = \sup_{x \in [0,1]} | f(x) |, \tag{7.5}$$

the supremum norm.

Another example is the L_2 norm

$$\| f \|_2 = \left(\int_0^1 \mathrm{d}x \, | f(x) |^2 \right)^{1/2}. \tag{7.6}$$

The integral over a continuous function on a finite range is well defined.

A linear space $\mathcal{L}$ with norm $\| . \|$ has a distance function. It is therefore a topological space as well which is usually denoted by $(\mathcal{L}, \| . \|)$ or simply by $\mathcal{L}$ if it is clear from the context which norm is assumed. As we have just seen, there may be more than one norm and consequently more than one topology for one and the same linear space.

7.1.4 Linear Space with Scalar Product

Let us further specialize. Ω is a linear space $\mathcal{H}$ with scalar product. Any pair is associated with a real or complex number (y, x) obeying

$$\begin{aligned}(y, x) &= (x, y)^* \\ (y, \lambda_1 x_1 + \lambda_2 x_2) &= \lambda_1 (y, x_1) + \lambda_2 (y, x_2) \\ (x, x) &\geq 0 \\ (x, x) &= 0 \text{ implies } x = 0.\end{aligned} \tag{7.7}$$

This applies to a complex linear space. The rules for a real linear space are the same except for the first line which then reads $(y, x) = (x, y)$.

Evidently, $|| x || = \sqrt{(x, x)}$ is a norm. Every linear space with scalar product is a linear space with norm, therefore a metric space and therefore a topological space.

As an example we again refer to the space $\mathcal{C}$ of complex valued continuous function living on $[0, 1]$. We declare a scalar product by

$$(g, f) = \int_0^1 \mathrm{d}x \, g^*(x) \, f(x). \tag{7.8}$$

Obviously this expression fulfills all requirements (7.7). Hence, $|| f ||$ is a norm which we know already as $|| f ||_2$ of (7.6).

7.1.5 Convergent Sequences

Recall the hierarchy of spaces. A linear space with scalar product is also a normed linear space. A linear space with norm is a metric space as well. And every metric space is a topological space.

We here focus on a metric space (Ω, d) with its natural topology as defined by the distance $d = d(x, y)$. With this tool we may check whether a sequence converges.

The sequence $x_1, x_2, \dots$ of points in Ω converges towards $x \in \Omega$ if $d(x, x_j) \to 0$ with $j \to \infty$. In other words, for any $\epsilon > 0$ there is an index n such that $d(x, x_j) \leq \epsilon$ for all $j \geq n$. x is the limit of the convergent sequence.

A subset $A \subset \Omega$ is closed if and only if every convergent sequence in A has a limit in A. This is an important statement since it allows to formulate topological concepts in terms of convergent sequences.

Here is the proof.

Let $a_1, a_2, \dots$ be a convergent sequence in A and assume A to be a closed set. Assume the limit a is not in A. It then belongs to $\Omega \setminus A$, and this set is open. There is a certain ball $K_\epsilon(a)$ with $\epsilon > 0$ which is contained entirely within $\Omega \setminus A$. It follows that $d(a, a_j) \geq \epsilon$ for all indexes j. This is a contradiction. We conclude that the assumption $a \notin A$ was false or that $a \in A$. Convergent sequences in a closed set have a limit in that set.

Now the converse assertion. Let A be a set such that any convergent sequence in A has a limit in A. Assume that A be not closed, i.e. $\Omega \setminus A$ be not open. Then there is a point y in $\Omega \setminus A$ such that the intersection of A with an arbitrary ball $K_\epsilon(y)$ is not empty. For $j = 1, 2, \ldots$ we define $B_j = K_{1/j}(y)$. Let us select points a_j in $B_j \cap A$. The sequence $a_1, a_2, \ldots$, on the one hand, is in A. On the other hand, it converges towards y, a point not in A. Again, this is a contradiction which proves that A is closed.

The natural topology of a metric space Ω may be defined by the convergence behavior of sequences. A subset $A \subset \Omega$ is closed if an arbitrary convergent sequence $a_1, a_2 \ldots$ of points in A converges towards a limit in A. A subset $B \subset \Omega$ is open if it is the complement of a closed set.

For metric spaces and for linear spaces with norm or scalar product one may run through topology by studying the convergence behavior of sequences. What refers to open or closed may be translated into statements about converging sequences.

7.1.6 Continuity

Let (Ω, d) and (Ω', d') be metric spaces and therefore topological spaces. $d(x_1, y_1)$ measures the distance in Ω and $d'(y_1, y_2)$ in Ω'. We focus on a mapping $f : \Omega \to \Omega'$. The mapping is continuous if the inverse image of an arbitrary open subset is open. The following formulation is equivalent: the inverse mapping of an arbitrary subset is closed. Both characterize a continuous function.

We now want to translate this topological definition of continuity into convergent sequences parlance.

A function $f : \Omega \to \Omega'$ is sequence-continuous if an arbitrary convergent sequence $x_1, x_2, \ldots$ is mapped into a convergent sequence:

$$\lim_{j\to\infty} f(x_j) = f(x) \text{ where } x = \lim_{j\to\infty} x_j. \tag{7.9}$$

Let f be a sequence-continuous function. Choose an arbitrary closed set $B \subset \Omega$. The corresponding inverse image is $A = f^{-1}(B)$. In case A is empty it is closed. If not, we pick an arbitrary converging sequence $x_1, x_2, \ldots$ of points in A the limit of which is denoted by x. Since f is sequence-continuous we know $\lim f(x_j) = f(x)$. B is closed, therefore $f(x)$ is in B as well and $x \in A$, i.e. A is closed. The inverse image A of an arbitrary closed set B is closed. f is topology-continuous.

The converse statement is also true.

Select an arbitrary converging sequence $x_1, x_2, \ldots$ with limit x. Let $\epsilon > 0$ be an arbitrary positive number. The ball $K_\epsilon(f(x))$ is an open set with respect to (Ω', d'). Because f shall be topology-continuous, the inverse image $f^{-1}(K_\epsilon(f(x)))$ is also open with respect to (Ω, d). There is a positive number δ so that

$$x \in K_\delta(x) \cup f^{-1}(K_\epsilon(f(x))) \tag{7.10}$$

holds true. Because of $x_j \to x$ there is an index n such that $x_j \in K_\delta(x)$ is true for all $j \geq n$. Consequently,

$$f(x_j) \in f(K_\delta(x)) \subset K_\epsilon(f(x)) \tag{7.11}$$

holds true for all $j \geq n$. We conclude that f is sequence-continuous.

To sum it up: topology-continuous and sequence continuous is the same for functions mapping from a metric space to another metric space. Consequently this is also true for normed linear spaces and for linear spaces with scalar product.

7.1.7 Cauchy Sequences and Completeness

The following remarks pertain to a metric space (Ω, d), therefore in particular to a linear space with norm or even with scalar product.

Consider a sequence $x_j \in \Omega$ which get closer and closer. For any $\epsilon > 0$ there is an index n such that

$$d(x_j, x_k) \leq \epsilon \text{ for all } j, k \geq n. \tag{7.12}$$

Equation (7.12) characterizes a Cauchy-convergent sequence. Cauchy-convergent sequences may not have a limit.

The basic set Ω is complete if any Cauchy-convergent sequence also converges in Ω

Incidentally, this is not a strong requirement, as we shall show now.

Two sequences $x_1, x_2, \dots$ and $y_1, y_2, \dots$ are equivalent if for any $\epsilon > 0$ there is an index n such that

$$d(x_j, y_j) \leq \epsilon \text{ for all } j \geq n \tag{7.13}$$

holds true.

We write $x \equiv y$ for equivalent sequences. Clearly, $x \equiv x$. Also, $x \equiv y$ implies $y \equiv x$. And $x \equiv y$, $y \equiv z$ imply $x \equiv z$. One therefore may speak of the class of equivalent sequences. The elements x of Ω are identified with sequences $x, x, \dots$, they belong to a certain class of equivalent sequences.

The set of old equivalence classes is augmented by the equivalence classes of Cauchy convergent sequences. We denote this set by $\bar{\Omega}$. Any Cauchy-convergent sequence now converges in $\bar{\Omega}$. The set $\bar{\Omega}$ is complete. The old distance function may be extended to a distance function for $\bar{\Omega}$ such that the previously defined distances remain the same.

The process of completion is best illustrated by an example. Consider the set $\mathbb{Q}$ of rational numbers. Not all Cauchy-convergent sequences of rational numbers converge. There are sequences which would converge towards $\sqrt{2}$, but this number does not exist so far. One simply augments the set of rational numbers by equivalence classes of Cauchy-convergent sequences and obtains $\bar{\mathbb{Q}} = \mathbb{R}$, the set of real numbers. The rules how to calculate with rational numbers determine the rules for real numbers.

A complete linear space with norm is called a Banach space. And most important for this book: a Hilbert space is a linear space with scalar product which is complete in its natural topology.

7.2 The Lebesgue Integral

Most integrals in this book are of the Lebesgue, and not Riemann type. In brief, the difference is a follows. Assume you have a heap of coins, 1, 2, 5, 10, …cents. You want to determine its sum. One way is to lay them out in a line, and you sum its values. That corresponds to Riemann's approach to integration. The Lebesgue way of doing so is to form sub-heaps of equally valued coins, one for the 1 cent pieces, one for the 2 cent pieces, and so on. For each value v there are n_v coins. The total amount then is calculated as the sum over all values v, each contributing with $v\, n_v$. For a finite number of coins, both methods should yield the same result. But for an infinite number of infinitesimal contributions, there are differences. For most calculations, these differences are irrelevant. We discuss the Lebesgue integral here because it is involved in the definition of square-integrable functions, it is central to Hilbert spaces and its linear operators.

7.2.1 Measure Spaces

Think of a non-empty set of points Ω. Associated with Ω is a σ-algebra $\mathfrak{M}$ of subsets which are called measurable:

- The empty set is measurable, $\emptyset \in \mathfrak{M}$
- The complement of a measurable set is measurable. $A \in \mathfrak{M}$ implies $\Omega \setminus A \in \mathfrak{M}$.
- A countable union of measurable sets is measurable. If $A_1, A_2, \ldots \in \mathfrak{M}$ then $A_1 \cup A_2 \cdots \in \mathfrak{M}$.

Moreover, there is a measure μ which associates a non-negative real number $\mu(A)$ with each measurable set A, or plus $+\infty$. The measure functional shall obeys

- The empty set has measure zero, $\mu(\emptyset) = 0$.
- If the measurable set A is the union of disjoint measurable sets A_j, their measures add. For $A_j \cap A_k = \emptyset$ where $j \neq k$ we demand

$$\mu(A_1 \cup A_2 \cup \ldots) = \mu(A_1) + \mu(A_2) + \cdots . \tag{7.14}$$

The triple $(\Omega, \mathfrak{M}, \mu)$ is called a measure space. There is a set Ω of points, a σ-algebra $\mathfrak{M}$ of measurable subsets, and an additive measure $\mu = \mu(A)$ which assigns a non-negative real number to each measurable set A. The measure is monotonous. For $A, B \in \mathfrak{M}$ and $A \subset B$ one finds $\mu(A) \leq \mu(B)$. The larger the set, the larger its measure. The measure functional is sub-additive:

$$\mu(A_1 \cup A_2 \cup \ldots) \leq \mu(A_1) + \mu(A_2) + \ldots \tag{7.15}$$

7.2.1.1 Borel Measure

Let us assume that the points are real numbers, $\Omega = \mathbb{R}$. The open intervals

$$(a, b) = \{x \in \mathbb{R} \,|\, a < x < b\} \tag{7.16}$$

with $a \leq b$ are measurable, the measure is $b - a$. The smallest σ-algebra containing these intervals is denoted by $\mathfrak{B}$, its members are called Borel sets. There is a natural measure μ_{B}, and the triple $(\mathbb{R}, \mathfrak{B}, \mu_{\mathrm{B}})$ is a measurable space. From now on we shall only consider real numbers, its Borel sets and the Borel measure. Generalizations to $\mathbb{R}^n$ or $\mathbb{C}$ are obvious.

It is not difficult to show that, for $a \leq b$, not only (a, b), but also $[a, b)$, $(a, b]$ and $[a, b]$ are Borel sets. They all have the same measure, namely $b - a$. Therefore, a set containing a single number has measure zero, and the same applies for a set containing a countable number of points only. For example, the set $\mathbb{Q}$ of rational numbers is a set of measure zero.

7.2.2 *Measurable Functions*

Consider two measurable spaces $(\Omega', \mathfrak{M}', \mu')$ and $(\Omega'', \mathfrak{M}'', \mu'')$. We speak about a mapping $f : \Omega' \to \Omega''$, or a function. For a measurable set $A'' \in \mathfrak{M}''$ we define by

$$A' = f^{-1}(A'') = \{x \in \Omega' \,|\, f(x) \in A''\}, \tag{7.17}$$

the inverse image of A. A function f is measurable if for each measurable set $A'' \in \mathfrak{M}''$ the inverse image $A' = f^{-1}(A'')$ belongs to $\mathfrak{M}'$, i. e. is measurable.

The property to be measurable survives the most common operation on functions:

- If f_1, f_2 are measurable functions, then $\alpha_1 f_1 + \alpha_2 f_2$ is measurable. Likewise, the product $f_1 f_2$ is measurable. This is also true for countable sums or products.
- A point-wise convergent series $f_1, f_2, \ldots$ of measurable functions is measurable.
- For any series $f_1, f_2, \ldots$ of measurable functions the supremum $f(x) = \sup_j f_j(x)$ is measurable.
- The composition $h(x) = g(f(x))$ is measurable if f and g are.

- If $|f|$ is measurable, so is f, and vice versa.

Specializing to real valued function of a real argument (Borel space), a function f is measurable if the inverse image of the interval (a, b) is measurable.

For example, the constant function $f(x) = c$ is measurable. Choose an interval $A = (a, b)$. If $c \in A$ then $f^{-1}(A) = \mathbb{R}$ which, as a complement of the empty set, is measurable. If $c \notin A$ then $f^{-1}(A) = \emptyset$ which is likewise measurable.

7.2.3 The Lebesgue Integral

We have a measure space $(\Omega, \mathfrak{M}, \mu)$ and the Borel space $(\mathbb{R}, \mathfrak{B}, \mu_B)$ in mind. Consider a measurable function $f : \Omega \to \mathbb{R}$.

Let us begin with defining the Lebesgue integral for non-negative functions, $f(x) \geq 0$ for all $x \in \Omega$. Assume a positive number h and define representative points $y_j = jh$ where j is an integer. Associated with each representative point is an interval $Y_j = [y_j, y_j + h)$. Hence

$$\mathbb{R}_+ = [0, \infty) = \bigcup_{j=0}^{\infty} Y_j. \tag{7.18}$$

Each Y_j is a borel set. Because f is assumed to be measurable, the inverse image $X_j = f^{-1}(Y_j)$ is a measurable set with measure μ_j. The following expression

$$I_h = \sum_{j=0}^{\infty} \mu_j y_j \tag{7.19}$$

is a lower estimate of the integral. It is an infinite sum of non-negative contributions which converges in any case, either towards a finite value or towards infinity. Now, I_h increases if h decreases, and there is a limit for $h \to 0$ which we denote by

$$\int \mathrm{d}\mu(f)\, y = \int \mathrm{d}x\, f(x). \tag{7.20}$$

Its value may be infinite.

Now, any real-valued function $f : \Omega \to \mathbb{R}$ may be split into its positive and negative contributions. We denote by Ω_+ the set of points where $f(x) > 0$, by Ω_- if $f(x) < 0$ and by Ω_0 the set where $f(x) = 0$. Clearly, $\Omega = \Omega_+ \cup \Omega_- \cup \Omega_0$. We define a function f_+ which coincides with f on Ω_+ and vanishes otherwise. Likewise, f_- is defined as $-f$ on Ω_-. Both functions, f_+ and f_- are positive and measurable. The Lebesgue integral of f is defined as

$$\int \mathrm{d}\mu(f)\, y = \int \mathrm{d}x\, f(x) = \int \mathrm{d}x\, f_+(x) - \int \mathrm{d}x\, f_-(x). \tag{7.21}$$

If $f : \Omega \to \mathbb{C}$ is a complex-valued function, it may be decomposed into its real and imaginary part. It is measurable if the real part $\mathrm{Re} f$ and the imaginary part $\mathrm{Im} f$ are both measurable. The Lebesgue integral is defined as

$$\int \mathrm{d}\mu(f)\, y = \int \mathrm{d}x\, f(x) = \int \mathrm{d}x\, \Re f(x) + \mathrm{i} \int \mathrm{d}x\, \mathrm{Im} f(x). \tag{7.22}$$

7.2.4 Function Spaces

Complex valued functions can be added and multiplied by complex scalars in a natural way. They form a linear space. Often more structure is required, such as a scalar product. As an example we discuss the Hilbert space

$$\mathcal{L}_2(\Omega) = \{f : \Omega \to \mathbb{C} \mid \int \mathrm{d}x\, |f(x)|^2 < \infty\}. \tag{7.23}$$

This seemingly clear definition has to be read within the proper context:

- An integral is involved, therefore the functions f must be measurable. Here the Lebesgue integral is understood.
- Ω therefore is the basic set of a measure space $(\Omega, \mathfrak{M}, \mu)$.
- The set of complex numbers $\mathbb{C}$ is assumed to be equipped with the Borel measure.
- In fact, f stands for any member of its equivalence class. Two functions are equivalent if they differ on a set of measure zero only.

The latter remark must be explained.

Consider two functions $f_1, f_2 : \Omega \to \mathbb{C}$. It is not difficult to show that the set

$$A = \{x \in \Omega \mid f_1(x) \neq f_2(x)\} \tag{7.24}$$

is measurable. If its measure $\mu(A)$ vanishes one says that the two functions coincide almost everywhere. They are equivalent with respect to integration, $f_1 \equiv f_2$. Since

- $f \equiv f$
- $f_1 \equiv f_2$ implies $f_2 \equiv f_1$
- $f_1 \equiv f_2$ and $f_2 \equiv f_3$ implies $f_1 \equiv f_3$

hold true, a set of functions which coincide almost everywhere form an equivalence class. Therefore, the f in (7.23) is not a particular function, but a representative of its equivalence class.

The space of square-integrable functions $\mathcal{L}_2(\Omega)$ allows for the scalar product

$$(g, f) = \int \mathrm{d}x\, g^*(x) f(x). \tag{7.25}$$

For $f, g \in \mathcal{L}_2(\Omega)$ it is well defined as guaranteed by the Cauchy–Schwarz inequality

$$\int \mathrm{d}x\, |g^*(x) f(x)| \leq \sqrt{\int \mathrm{d}x\, |g(x)|^2}\, \sqrt{\int \mathrm{d}x\, |f(x)|^2}. \tag{7.26}$$

Therefore, the scalar product (7.25) induces a norm, and this norm a topology. The space $\mathcal{L}_2(\Omega)$ is complete: any Cauchy series $f_1, f_2, \ldots$ has a limit in $\mathcal{L}_2(\Omega)$. Therefore, it is a Hilbert space.

7.3 On Probabilities

In this section we explain the modern approach to probability. Probability theory by is a well established branch of mathematics. However, it has nothing to say to which situations it is applicable, which events are truly governed by chance. For a mathematical foundation, a base set Ω of is required. This may be the phase space of a large classical system. However, the laws of probability never refer to this set explicitly. The observables of a quantum-physical system behave as random variables with one important difference. Different observables do not commute in general, unlike in classical probability theory. This section is intended to familiarize the reader with the vocabulary of probability theory: event, probability, and, or, not, random variable and functions thereof, probability distribution, probability density, independent, law of large numbers, central limit theorem and more. It is an overview, not a treatise on probability theory.

7.3.1 Probability Spaces

Consider a usually huge set Ω of points. There is a σ-algebra $\mathfrak{E}$ of subsets $E \subset \Omega$ which are called events:

- The empty set $\emptyset$ is an event.
- If E is an event, its complement $\Omega \setminus E$ is also an event.
- A countable union $\bigcup_j E_j$ of events $E_1, E_2, \ldots$ is also an event.

If E and F are events, then $E \cup F$ is the event 'E or F'. Likewise $E \cap F$ is the event 'E and F'. $\Omega \setminus E$ is interpreted as 'not E'. Note that $\Omega \in \mathfrak{E}$ is an event, it stands for 'always true' or certain such as $\emptyset \in \mathfrak{E}$ means 'never true', or impossible. We have just given a precise meaning to otherwise vaguely defined words: *or, and, not, impossible, certain.*

There is also a probability function $\mathrm{pr} = \mathrm{pr}\,(E)$ which assigns to each event a real number in [0, 1]. The probability function obeys

- $\mathrm{pr}\,(\emptyset) = 0$
- $\mathrm{pr}\,(\Omega) = 1$
- $\mathrm{pr}\,(E_1 \cup E_2 \cup \ldots) = \mathrm{pr}\,(E_1) + \mathrm{pr}\,(E_2) + \ldots$ if $E_j \cap E_k = \emptyset$ for $j \neq k$

The first item says that the impossible event occurs with probability zero. The second item states that the certain event occurs with probability one. The last line says that the probabilities of events add upon the 'or' operation, provided that the events are disjoint. E and F are disjoint in the sense 'either E or F, but not both'. A triple $(\Omega, \mathfrak{E}, \text{pr})$ is a probability space. Formally, a probability space is a measure space with the additional requirement that $\text{pr}(\Omega) = 1$.

7.3.2 Random Variables

Consider a real valued function $X : \Omega \to \mathbb{R}$. Ω shall be the base set of a probability space $(\Omega, \mathfrak{E}, \text{pr})$. If for any Borel set $A \subset \mathbb{R}$ the inverse image $X^{-1}(A)$ is an event, we speak of a random variable. See Sect. 7.2 on the Lebesgue integral for the definition of the Borel space $(\mathbb{R}, \mathfrak{B}, \mu_B)$. It is sufficient to show that

$$\{\omega \in \Omega \mid X(\omega) \in (a, b)\} \in \mathfrak{E} \tag{7.27}$$

holds true for arbitrary intervals (a, b). A random variable is called so because it is a variable the value of which is or seems to be governed by chance. If you measure X, dice are thrown to determine a value ω, and $X(\omega)$ is returned.

7.3.2.1 Functions of Random Variables

If α is a real number and X a random variable, then αX is a random variable as well. If X and Y are random variables, the sum $X + Y$ is also a random variable. The same is true for $XY = YX$, since multiplication of real numbers is commutative. The observables of a quantum-physical system behave as random variables, but not in this respect, they do not commute in general.

Let $f : \mathbb{R} \to \mathbb{R}$ be a Borel measurable function. Then $\omega \to f(X(\omega))$ defines a random variable $f(X)$. Indeed, for any Borel set A the set $B = f^{-1}(A)$ is also a Borel set, and $X^{-1}(B)$ is an event.

7.3.2.2 Probability Distributions

Let $X : \Omega \to \mathbb{R}$ be a measurable function. The set

$$(X \le x) = \{\omega \in \Omega \mid X(\omega) \le x\} \tag{7.28}$$

is an event since $(X \le x)$ is the inverse image of the Borel set $(-\infty, x]$. It occurs with probability

$$P(X; x) = \text{pr}(X \le x). \tag{7.29}$$

The probability distribution $P = P(X; x)$ of the random variable X is characterized by

- $P(-\infty) = 0$
- $P(+\infty) = 1$
- $P(x_2) \geq P(x_1)$ if $x_2 \geq x_1$

The probability distribution of a random variable is a monotonously increasing function $P : \mathbb{R} \rightarrow [0, 1]$. It is absolutely continuous in the sense of

$$P(x) = \int_{-\infty}^{x} \mathrm{d}s\, p(s) \tag{7.30}$$

where $p = p(s)$ is a non-negative Borel-measurable function, the probability density. Mind the notation: pr (E) is the probabilty of an event E. By $P(X; x)$ we denote the probability that the random variable has a value less or equal to x. Its derivative is $p(X; x) = P'(X; x)$.

7.3.2.3 Expectation Values

Let X be a random variable with probability distribution $P = P(X; x)$. Its probability density shall be denoted by $p = P'(x)$. Recall $p(x) \geq 0$ and

$$\int \mathrm{d}s\, p(s) = 1, \tag{7.31}$$

where the integral extends over $\mathbb{R}$, from $-\infty$ to $+\infty$. The expectation value of the random variable $f(X)$ is declared as

$$\langle f(X) \rangle = \int \mathrm{d}s\, p(X; s)\, f(s). \tag{7.32}$$

Note that $f : \mathbb{R} \rightarrow \mathbb{R}$ is a Borel measurable function and that the details of the random variable X are described by p, its probability density. The base set Ω does no more appear explicitly.

7.3.2.4 Joint Distributions

Let X and Y be random variables. That X has a value not exceeding x is an event, $(X \leq x)$, just as $(Y \leq y)$. It follows that $(X \leq x)$ <u>and</u> $(Y \leq y)$ is also an event, and we denote its probability by

$$P(X, Y; x, y) = \text{pr}\,((X \leq x) \cap (Y \leq y)). \tag{7.33}$$

The following statements

$$P(X, Y; x, \infty) = P(X; x) \text{ and } P(X, Y; \infty, y) = P(Y; y) \tag{7.34}$$

are obvious. Two random variables X and Y are said to be independent if

$$P(X, Y; x, y) = P(X; x)\, P(Y; y) \tag{7.35}$$

holds true for all pairs (x, y) of real numbers. The counterpart of (7.30) is

$$P(X, Y; x, y) = \int_{-\infty}^{x} \mathrm{d}s \int_{-\infty}^{y} \mathrm{d}t\; p(X, Y; s, t). \tag{7.36}$$

$p = p(X, Y; x, y)$ is the joint probability density, a non-negative measurable function of two real arguments. For independent random variables X and Y the joint probability density factorizes:

$$p(X, Y; x, y) = p(X; x)\, p(Y; y). \tag{7.37}$$

Equations (7.35) and (7.37) say the same: X and Y are independent random variables.

7.3.2.5 Generating Function

Let X be a random variable and $p(X; x)$ its probability density. We define the characteristic function by

$$\pi(X; \lambda) = \int \mathrm{d}x\; \mathrm{e}^{\,\mathrm{i}x\lambda}\; p(X; x), \tag{7.38}$$

i.e. its Fourier transform. It allows to calculate the momenta of X in a simple way:

$$\langle X^k \rangle = \int \mathrm{d}x\; p(X; x)\, x^k = (-\mathrm{i})^k \pi^{(k)}(X; 0), \tag{7.39}$$

where $f^{(k)}$ denotes the k-th derivative of a function f. Moreover, the characteristic functions of two random variables X and Y simply multiply if the random variables are added. The probability density of $X + Y$ is

$$p(X + Y, s) = \int \mathrm{d}t\; p(X; t)\, p(Y; s - t), \tag{7.40}$$

a convolution. Since the Fourier transform of a convolution is the same as the product of the Fourier transforms, we may write

$$\pi(X+Y;\lambda) = \pi(X;\lambda)\,\pi(Y;\lambda). \tag{7.41}$$

A very useful result, as we shall see soon.

7.3.3 *Law of Large Numbers and Central Limit Theorem*

We discuss two important results of probability theory. The law of large numbers guarantees that measuring a certain observable again and again will result in a stable mean value which coincides with the expectation value. The central value theorem states that deviations from the mean value, if weighted by the inverse square root of the numbers of repetitions, are normally distributed.

7.3.3.1 Law of Large Numbers

Think of a series of independent measurements $X_1, X_2, \dots$ of one and the same property X. X_j and X_k are independent, and the probability densities $p = p(X_j; x)$ do not depend on the index j. We define by

$$M_n = \frac{X_1 + X_2 + \dots + X_n}{n} \tag{7.42}$$

the mean of the first n measurements, a random variable. Its characteristic function is

$$\pi(M_n;\lambda) = \prod_{j=1}^{n} \pi\left(\frac{1}{n}X_j;\lambda\right) = \pi\left(\frac{\lambda}{n}\right)^n. \tag{7.43}$$

The right hand side evaluates to

$$\left(1 + \frac{i\lambda}{n}\langle X\rangle + \frac{1}{2}\left(\frac{i\lambda}{n}\right)^2 \langle X^2\rangle + \cdots\right)^n \tag{7.44}$$

the limit of which is

$$\lim_{n\to\infty} \pi(M_n,\lambda) = \pi_\infty(\lambda) = e^{i\lambda\langle X\rangle}. \tag{7.45}$$

Its Fourier back-transform gives the associated probability density

$$p_\infty(x) = \delta(\langle X\rangle - x). \tag{7.46}$$

This is the so-called law of large numbers. If the measurement of a random variable X is repeated again and again, the mean value will converge towards the expectation value better and better.

7.3.3.2 Central Limit Theorem

Now assume that X is a deviation, or fluctuation: $\langle X \rangle = 0$. We are concerned with

$$Q_n = \frac{X_1 + X_2 + \cdots + X_n}{\sqrt{n}}. \tag{7.47}$$

The X_j are independent deviations from the mean value when measuring X repeatedly. Equation (7.44) therefore reads

$$\pi(Q_n; \lambda) = \left(1 + \frac{1}{2}\left(\frac{\mathrm{i}\lambda}{\sqrt{n}}\right)^2 \langle X^2 \rangle + \cdots\right)^n. \tag{7.48}$$

With $n \to \infty$ this expression converges towards

$$\pi_\infty(\lambda) = \mathrm{e}^{-\lambda^2 \langle X^2 \rangle / 2}. \tag{7.49}$$

The associated probability density is

$$p_\infty(x) = \frac{1}{\sqrt{2\pi\sigma^2}} \mathrm{e}^{-x^2/2\sigma^2}, \tag{7.50}$$

where $\sigma^2 = \langle X^2 \rangle - \langle X \rangle^2$ denotes the variance of the random variable X.

7.4 Generalized Functions

Some objects of theoretical physics seem to be functions, but they are not. A well known example is Dirac's δ-function which, however, is not function. In this section we sketch how such generalized functions, or distributions, are defined and dealt with. In particular, any generalized function has a Fourier transform. The set of generalized functions comprises a large set of ordinary functions which explains why distributions are also called generalized functions.

7.4.1 Test Functions

A test function $t : \mathbb{R} \to \mathbb{C}$ is a particularly well behaved function. It can be differentiated arbitrarily often and falls off at infinity faster than any negative power of its argument. More precisely, for a test function t all derivatives $t^{(m)}$ are continuous, and

$$|| t ||_{m,n} = \sup_{x \in \mathbb{R}} |x^n t^{(m)}(x)| \tag{7.51}$$

is finite for all orders $m = 0, 1, \ldots$ and all powers $n = 0, 1, \ldots$ The Gaussian

$$t(x) = a\,\mathrm{e}^{-(x-b)^2/2\sigma^2} \tag{7.52}$$

is a well known example. Test functions can be added and multiplied by complex scalars, they form a linear space $\mathcal{S}$. A sequence $t_1, t_2, \ldots$ converges to a test function t if

$$\lim_{j \to \infty} || t - t_j ||_{m,n} = 0 \tag{7.53}$$

for all $m, n \in \mathbb{N}$. We shall write $t_j \to t$ in this case.

7.4.2 Distributions

A continuous linear functional $\Phi : \mathcal{S} \to \mathbb{C}$ is called a distribution, or generalized function. This means

$$\Phi(z_1 t_1 + z_2 t_2) = z_1 \Phi(t_1) + z_2 \Phi(t_2) \tag{7.54}$$

for complex numbers z_1, z_2 and test functions t_1, t_2. The linear functional must also be continuous, i.e.

$$\lim_j \Phi(t_j) = \Phi(t) \text{ for } t_j \to t. \tag{7.55}$$

A well behaved function $f = f(x)$ induces a distribution via

$$\Phi(t) = \int \mathrm{d}x\, t(x) f(x). \tag{7.56}$$

By running through all test functions you may reconstruct the properties of f in detail. But even if the distribution is not generated by a function, one writes it as (7.56) where $f = f(x)$ is merely a symbol for the distribution.

Take the Dirac 'function' $\delta(x)$ as an example. It is defined by

$$\Phi(t) = t(0) = \int \mathrm{d}x\, t(x)\, \delta(x). \tag{7.57}$$

This functional is obviously linear. It is also continuous since (7.53) assures point-wise convergence of test functions ($m = n = 0$). However, $\delta(x)$ is certainly not a function.

A function $f : \mathbb{R} \to \mathbb{C}$ is locally integrable if the integral over its absolute value

$$\int_a^b \mathrm{d}x\, |f(x)| \tag{7.58}$$

is well defined for any finite interval $a \le x \le b$. It is weakly growing if there is a natural number n such that

$$K = \sup_x \frac{|f(x)|}{1 + |x|^n} < \infty. \tag{7.59}$$

Put otherwise, it does not grow faster than a certain power of its argument at infinity.

Locally integrable weakly growing functions f induce distribution by

$$\Phi(t) = \int \mathrm{d}x\, t(x)\, f(x). \tag{7.60}$$

Clearly, the functional $\Phi : \mathcal{S} \to \mathbb{C}$ is linear; this is a property of the integral. It remains to show that the functional is continuous. We write

$$\Phi(t) = \int \mathrm{d}x\, t(x) \frac{f(x)}{1 + |x|^n}(1 + |x|^n) \tag{7.61}$$

which is bounded by

$$|\Phi(t)| \le K \sum_{k=0}^{n} c_k \|\, t\, \|_{0,k}. \tag{7.62}$$

Consequently, $\Phi(t_j) \to \Phi(0)$ with $t_j \to 0$. To sum it up: a locally integrable weakly growing function induces a distribution by the expression (7.60).

7.4.3 Derivatives

Distributions can be differentiated. The derivative Φ' of a distribution Φ is defined by

$$\Phi'(t) = -\Phi(t'). \tag{7.63}$$

$t \to \Phi'$ is obviously a linear functional. It is also continuous because $t'_j \to t'$ if $t_j \to t$ since (7.53) assures point-wise convergence of test function derivatives

($m = 1, n = 0$). Hence Φ' is a distribution. If Φ is represented by the symbol f, we associate with its derivative the symbol f' according to

$$\Phi'(t) = -\int \mathrm{d}x\, t'(x)\, f(x) = \int \mathrm{d}x\, t(x)\, f'(x). \tag{7.64}$$

In case the distribution is generated by a differentiable function f, then f' is its common derivative.

7.4.4 *Fourier Transforms*

We show that Fourier transforming a test function results in another test function. Moreover, the operation is not only linear and invertible, it is also continuous. These facts allow to define the Fourier transform of an arbitrary generalized function.

7.4.4.1 Fourier Transformed Test Functions

The Fourier transform of a test function t, namely

$$\hat{t}(y) = \int \mathrm{d}x\, \mathrm{e}^{\,\mathrm{i}yx}\, t(x) \tag{7.65}$$

is a test function as well. Its mth derivative

$$\hat{t}^{(m)}(y) = (\mathrm{i})^m \int \mathrm{d}x\, \mathrm{e}^{\,\mathrm{i}yx}\, t(x)\, x^m \tag{7.66}$$

is well defined. Moreover,

$$\left| y^n \hat{t}^{(m)}(y) \right| = \left| \int \mathrm{d}x\, \mathrm{e}^{\,\mathrm{i}yx}\, \frac{\mathrm{d}^n}{\mathrm{d}x^n} x^m t(x) \right|, \tag{7.67}$$

so that

$$||\, \hat{t}\, ||_{m,n} \le \int \mathrm{d}x\, \left| \frac{\mathrm{d}^n}{\mathrm{d}x^n} x^m t(x) \right| < \infty \tag{7.68}$$

is guaranteed. To summarize, the Fourier transform maps the space of test functions into itself. Even better. Because of

$$t(x) = \int \frac{\mathrm{d}y}{2\pi}\, \mathrm{e}^{\,-\mathrm{i}xy}\, \hat{t}(y) \tag{7.69}$$

each test function t is the Fourier transform of a Fourier transformed test function $\hat{t}$. We conclude that the Fourier transform maps the space of test functions onto itself. $t \to \hat{t}$ is invertible.

7.4.4.2 Continuity

Since the mapping $t \to \hat{t}$ is obviously linear, we only have to show that $t_j \to 0$ implies $\hat{t}_j \to 0$. Let us have a closer look at (7.68). The right hand side is a finite sum over terms

$$\int \mathrm{d}x \, \left| x^p t^{(q)}(x) \right| = ||\, t \,||_{q,p} \tag{7.70}$$

all of which vanish with $t_j \to 0$. We conclude that the operation of Fourier transforming test functions is not only linear and invertible, but also continuous. Close-by test functions have close-by Fourier transforms.

7.4.4.3 Fourier Transform

Let $\Phi = \Phi(t)$ be a distribution, a continuous linear functional which associates with each test function $t \in \mathcal{S}$ a complex number. Recall that we write

$$\Phi(t) = \int \mathrm{d}x \, t(x) \, f(x) \tag{7.71}$$

where f is either a function $f = f(x)$ proper or merely a symbol characterizing the distribution, such as $\delta = \delta(x)$ for the Dirac distribution. The Fourier transform $\hat{\boldsymbol{\Phi}}$ of a distribution Φ is defined as

$$\hat{\boldsymbol{\Phi}}(t) = \Phi(\hat{t}\,). \tag{7.72}$$

If Φ is represented by a well-behaved function f, then

$$\hat{\boldsymbol{\Phi}}(t) = \int \mathrm{d}y \, f(y) \int \mathrm{d}x \, \mathrm{e}^{\mathrm{i}xy} \, t(x) = \int \mathrm{d}x \, t(x) \int \mathrm{d}y \, \mathrm{e}^{\mathrm{i}xy} \, f(y) \tag{7.73}$$

holds true saying that the Fourier transformed distribution $\hat{\boldsymbol{\Phi}}$ is represented by $\hat{f}$.

7.4.4.4 Examples

Our first example is the unit distribution generated by $1(x) = 1$. It is locally integrable and weakly growing. Its derivative vanishes, and applying it to a test function returns its integral.

Our second example is the Heaviside jump function

$$\theta(x) = \begin{cases} 0 & \text{for } x < 0 \\ \frac{1}{2} & \text{for } x = 0 \\ 1 & \text{for } x > 0 \end{cases} \tag{7.74}$$

It is locally integrable and weakly growing and therefore describes a distribution, namely

$$\Phi(t) = \int \mathrm{d}x\, \theta(x) t(x) = \int_0^\infty \mathrm{d}x\, t(x). \tag{7.75}$$

Its derivative certainly is not a function.

The third example is the Dirac distribution:

$$\Phi(t) = \int \mathrm{d}x\, t(x)\, \delta(x) = t(0). \tag{7.76}$$

Because of

$$\int \mathrm{d}x\, \theta'(x)\, t(x) = -\int \mathrm{d}x\, \theta(x)\, t'(x) = t(0) \tag{7.77}$$

we conclude

$$\theta'(x) = \delta(x). \tag{7.78}$$

The Fourier transform of Dirac's distribution is determined as follows:

$$\int \mathrm{d}x\, t(x)\, \hat{\boldsymbol{\delta}}(x) = \int \mathrm{d}x\, \hat{\boldsymbol{t}}(x) \delta(x) = \hat{\boldsymbol{t}}(0) = \int \mathrm{d}x\, t(x), \tag{7.79}$$

which amounts to

$$\hat{\boldsymbol{\delta}} = 1. \tag{7.80}$$

The Fourier transform of the 1-function is

$$\int \mathrm{d}x\, t(x)\, \hat{\mathbf{1}}(x) = \int \mathrm{d}x\, \hat{\boldsymbol{t}}(x) = 2\pi t(0), \tag{7.81}$$

meaning

$$\hat{\mathbf{1}} = 2\pi\delta. \tag{7.82}$$

The Fourier transform of Heaviside's jump function jump function θ is formally defined by

$$\hat{\boldsymbol{\theta}}(y) = \int_0^\infty \mathrm{d}x\, \mathrm{e}^{ixy}. \tag{7.83}$$

Now, this integral does not converge. We therefore replace it with

Table 7.1 Some distributions, their derivatives and Fourier transforms δ, θ and ϵ are explained in the text. $t = t(x)$ is an arbitrary test function

$\Phi(t) = \int \mathrm{d}x\, \phi(x)\, t(x)$	$\phi(x)$	$\phi'(x)$	$\hat{\phi}(y)$
$\int_{-\infty}^{\infty} \mathrm{d}x\, t(x)$	1	0	$2\pi\delta(y)$
$\int_{0}^{\infty} t(x)$	$\theta(x)$	$\delta(x)$	$\frac{\mathrm{i}}{y + \mathrm{i}\epsilon}$
t(0)	$\delta(x)$	$\delta'(x)$	1

$$\hat{\theta}(y) = \int_0^\infty \mathrm{d}x\ \mathrm{e}^{\mathrm{i}x(y+\mathrm{i}\epsilon)} = \frac{\mathrm{i}}{y + \mathrm{i}\epsilon} \tag{7.84}$$

where $\epsilon > 0$ and the limit $\epsilon \to 0$ is understood. The backward Fourier transform is

$$\int \frac{\mathrm{d}y}{2\pi} \frac{\mathrm{i}}{y + \mathrm{i}\epsilon} \mathrm{e}^{-\mathrm{i}yx} . \tag{7.85}$$

The integrand is an analytic fuction with a pole like singularity at $y = -\mathrm{i}\epsilon$. For $x < 0$ one may add to the integration path an infinite half circle in the upper complex plane and contract it to the zero path without crossing a singularity. For $x < 0$, the expression (7.85) vanishes. For $x > 0$ an infinite half circle in the lower complex plane can be added. However, when contracting it, the pole at $y = -\mathrm{i}\epsilon$ is encountered. The residue is made up of factors $2\pi\mathrm{i}$, i, and the exponential of ϵx, and a factor -1 because the integral runs clockwise around the pole. Therefore the expression (7.85) yields 1 for $x > 0$. Hence, the back Fourier transform of the right hand side of (7.84) is in fact the Heaviside jump function θ. Thus we have shown

$$\hat{\theta}(y) = \frac{\mathrm{i}}{y + \mathrm{i}\epsilon} \tag{7.86}$$

where $\epsilon > 0$ and $\epsilon \to 0$ is meant. Whereas the jump function θ may be considered a function—albeit discontinuous—its derivative θ' and it Fourier transform $\hat{\theta}$ are not.

We summarize these findings in Table 7.1.

7.5 Linear Spaces

A linear space $\mathcal{L}$ is a set of objects—we call them vectors—which can be added and multiplied by scalars.

7.5.1 Scalars

If not stated otherwise, the scalars are complex numbers. They form the largest set of numbers in a chain $\mathbb{N} \subset \mathbb{Z} \subset \mathbb{Q} \subset \mathbb{R} \subset \mathbb{C}$. The set of real numbers is complete in the sense that every Cauchy sequence has a limit. Complex numbers contain in addition the symbolic number i with the property $i^2 = -1$. Every complex number $z \in \mathbb{C}$ may be uniquely characterized by $z = x + iy$ where x and y are real numbers. $x = \Re(z)$ is the real part, $y = \text{Im}(z)$ the imaginary part. The number $z^* = x - iy$ is the complex conjugate. Recall the expressions $z^* + z = 2\text{Re}(z)$ and $z^* - z = 2i\text{Im}(z)$. The norm $|z|$ of a complex number z is declared by $|z|^2 = z^*z = x^2 + y^2$. Convergence of sequences of complex numbers alway refers to this norm.

7.5.2 Vectors

So far to the scalars. The vectors are treated as points without reference to details of their nature. They can be added and multiplied by scalars according to the following rules. x, y, z are vectors and α, β denote scalars:

- Commutativity: $x + y = y + x$.
- Vector addition is associative: $(x + y) + z = x + (y + z)$
- Neutral element 0 of addition: $0 + x = x$
- Inverse element $-x$: $(-x) + x = x + (-x) = 0$
- Scalar multiplication is associative: $\alpha(\beta x) = (\alpha\beta)x$.
- Distributive law of scalar addition. $(\alpha + \beta)x = \alpha x + \beta x$.
- Distributive law of vector addition: $\alpha(x + y) = \alpha x + \beta y$.
- Neutral Elements of scalar-vector multiplication: $1x = x$

7.5.3 Linear Subspaces

Let $\mathcal{L}' \subset \mathcal{L}$ be a subset. If it is a linear space as well we call it a linear subspace, or a subspace for short. For $x, y \in \mathcal{L}'$ any linear combination $z = \alpha x + \beta y$ is in $\mathcal{L}'$ as well.

A set $\{x_1, x_2, \dots, x_n\}$ of vectors in $\mathcal{L}$ is linearly independent if the equation

$$\alpha_1 x_1 + \alpha_2 x_2 + \cdots + \alpha_n x_n = 0 \tag{7.87}$$

allows one solution only, namely $\alpha_1 = \alpha_2 = \cdots = \alpha_n = 0$.

$$\mathcal{L}' = \{x \in \mathcal{L} \mid x = \sum_{j=1}^{n} \alpha_j x_j \ \text{ with } \ \alpha_j \in \mathbb{C}\} \tag{7.88}$$

is a linear subspace of $\mathcal{L}$ which is spanned by vectors $x_1, x_2, \ldots, x_n$ in $\mathcal{L}$. We write $\mathcal{L}' = \text{span}\,\{x_1, x_2, \ldots, x_n\}$.

7.5.4 Dimension

The set $\{e_1, e_2, \ldots, e_n\}$ is a base if the vectors are linearly independent. Only then.

One may span the same linear space by linearly independent vector sets $\{e_1, e_2, \ldots, e_n\}$ and by $\{f_1, f_2, \ldots, f_m\}$. This is possible only for $m = n$. The number of linearly independent vectors which span a linear subspace is its dimension.

Assume one can find a countably infinite set of vectors in $\mathcal{L}$ such that the vectors of any finite subset are linearly independent. In this case we say that the linear space $\mathcal{L}$ has the dimension countably infinite (or $\aleph_0$, aleph naught).

There are linear spaces with larger dimensions which we shall not discuss. We simply speak of an infinite-dimensional linear space.

Note that according to our definition the linear space $\mathcal{L}$ itself is a linear subspace as well. The smallest subspace is the trivial space span $\{0\}$ which contains the null vector only. Its dimension is zero.

7.5.5 Linear Mappings

Let $\mathcal{L}_1$ and $\mathcal{L}_2$ be two linear spaces. A mapping $L : \mathcal{L}_1 \to \mathcal{L}_2$ is linear if any linear combination is mapped into the corresponding linear combination. That is:

$$L(\alpha x + \beta y) = \alpha L(x) + \beta L(y). \tag{7.89}$$

Here $\alpha, \beta \in \mathbb{C}$ are arbitrary scalars, $x, y \in \mathcal{L}_1$ and $L(x), L(y) \in \mathcal{L}_2$.

We shall later discuss in particular mappings of a linear space $\mathcal{L}$ into itself. In this case we will use operator notation, $L(x) = Lx$, i.e. omit the parentheses. A linear combination is mapped as $L(\alpha x + \beta y) = \alpha Lx + \beta Ly$.

7.5.6 Ring of Linear Operators

Denote by $\mathfrak{R}(\mathcal{L})$ the set of linear operators mapping $\mathcal{L}$ into itself. For any linear operator and any complex number α one may define αL by $(\alpha L)x = \alpha Lx$, for all $x \in \mathcal{L}$. In the same way one may declare the sum of two linear operators by $(L_1 + L_2)x = L_1 x + L_2 x$.

Applying one linear operator after the other, one defines the product $L_2 L_1$ by $(L_2 L_1)x = L_2(L_1 x)$ for each $x \in \mathcal{L}$. The linear operators thus form a linear space with an additional structure, namely multiplication. Note that $L_2 L_1$ and $L_1 L_2$ may be different operators. Multiplication is not commutative. Multiplication with scalars, addition obey the usual rules.

The set $\mathfrak{R}(\mathcal{L})$ is a ring. It contains a neutral element with respect to addition (the zero-operator 0 which maps each vector to to the zero vector) and multiplication, namely the one-operator which maps an arbitrary vector to itself, $Ix = x$. Note however, that there is no division for a ring. The linear operators do not necessarily have an inverse, i.e. $XL = I$ in general has no solution $X = L^{-1}$.

7.6 Hilbert Spaces

The vectors of a linear space may be added and multiplied by scalars, i.e. by complex numbers. Let us add more structure. We demand that for any pair of vectors x, y there is a scalar product (x, y). Its obeys the following rules (x and y are vectors, α denotes a scalar, a complex number):

$$\begin{aligned}(x, \alpha y) &= \alpha\,(x, y)\\ (x, y+z) &= (x, y) + (x, z)\\ (x, y) &= (y, x)^*\\ (x, x) &\geq 0\\ (x, x) &= 0 \text{ if and only if } x = 0.\end{aligned} \tag{7.90}$$

The scalar product of a vector x with itself is never negative. We may therefore declare the norm of a vector by

$$||\, x\, || = \sqrt{(x, x)}. \tag{7.91}$$

Equation (7.91) defines a norm, in fact.

The rule

$$||\, \lambda x\, || = |\, \lambda\, |\, ||\, x\, || \tag{7.92}$$

is evident. The Cauchy–Schwarz inequality says

$$|\, (x, y)\, | \leq ||\, x\, ||\, ||\, y\, ||. \tag{7.93}$$

For $y = 0$ this is true, therefore we assume $y \neq 0$. For an arbitrary complex number α we calculate

$$0 \leq (x - \alpha y, x - \alpha y) = (x, x) + |\, \alpha\, |^2 (y, y) - \alpha^*(y, x) - \alpha(x, y). \tag{7.94}$$

Choose $\alpha = (x, y)/(y, y)$ and multiply by (y, y). The result turns out to be $0 \leq (x, x)\,(y, y) - (x, y)(y, x)$, the same as (7.93). We thus have proven the Cauchy-Cauchy–Schwarz inequality. We write

$$||\,x+y\,||^2 = ||\,x\,||^2 + ||\,y\,||^2 + 2\Re(x, y) \le ||\,x\,||^2 + ||\,y\,||^2 + 2|\,(x, y)\,|, \quad (7.95)$$

and with the Cauchy–Schwarz inequality

$$||\,x+y\,||^2 \le ||\,x\,||^2 + ||\,y\,||^2 + 2||\,x\,||||\,y\,||, \quad (7.96)$$

which is the triangle inequality

$$||\,x+y\,|| \le ||\,x\,|| + ||\,y\,||. \quad (7.97)$$

For a linear space with scalar product, a norm $||\,x\,||$ is defined which satisfies the requirements (7.92) and (7.97). We therefore may speak of convergent and Cauchy-convergent sequences. A linear space with scalar product is called a Hilbert space $\mathcal{H}$ if arbitrary Cauchy-convergent sequences of vectors have a limit in $\mathcal{H}$.

This is not at all evident. Just consider the linear space of complex valued continuous functions living on $[-1, 1]$. The scalar product shall be

$$(g, f) = \int_{-1}^{1} \mathrm{d}x\, g^*(x)\, f(x). \quad (7.98)$$

In Fig. 7.1 we show a family of such functions which, with $\epsilon \to 0$, converges towards the jump function which is not continuous.

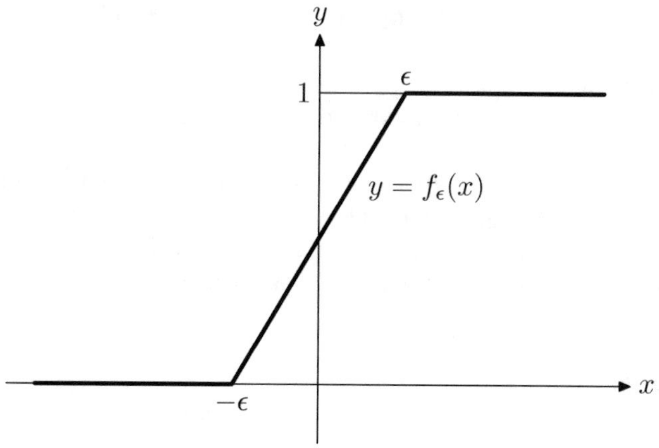

Fig. 7.1 A Cauchy convergent sequence of continuous functions. The corresponding linear space with standard scalar product is not complete. It is not a Hilbert space

7.6.1 Operator Norm

Consider a bounded operator L. There is a positive constant κ, an upper bound, such that $|| Lx || \le \kappa || x ||$ holds true for alle $x \in \mathcal{H}$. Bounded linear operators are continuous, and the converse is also true.

We may define the operator norm $\| L \|$ as the smallest upper bound:

$$\| L \| = \inf_{|| x ||=1} || Lx ||. \tag{7.99}$$

Equation (7.99) is a norm of for the ring of linear operators:

$$\begin{aligned} \| \alpha L \| &= | \alpha | \| L \| \\ \| L_1 + L_2 \| &\le \| L_1 \| + \| L_2 \| \\ \| L_2 L_1 \| &\le \| L_2 \| \| L_1 \|. \end{aligned} \tag{7.100}$$

The first equation is obvious. The second is a consequence of the triangle inequality for vectors. For an arbitrary normalized x we may write

$$|| (L_1 + L_2)x || \le || L_1 x || + || L_2 x || \le \| L_1 \| + \| L_2 \|, \tag{7.101}$$

and by taking the infimum over all normalized vectors one arrives at the operator triangle inequality, i.e. the second line of (7.100). To prove the third line, choose an arbitrary normalized vector x. One estimates

$$|| L_2 L_1 x || \le \| L_2 \| \, || L_1 x || \le \| L_2 \| \| L_1 \| \tag{7.102}$$

and takes the infimum over all such x.

Note that the ring of linear operators has a unit element. Obviously its ring norm is unity, $\| I \| = 1$.

Bounded operators, i.e. operators with a finite norm, are continuous. Just consider a convergent sequence $x_1, x_2, \dots$ the limit of which is x. Because of $|| Lx - Lx_j || \le \| L \| \, || x - x_j ||$ the sequence images converge toward the image of the limit. Therefore, the mapping $x \to Lx$ is continuous if L is bounded.

7.6.2 Adjoint Operator

We concentrate on continuous linear functionals, i.e. on linear mappings $\phi : \mathcal{H} \to \mathbb{C}$. Such functionals map vectors into complex numbers.

Choose a vector $y \in \mathcal{H}$. The scalar product $x \to \phi(x) = (y, x) \in \mathbb{C}$ is a linear functional. If $x_1, x_2, \dots$ converges to x then (y, x_j) converges to (y, x). Therefore, the functional defined above is continuous. The Riesz representation theorem states that the converse is also true: every continuous linear functional on $\mathcal{H}$ may be written as a scalar product with a suitable vector.

This theorem is far from trivial, and we shall refrain from proving it here.

Pick up a vector y and consider the mapping $x \to (y, Lx)$. It is linear and continuous. Continuous, because for any Cauchy-convergent sequence x_j with limit x the following inequality

$$| (y, Lx - Lx_j | \le || y |||| L(x - x_j) || \le || y ||\| L \||| x - x_j || \tag{7.103}$$

holds true.

According to the Riesz representation theorem there is a vector $z \in \mathcal{H}$ such that $(y, Lx) = (z, x)$ is true for all vectors x. It is a simple exercise to show that this z depends linearly on y. We may therefore write $z = L^\star y$ with a certain linear operator $L^\star$. One calls it the adjoint operator.

Every bounded operator L is associated with an adjoint operator $L^\star$ such that

$$(y, Lx) = (L^\star y, x) \tag{7.104}$$

holds for all vectors $x, y \in \mathcal{H}$.

7.7 Projection Operators

Let $\mathcal{L} \subset \mathcal{H}$ be a linear subspace of a Hilbert space $\mathcal{H}$. Note that we do not demand that $\mathcal{L}$ be closed. It need not be a Hilbert space itself.

Two vectors $x, y \in \mathcal{H}$ are orthogonal if their scalar product vanishes, $(y, x) = 0$. This can be extended to linear subspaces. The subspaces $\mathcal{L}_1$ and $\mathcal{L}_2$ are orthogonal if $(y, x) = 0$ for all $x \in \mathcal{L}_1$ and for all $y \in \mathcal{L}_2$. We write $(\mathcal{L}_2, \mathcal{L}_1) = 0$ in this case.

As a base for the linear space $\mathcal{L}$ one chooses a system $\{e_1, e_2, \dots, e_n\}$ of mutually orthogonal unit vectors. Such an orthonormal system is characterized by

$$(e_k, e_j) = \delta_{kj}. \tag{7.105}$$

Any vector y in the subspace $\mathcal{L} = \text{span}\,\{e_1, e_2, \dots, e_n\}$ may be expanded as

$$y = \sum_{j=1}^{n} \alpha_j e_j \;\text{ with }\; \alpha_j = (y, e_j). \tag{7.106}$$

7.7.1 *Projectors*

Let $x \in \mathcal{H}$ be an arbitrary vector. Its projection y on the subspace $\mathcal{L} \subset \mathcal{H}$ is defined by

$$y = \sum_{j=1}^{n} (x, e_j) e_j. \tag{7.107}$$

Because of $(y, e_j) = (x, e_j)$ we conclude that $x - y$ is orthogonal to all vectors in $\mathcal{L}$. We have thus decomposed $x \in \mathcal{H}$ into a vector $y \in \mathcal{L}$ and a rest $x - y$, which is orthogonal to that subspace.

Now, y depends linearly on x, and we may write $y = \Pi x$. The projector Π is characterized by

$$\Pi^2 = \Pi \text{ and } \Pi^\star = \Pi. \tag{7.108}$$

Because of $\Pi e_j = e_j$ the first assertion is evident. The second—that the projector is self-adjoint—may be verified by comparing the expression $(z, \Pi x) = \sum_j (e_j, x)(z, e_j)$ with $(\Pi z, x) = \sum_j (e_j, z)^* (e_j, x)$; they are the same.

Every linear subspace $\mathcal{L} \subset \mathcal{H}$ is described by a projector Π, and vice versa. Therefore, $\mathcal{L} = \Pi \mathcal{H}$ is an appropriate notation.

The dimension of a projector Π is the dimension of the subspace $\Pi \mathcal{H}$ it projects on. We can show this directly only for finite-dimensional subspaces, the statement however remains correct for a closed subspace. Each convergent sequence of vectors in the subspace then has a limit in the subspace.

An operator M' is smaller or equal to an operator M'' if

$$(x, M'x) \le (x, M''x) \text{ for all } x \in \mathcal{H} \tag{7.109}$$

is true. In this sense $0 \le \Pi \le I$ is true. A projector is non-negative and bounded from above by the unit operator.

Let us remove a deficiency of definition (7.107). The subspace under discussion $\mathcal{L} = \text{span}\,\{e_1, e_2, \dots, e_n\}$ can also be spanned by another orthonormal set, $\mathcal{L} = \text{span}\,\{\bar{e}_1, \bar{e}_2, \dots, \bar{e}_n\}$. We may write

$$e_j = \sum_k u_{jk} \bar{e}_k, \tag{7.110}$$

with

$$\delta_{ij} = (e_i, e_j) = \sum_{kl} u_{ik}{}^* u_{jl} (\bar{e}_k, \bar{e}_l) = \sum_k u_{jk} u_{ik}{}^*, \tag{7.111}$$

i.e. $uu^\star = I$ for the $n \times n$ matrix u. It is invertible which implies $u^\star u = I$. Therefore:

$$\Pi x = \sum_i (e_i, x) e_i = \sum_{ij} u_{ij}{}^* (\bar{e}_j, x) \sum_k u_{ik} \bar{e}_k = \sum_j (\bar{e}_j, x) \bar{e}_j. \tag{7.112}$$

We have made use of $\sum_i u_{ij}{}^* u_{ik} = \delta_{jk}$.

Equation (7.112) says that it is irrelevant which base for $\mathcal{L}$ serves to calculate the projector Π onto $\mathcal{L}$.

7.7.2 Decomposition of Unity

Let Π_1 be a projector: $\Pi_1^2 = \Pi_1$ and $\Pi_1{}^\star = \Pi_1$. Define $\Pi_2 = I - \Pi_1$. We calculate $\Pi_2^2 = \Pi_2$ and $\Pi_2{}^\star = \Pi_2$. With other words, Π_2 is a projector as well. For arbitrary $x \in \mathcal{H}$ we find $(\Pi_2 x, \Pi_1 x) = 0$. The linear subspaces $\mathcal{L}_1 = \Pi_1\mathcal{H}$ and $\mathcal{L}_2 = \Pi_2\mathcal{H}$ are obviously orthogonal, $(\Pi_1\mathcal{H}, \Pi_2\mathcal{H}) = 0$. This statement is equivalent with $\Pi_2\Pi_1 = 0$ or with $\Pi_1\Pi_2 = 0$. The projectors are orthogonal to each other.

We have just described a decomposition $I = \Pi_1 + \Pi_2$ of unity into two mutually orthogonal projectors. This corresponds to a decomposition of the entire Hilbert space into two orthogonal subspaces.

In $\mathbb{R}^3$ one subspace might be a plane including the origin, the other a perpendicular line through the origin. An arbitrary vector is the sum of two contribution, one a projection onto the plane, the other one the projection onto the line.

What has been said about two projectors may be generalized to arbitrarily many. We speak of a decomposition of unity if there are mutually orthogonal projectors $\Pi_1, \Pi_2, \ldots$ adding up to I. With other words

- projectors $\Pi_j = \Pi_j{}^\star = \Pi_j^2$,
- mutually orthogonal, $\Pi_j\Pi_k = 0$ if $j \neq k$,
- add up to $I = \Pi_1 + \Pi_2 + \cdots$

The concept of a decomposition of unity into mutually orthogonal projectors is helpful for characterizing normal operators.

7.8 Normal Operators

An operator N is called normal if it commutes with its own adjoint,

$$NN^\star = N^\star N. \tag{7.113}$$

A normal operator may always be written as a sum of multiples of mutually orthogonal projection operators Π_j which form a decomposition of unity. The factors by which the projectors are multiplied are the eigenvalues of N. The projectors project on linear subspaces $\mathcal{L}_j = \Pi_j\mathcal{H}$, and the normal operators map the subspace into itself by simply re-scaling each vector by the eigenvalue.

7.8.1 Spectral Decomposition

Consider a decomposition of unity into mutually orthogonal projectors,

$$I = \sum_j \Pi_j \quad \text{where} \quad \Pi_j\Pi_k = 0 \quad \text{for} \quad j \neq k. \tag{7.114}$$

Let $\nu_1, \nu_2, \ldots$ be a sequence of complex numbers. We define a linear operator N by

$$N = \sum_j \nu_j \Pi_j. \tag{7.115}$$

Its adjoint is

$$N^\star = \sum_j \nu_j{}^* \Pi_j. \tag{7.116}$$

Calculate $NN^\star$ and $N^\star N$. Both expressions yield

$$NN^\star = N^\star N = \sum_j |\nu_j|^2 \Pi_j. \tag{7.117}$$

It follows that the N of (7.115) is normal.

The converse is true as well. However we will treat a finite-dimensional Hilbert space only, $\mathcal{H} = \mathbb{C}^n$ say. Operators then are represented by complex $n \times n$ matrices. Adjoining now means complex conjugating and interchanging the role of rows and columns.

The eigenvalue equation

$$Nf = \nu f \tag{7.118}$$

has non-vanishing solution vectors f if, and only if the characteristic polynomial

$$\chi(\nu) = \det(N - \nu I) = 0 \tag{7.119}$$

vanishes. There is alway at least one zero if $\nu \in \mathbb{C}$ is allowed, so the fundamental theorem of algebra. After all, the characteristic polynomial cannot be constant for $n \geq 1$.

Let us denote by $\mathcal{L}$ the eigenspace for a solution ν of (7.119),

$$\mathcal{L} = \{f \in \mathcal{H} \mid Nf = \nu f\}. \tag{7.120}$$

Up to now N was an arbitrary operator. We now make us of N being normal. For $f \in \mathcal{L}$ one calculates

$$NN^\star f = N^\star N f = \nu N^\star f, \tag{7.121}$$

i.e.

$$N^\star \mathcal{L} \subset \mathcal{L}. \tag{7.122}$$

For all $f, g \in \mathcal{L}$ one has

$$0 = (g, (N - \nu I) f) = ((N^\star - \nu^* I) g, f), \tag{7.123}$$

meaning that $\mathcal{L}$ is also an eigenspace of $N^\star$ with eigenvalue ν^*.

We characterize the eigenspace $\mathcal{L}$ by the projector Π. On $\mathcal{L} = \Pi\mathcal{H}$ the operator N acts as νI and $N^\star$ as $\nu^* \Pi$.

$\mathcal{L}_\perp = (I - \Pi)\mathcal{H}$ is the subspace orthogonal to $\mathcal{L}$. For $g \in \mathcal{L}$ and $f \in \mathcal{L}_\perp$ we calculate

$$(g, Nf) = (N^\star g, f) = (\nu^* g, f) = \nu(g, f) = 0, \tag{7.124}$$

i.e. $N\mathcal{L}_\perp \subset \mathcal{L}_\perp$. N maps the subspace $\mathcal{L}_\perp$ (which is orthogonal to $\mathcal{L}$) into itself. The same is true for $N^\star$. Restricted to $\mathcal{L}_\perp$ the operator N is normal as well.

Hence, one can repeat the above procedure once more, with $\mathcal{L}_\perp$ instead of $\mathcal{H}$. However, the dimension is lower than before. Therefore, after finitely many repetitions one arrives at the trivial zero space. Our goal has been reached. The normal operator N may be represented as

$$N = \sum_j \nu_j \Pi_j \text{ and } N^\star = \sum_j \nu_j{}^* \Pi_j, \tag{7.125}$$

with a decomposition $I = \sum_j \Pi_j$ into mutually orthogonal projectors Π.

7.8.2 Unitary Operators

Consider a mapping U of $\mathcal{H}$ onto itself which preserves the scalar product. For all $f, g \in \mathcal{H}$ we demand

$$(g, f) = (Ug, Uf) = (g, U^\star U f). \tag{7.126}$$

We conclude

$$U^\star U = I. \tag{7.127}$$

The equation $Uf_1 = Uf_2$ implies $f_1 = f_2$. With $g = f_2 - f_1$ the equation $Ug = 0$ has to be solved, i.e. $|| Ug || = || g || = 0$ or $g = 0$. We conclude that the mapping $U : \mathcal{H} \to \mathcal{H}$ can be inverted. From (7.127) one deduces $U^\star = U^{-1}$ and

$$UU^\star = I. \tag{7.128}$$

One compares (7.127) with (7.128) and concludes that a unitary operator is normal. It can be represented in the usual way as

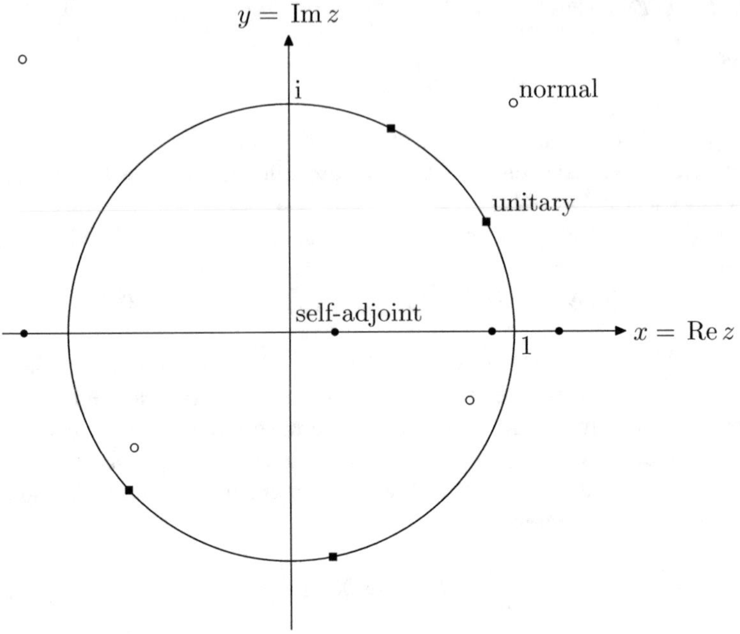

Fig. 7.2 Normal operators The spectrum of normal operators, schematic. Values in the complex plane (open circles) stand for a general normal operator. Full squares are situated at the unit circle; they are the eigenvalues of a unitary operator. A self-adjoint operator has eigenvalues on the real axis (full circles). The operator is positive if its eigenvalues lie on the positive x-axis.

$$U = \sum_j u_j \Pi_j, \tag{7.129}$$

and the eigenvalues u_j are phase factors, i.e.

$$|\, u_j \,| = 1. \tag{7.130}$$

The eigenvalues of a unitary operator lie on the unit circle of the complex plane, see Fig. 7.2.

7.8.3 Self-Adjoint Operators

A self-adjoint operator is adjoint to itself,

$$A^\star = A. \tag{7.131}$$

Therefore, the operator is normal. There is a decomposition of unity into mutually orthogonal projectors Π_j such that

$$A = \sum_j a_j \Pi_j \quad \text{with} \quad a_j = a_j{}^* \tag{7.132}$$

holds true. A self-adjoint operator has real eigenvalues, as indicated in Fig. 7.2.

7.8.4 *Positive Operators*

There are two definitions of a positive[2] operator P. One says that arbitrary diagonal elements (x, Px) are always positive. The other definition asserts that one may write $P = B^\star B$ with a certain operator B.

From the second definition it follows that P is self-adjoint. It may be written as

$$P = \sum_j p_j \Pi_j \quad \text{with} \quad p_j \geq 0. \tag{7.133}$$

A positive operator has positive eigenvalues. We find for an arbitrary vector $x \in \mathcal{H}$ that

$$(x, Px) = \sum_j p_j (x, \Pi_j x) = \sum_j p_j (\Pi_j x, \Pi_j x) \tag{7.134}$$

is a sum of positive contributions and therefore never negative. This is the first of the two definitions for a positive operator.

7.8.5 *Probability Operators*

A probability operator ϱ is positive (again in the sense of non-negative) with a trace of one. The trace of a projector is its dimension, i.e. the dimension of the linear space $\mathcal{L} = \Pi\mathcal{H}$. We therefore may write

$$\varrho = \sum_j w_j \Pi_j \quad \text{with} \quad w_j \geq 0 \quad \text{and} \quad \text{Tr}\, \varrho = \sum_j w_j \dim(\Pi_j). \tag{7.135}$$

It should be clear from this definition that not every non-negative operator has a finite trace.

[2]Here in the sense of non-negative.

7.9 Operator Functions

There are two possibilities to declare functions of an operator. The first one makes use of the algebraic properties of the operator ring. Any polynomial of an operator is well defined, and this may be extended to convergent power series. The second definition is applicable to normal operators only. The decomposition of unity into mutually orthogonal projectors remains unchanged, but the eigenvalues are mapped by the function. In cases where both definitions make sense both yield identical results.

We restrict the discussion to bounded operators, thereby avoiding difficult questions of the domain of definition. Non-continuous linear mappings must be dealt with as the case arises.

7.9.1 Power Series

Linear operators may be multiplied by scalars, added, and multiplied. With these operations one can define arbitrary polynomials.

Consider a power series $f(x) = c_0 + c_1 z + c_2 z^2 + \cdots$ The variable z and the coefficients c_k are complex numbers. Such a power series converges in the interior of a circle of radius

$$r = \frac{1}{\limsup_{k\to\infty} \sqrt[k]{|\, c_k \,|}}, \tag{7.136}$$

i.e. for $|\, z \,| < r$.

Let us now define the power series for an operator L:

$$f(L) = \sum_{k=0}^{\infty} c_k L^k = I + c_1 L + c_2 L^2 + \cdots \tag{7.137}$$

We estimate its operator norm. Applying the rules for αL, $L_1 + L_2$ and $L_2 L_1$ we arrive at

$$\| \, f(L) \, \| = \| \sum_{k=0}^{\infty} c_k L^k \, \| \le \sum_{k=0}^{\infty} |\, c_k \,| \, \| \, L \, \|^k. \tag{7.138}$$

The power series of an operator converges if its norm is smaller than the radius of convergence, $\| \, L \, \| < r$.

7.9.2 Normal Operator

Assume now that L is normal. One may therefore write

$$L = \sum_j \lambda_j \Pi_j \tag{7.139}$$

where the Π_j are a sequence of mutually orthogonal projectors summing up to $\Pi_1 + \Pi_2 + \cdots = I$. The eigenvalues λ_j are complex numbers in general, lie on the unit circle for unitary operators or are real numbers for self-adjoint operators.

Let $f : \mathbb{C} \to \mathbb{C}$ an arbitrary complex valued function of a complex variable. One declares

$$f(L) = \sum_j f(\lambda_j) \Pi_j, \tag{7.140}$$

with the same decomposition of unity. $f(L)$ has the same eigenspaces as L, but the eigenvalues are mapped by our function f.

7.9.3 Comparison

Both definitions for operator valued functions are plausible. The power series method works if the function can be represented as a power series in a restricted domain. The operator is arbitrary but must be small enough. The spectral method is applicable to normal operators only, but the function may be arbitrary.

However there are cases where both definitions are applicable. Do the results agree?

Let f be a convergent power series and L a normal operator the norm of which is smaller than the radius of convergence. One calculates

$$\sum_j f(\lambda_j)\, \Pi_j = \sum_j \sum_k c_k\, \lambda_j^k\, \Pi_j = \sum_k c_k \sum_j \lambda_j^k\, \Pi_j = \sum_k c_k\, L^k. \tag{7.141}$$

Clearly, both definitions are equivalent. We have re-summed which is allowed since the power series converges absolutely.

7.9.4 Example

Consider the three Pauli matrices

$$\sigma_1 = \begin{pmatrix} 0 & 1 \\ 1 & 0 \end{pmatrix}\ ,\ \sigma_2 = \begin{pmatrix} 0 & -\mathrm{i} \\ I & 0 \end{pmatrix} \text{ and } \sigma_3 = \begin{pmatrix} 1 & 0 \\ 0 & -1 \end{pmatrix}. \tag{7.142}$$

They are operators in the Hilbert space $\mathbb{C}^2$. All three operators are self-adjoint since $\sigma_{jk} = {\sigma_{kj}}^*$.

σ_3 may be written as

$$\sigma_3 = \Pi_+ - \Pi_- = \begin{pmatrix} 1 & 0 \\ 0 & 0 \end{pmatrix} - \begin{pmatrix} 0 & 0 \\ 0 & 1 \end{pmatrix} . \tag{7.143}$$

The two matrices are orthogonal projectors. Therefore, σ_3 has the two eigenvalues $+1$ and -1.

Let us calculate $U = \mathrm{e}^{\mathrm{i}\phi\sigma_3}$. An easy job with the spectral decomposition (7.143):

$$U = \mathrm{e}^{\mathrm{i}\phi} \begin{pmatrix} 1 & 0 \\ 0 & 0 \end{pmatrix} + \mathrm{e}^{-\mathrm{i}\phi} \begin{pmatrix} 0 & 0 \\ 0 & 1 \end{pmatrix} = \begin{pmatrix} \mathrm{e}^{\mathrm{i}\phi} & 0 \\ 0 & \mathrm{e}^{-\mathrm{i}\phi} \end{pmatrix} . \tag{7.144}$$

Now we take into account that the exponential function is an always convergent power series. One obtains

$$U = I + \frac{\mathrm{i}\phi}{1!}\sigma_3 + \frac{(\mathrm{i}\phi)^2}{2!}\sigma_3^2 + \cdots . \tag{7.145}$$

Since $\sigma_3^2 = I$ this reads

$$U = \cos\phi \, I + \mathrm{i} \sin\phi \, \sigma_3 = \begin{pmatrix} \mathrm{e}^{\mathrm{i}\phi} & 0 \\ 0 & \mathrm{e}^{-\mathrm{i}\phi} \end{pmatrix} . \tag{7.146}$$

Very satisfying!

7.10 Translations

The continuum $x \in \mathbb{R}^3$ causes problems because the corresponding operator X is not bounded. We therefore treat locations as positions on a circular ring of radius R. The idealization of an infinite x-axis is approximated better and better by sending R to infinity. Translational invariance is achieved by insisting on periodic functions.

In this and in the following subsections we again shall set $\hbar = 1$. Planck's quantum of action is not a constant of nature; it corrects for the conceptual clash between the particle and the wave aspect of matter and is irrelevant for the mathematics behind.

7.10.1 Periodic Boundary Conditions

Let us begin our investigation for the case $R = 1$. We consider square-integrable complex valued functions on $\Omega = [-\pi, \pi]$. Square-integrable in the sense of the Lebesgue integral, see Sect. 7.2 on the Lebesgue Integral. The points x and $x + 2\pi$ are to be identified. Our Hilbert space is

$$\mathcal{H} = \{f : [-\pi, \pi] \to \mathbb{C} \mid f(x) = f(x + 2\pi) \text{ and } \int_{-\pi}^{\pi} \mathrm{d}x \mid f(x) \mid^2 \}. \tag{7.147}$$

Cauchy-convergent sequences of periodic functions converge towards periodic functions, and $\mathcal{H}$ is therefore complete.

The scalar product is translationally invariant:

$$\int_{-\pi}^{\pi} \mathrm{d}x\, g^*(x)\, f(x) = \int_{-\pi}^{\pi} \mathrm{d}x\, g^*(x + a)\, f(x + a). \tag{7.148}$$

The translation

$$f \to f_a = U_a f \text{ where } f_a(x) = f(x + a) \tag{7.149}$$

leaves all scalar products unchanged. The U_a are a family of unitary operators indexed by the translation a.

We expand f_a into a Taylor series:

$$(U_a f)(x) = f(x + a) = f(x) + \frac{a}{1!} f'(x) + \frac{a^2}{2!} f''(x) + \cdots \tag{7.150}$$

and recognize the exponential function. With the operator

$$P = -\mathrm{i}\frac{\mathrm{d}}{\mathrm{d}x} \tag{7.151}$$

we may rewrite (7.150) into

$$U_a = \mathrm{e}^{\mathrm{i}aP}. \tag{7.152}$$

7.10.2 Domain of Definition

The location operator

$$(Xf)(x) = x\, f(x) \tag{7.153}$$

is defined on the entire Hilbert space of (7.147) because the values in $[-\pi, \pi]$ are bounded. However, the momentum operator P which generates the translation operators U_a, cannot be defined everywhere since a square-integrable function is not necessarily differentiable.

For an arbitrary $f \in \mathcal{H}$ we define

$$F(x) = c + \int_{-\pi}^{x} \mathrm{d}s\, f(s). \tag{7.154}$$

Such a function F is called absolutely continuous. On the one hand it is continuous and therefore—on a finite interval—square-integrable. On the other hand, it is differentiable as defined by

$$F'(x) = f(x), \tag{7.155}$$

the result being square-integrable. Differentiable functions are smoother than merely continuous functions.

We define $\mathcal{D} \subset \mathcal{H}$ as the set of all absolutely continuous, periodic and square-integrable functions. $\mathcal{D}$ is a linear subspace. The momentum operator $P : \mathcal{D} \rightarrow \mathcal{H}$ maps absolutely continuous and periodic functions into square-integrable periodic functions.

7.10.3 Selfadjointness

Let us now calculate the adjoint operator $P^\star$. For each function G we must find a square-integrable g which causes $(G, PF) = (g, F)$ for all $F \in \mathcal{D}$. This means

$$-\mathrm{i} \int_{-\pi}^{\pi} \mathrm{d}x\, G^*(x) F'(x) = -\mathrm{i}\Delta + \mathrm{i} \int_{-\pi}^{\pi} \mathrm{d}x\, G'^*(x) F(x), \tag{7.156}$$

where $\Delta = G^*(\pi)F(\pi) - G^*(-\pi)F(-\pi)$. The function G must be periodic, $G(\pi) = G(-\pi)$, and Δ vanishes. At the same time, the function G has to be differentiated, it must be absolutely continuous. Thus we have shown that $P^\star$ should be defined on $\mathcal{D}$ just as P. So P and $P^\star$ have the same domain of definition and there they do the same. P is self-adjoint.

Clearly, if $\mathcal{D}$ would be shrunk the domain of definition for $P^\star$ would grow, and vice versa. We just have found the subspace $\mathcal{D}$ which serves as a domain of definition for both P and $P^\star$.

7.10.4 Spectral Decomposition

We look for the eigenfunctions of the momentum operator. The eigenvalue equation

$$Pe = -\mathrm{i}e' = p\,e \tag{7.157}$$

has the following solutions in $\mathcal{D}$:

$$e_j(x) = \frac{1}{\sqrt{2\pi}}\,\mathrm{e}^{\mathrm{i}jx}. \tag{7.158}$$

The corresponding eigenvalues are

$$p_j = j. \tag{7.159}$$

If we chose a ring of radius R (instead of $R = 1$), the eigenfunctions change to

$$e_j(x) = \frac{1}{\sqrt{2\pi R}}\,\mathrm{e}^{\mathrm{i}jx/R}, \tag{7.160}$$

the eigenvalues to

$$p_j = \frac{j}{R}. \tag{7.161}$$

We see that for $R \to \infty$ every real number is among the eigenvalues. However, the solutions

$$e_p \propto \mathrm{e}^{\mathrm{i}px} \tag{7.162}$$

of (7.157) are not eigenfunctions because they are not square-integrable according to

$$\int_{-\infty}^{\infty} \mathrm{d}x \mid f(x) \mid^2 < \infty. \tag{7.163}$$

One speaks of quasi-eigenfunctions, and the sum over the index j labeling the eigenvalues must be replaced by an integral.

There is a projector valued measure $\mathrm{d}\Pi_p$ with

$$I = \int \mathrm{d}\Pi_p \tag{7.164}$$

such that the linear momentum operator is represented by

$$P = \int \mathrm{d}\Pi_p\, p. \tag{7.165}$$

The concept of 'mutually orthogonal' and of the integrals has to be explained. We shall stop here because the substantial effort does not really lead to new insight.

7.11 Fourier Transform

The result of the preceding subsection are so important that we should discuss them in greater detail.

We consider periodic square-integrable functions on a finite interval,

$$\mathcal{H} = \{f : [-\pi, \pi] \to \mathbb{C} \mid f(x + 2\pi) = f(x); \int_{-\pi}^{\pi} \mathrm{d}x \mid f(x) \mid^2 < \infty\}. \quad (7.166)$$

The translation $f \to f_a = U_a f$ is defined by $f_a(x) = f(x + a)$. The unitary translation operators may be written as

$$U_a = \mathrm{e}^{iaP} \quad \text{with } P = -\mathrm{i}\frac{\mathrm{d}}{\mathrm{d}x}. \quad (7.167)$$

The momentum operator, if defined for periodic and absolutely continuous functions, is self-adjoint.

7.11.1 *Fourier Series*

The normalized eigenfunctions of P are

$$e_j(x) = \frac{1}{\sqrt{2\pi}} \mathrm{e}^{\mathrm{i}jx} \quad (7.168)$$

for $j \in \mathbb{Z}$. Since (7.168) are all eigenfunctions of a self-adjoint operator, namely P, the e_j form a base; they span the entire Hilbert space.

Every periodic square-integrable function $f \in \mathcal{H}$ may be expanded into a Fourier series:

$$f(x) = \sum_{j\in\mathbb{Z}} \hat{f}_j\, e_j(x) = \frac{1}{\sqrt{2\pi}} \sum_{j\in\mathbb{Z}} \hat{f}_j\, \mathrm{e}^{\mathrm{i}jx}. \quad (7.169)$$

Even better, we know how the Fourier coefficient $\hat{f}_j$ must be calculated:

$$\hat{f}_j = (e_j, f) = \frac{1}{\sqrt{2\pi}} \int_{-\pi}^{\pi} \mathrm{d}x\, \mathrm{e}^{-\mathrm{i}jx} f(x). \quad (7.170)$$

One easily can show Parseval's identity

$$(f, f) = \int_{-\pi}^{\pi} \mathrm{d}x \mid f(x) \mid^2 = \sum_{j \in \mathbb{Z}} \mid \hat{f}_j \mid^2 \tag{7.171}$$

7.11.2 Fourier Expansion

Even on a large computer you cannot handle an infinite number of contributions to the expansion (7.169). One has to replace it by a finite sum. We may only approximate the arbitrary function by

$$f(x) \approx \sum_{\mid j \mid \leq n} \hat{f}_j \, e_j(x) + r_n(x), \tag{7.172}$$

with a certain rest $r_n(x)$. The rest is always orthogonal to the approximation. Therefore, the coefficients $\hat{f}_j$ of the approximation do not depend on the approximation order. If one improves the approximation by allowing for more terms, the previously calculated coefficients remain the same. In this respect the approximation by a finite instead of an infinite Fourier sum is optimal.

7.11.3 Fourier Integral

Now we direct our attention to periodic square-integrable functions which live on $[-\pi R, \pi R]$ and send R to infinity. This amounts to $\mathcal{H} = \mathcal{L}_2(\mathbb{R})$. Recall that we really speak about equivalence classes of functions. Two functions are equivalent if their definitions differ on a set D only the measure of which vanishes. One says that they are almost equal. We have dropped the requirement that the functions be periodic; this makes no sense for the entire real axis.

Instead of summing over eigenfunctions we must integrate over quasi-eigen functions:

$$f(x) = \int \frac{\mathrm{d}p}{2\pi} \, \hat{f}(p) \, \mathrm{e}^{\,ipx} \tag{7.173}$$

where

$$\hat{f}(p) = \int \mathrm{d}x \, f(x) \, \mathrm{e}^{\,-ipx} \, . \tag{7.174}$$

We call $\hat{f} = \hat{f}(p)$ the Fourier transform of $f = f(x)$. Evidently $f = f(x)$ in turn is the Fourier transform of $\hat{f} = \hat{f}(p)$, up to a change of sign in the exponent and the factor 2π. In this sense, the Fourier transform is an invertible operation. We have discussed this before in Sect. 7.4 on Generalized Functions.

Incidentally, Parseval's identity now reads

$$\int \mathrm{d}x \mid f(x) \mid^2 = \int \frac{\mathrm{d}p}{2\pi} \mid \hat{f}(p) \mid^2. \tag{7.175}$$

The Fourier transform $\hat{f} = \hat{f}(p)$ is square-integrable as well.

As an example we present the Fourier transform of the normalized Gaussian

$$f(x) = \frac{1}{\sqrt{2\pi}} \, \mathrm{e}^{-x^2/2}. \tag{7.176}$$

It is the probability distribution of a normally distributed random variable F with means $\langle F \rangle = 0$ and variance $\langle F^2 \rangle = 1$. See Sect. 7.3 on Probabilities. Its Fourier transform is

$$\hat{f}(p) = \mathrm{e}^{-p^2/2}. \tag{7.177}$$

$\hat{f}(0) = 1$ is a check whether we have normalized properly. The value of the Fourier transform at zero is the integral over the function.

7.11.4 Convolution Theorem

The convolution $h = g * f$ is defined by

$$h(x) = \int \mathrm{d}y \, g(x - y) \, f(y). \tag{7.178}$$

Because of

$$h(x) = \int \mathrm{d}y \int \frac{\mathrm{d}p}{2\pi} \hat{g}(p) \, \mathrm{e}^{ip(x-y)} \int \frac{\mathrm{d}q}{2\pi} \hat{f}(q) \, \mathrm{e}^{iqy} \tag{7.179}$$

we conclude—by interchanging integrations—

$$h(x) = \int \frac{\mathrm{d}p}{2\pi} \hat{g}(p) \, \hat{f}(p) \, \mathrm{e}^{ipx}. \tag{7.180}$$

We have made use of

$$\int \mathrm{d}y \, \mathrm{e}^{\mathrm{i}(q-p)x} = 2\pi\delta(q - p). \tag{7.181}$$

For details see Sect. 7.4 on Generalized Functions. The result of the above calculation is

$$\hat{h}(p) = \hat{g}(p)\,\hat{f}(p). \tag{7.182}$$

The Fourier transform of a convolutions is the product of the Fourier transforms of the factors. This is an important result, and we have used it on various occasions. Note that one of the factors in (7.178) may be a generalized function which always can be Fourier transformed.

7.11.4.1 Causal Functions

A causal function $f : \mathbb{R} \to \mathbb{C}$ vanishes for negative argument. Therefore, one may write[3]

$$f(t) = \theta(t)\, f(t). \tag{7.183}$$

$\theta = \theta(t)$ denotes Heaviside's step function: it vanishes for $t < 0$ and has the value 1 for positive t. By the convolution theorem, the Fourier transform $g = \hat{f}$ of (7.183) is the convolution of the respective Fourier transforms:

$$g(\omega) = \hat{f}(\omega) = \int \frac{\mathrm{d}u}{2\pi}\,\hat{\theta}(u)\, g(\omega - u). \tag{7.184}$$

The Fourier transform of the Heaviside function is to be found in Table 7.1. We insert it:

$$g(\omega) = \int \frac{\mathrm{d}u}{2\pi \mathrm{i}}\,\frac{g(u)}{u - \omega - \mathrm{i}\epsilon}. \tag{7.185}$$

$\epsilon > 0$ and $\epsilon \to 0$ at the end is silently understood.

The integral is along a way avoiding the pole at $\omega = u$, with weight 1/2 in the upper and with weight 1/2 in the lower complex u-plane. This amounts to the principal value integral. We may write

$$g(\omega) = PV \int \frac{\mathrm{d}u}{\pi}\,\frac{g(u)}{u - v}. \tag{7.186}$$

Note that the principal value integral is the limit

$$PV \int \mathrm{d}x \;\cdots = \left\{ \int_{-\infty}^{x-\epsilon} + \int_{x+\epsilon}^{\infty} \right\} \mathrm{d}x \;\ldots \tag{7.187}$$

---

[3]Causality usually refers to time, and we write t for the argument of the causal functions and ω for the argument of its Fourier transform.

Let us denote the real part by g' and the imaginary part by g''. One obtains

$$g'(\omega) = PV \int \frac{\mathrm{d}u}{\pi} \frac{g''(u)}{u - \omega} \tag{7.188}$$

and

$$g''(\omega) = PV \int \frac{\mathrm{d}u}{\pi} \frac{g'(u)}{\omega - u}. \tag{7.189}$$

We see that the imaginary part is a function of the real part, and vice versa. If the causal function f turns out to be real there are further restrictions on the Fourier transform g.

7.12 Position and Momentum

Space and time are of particular interest for physics. Therefore the location operator X and its associated linear momentum P play an important role. However, neither of them can be defined on the entire Hilbert space since they are unbounded. We have shown in a previous section how to restrict the domain of definition in such a way that the momentum becomes self-adjoint. We did so for a model space with the topology of a ring. The standard case where each of the three dimensions of space is modeled by the real line was not covered.

7.12.1 Test Functions

We focus on the Hilbert space $\mathcal{H} = \mathcal{L}_2(\mathbb{R})$ of complex valued Lebesgue square-integrable functions $f : \mathbb{R} \to \mathbb{C}$ defined on the real axis $\mathbb{R}$. As mentioned before, neither X nor P can be defined for all functions $f \in \mathcal{H}$.

Instead of all functions we consider test functions only. Test functions t are arbitrarily often differentiable and vanish so rapidly at infinity that even $|\, x \,|^n \, t(x)$ vanishes at infinity, for any natural number n. The set of all test functions is denoted by $\mathcal{S}$. Clearly, $\mathcal{S} \subset \mathcal{H}$. For details see Sect. 7.4 on Generalized Functions.

Although test functions are very well behaved, the linear space $\mathcal{S}$ is dense in $\mathcal{H}$ in the natural topology. Any square-integrable function f is arbitrarily close to a test function. For $\epsilon > 0$ there is always a $t \in \mathcal{S}$ such that $||\, f - t \,|| \le \epsilon$. The proof is not trivial, and we skip it here.

Test functions are very convienient. They are so brave that one may differentiate or multiply with the argument as often as one likes.

7.12.2 Canonical Commutation Rules

The location operator is well defined for test functions:

$$(Xt)(x) = x\,t(x) \quad \text{for} \quad t \in \mathcal{S}. \tag{7.190}$$

The momentum operator is declared by

$$(Pt)(x) = -\mathrm{i}t'(x) \quad \text{form } t \in \mathcal{S}, \tag{7.191}$$

it is likewise well defined for test functions. Both operators, X and P, map $\mathcal{S}$ into $\mathcal{S}$.

These two operators do not commute. For an arbitrary test function we calculate

$$XPt - PXt = -\mathrm{i}xt' + \mathrm{i}(t + xt') = \mathrm{i}t, \tag{7.192}$$

i.e.

$$[X, P] = \mathrm{i}I. \tag{7.193}$$

Equation (7.193) is the canonical commutation rule. It is called so because it expresses a fundamental law setup by high authorities (in the parlance of the catholic church).

It is fundamental in so far as it is stable against unitary transformations. Recall that unitary transformations send one complete orthonormal system into another one such that

$$(\bar{e}_j, \bar{e}_k) = (Ue_j, Ue_k) = (e_j, U^\star U e_k) = (e_j, e_k) \tag{7.194}$$

holds true. This is how vectors are to be transformed: $f \to \bar{f} = Uf$. Linear operators should be altered as $\bar{L} = ULU^\star$ such that

$$\bar{L}\bar{f} = \overline{Lf}. \tag{7.195}$$

Let us check what happens to the canonical commutation rules. With $\bar{X} = UXU^\star$ and $\bar{P} = UPU^\star$ we calculate

$$[\bar{X}, \bar{P}] = UXU^\star UPU^\star - UPU^\star UXU^\star = U[X, P]U^\star = \mathrm{i}\bar{I} = \mathrm{i}I. \tag{7.196}$$

As announced, the canonical commutation rules are stable with respect to unitary transformations.

7.12.3 Uncertainty Relation

Location and momentum cannot be diagonalized simultaneously. Representations $X = \sum_j x_j \Pi_j$ and $P = \sum_j p_j \Pi_j$ with a common decomposition of unity are impossible. Otherwise X and P would commute, and this is not the case.

Let us denote by ϱ the state which has been prepared. ϱ is a probability operator. We denote by $\langle X \rangle = \text{Tr}\, \varrho X$ the location expectation value and define the uncertainty $\sigma(X)$ by the variance

$$\sigma^2(X) = \langle X^2 \rangle - \langle X \rangle^2. \tag{7.197}$$

The momentum uncertainty $\sigma(P)$ is defined similarly.

The expression $(X + \mathrm{i}\alpha P)(X - \mathrm{i}\alpha P)$ is, for real α, of type $AA^\star$, hence never negative. Its expectation value is therefore also never negative,

$$\langle X^2 \rangle + \alpha^2 \langle P^2 \rangle + \alpha \geq 0. \tag{7.198}$$

Here we have replaced $-\mathrm{i}\alpha(XP - PX)$ by αI.

The left-hand side of (7.198) is smallest if we choose

$$2\alpha \langle P^2 \rangle + 1 = 0, \tag{7.199}$$

resulting in

$$\langle X^2 \rangle \geq \frac{1}{4\langle P^2 \rangle}. \tag{7.200}$$

This result is known as Heisenberg's uncertainty relation. One has to replace X by $X - \langle X \rangle$ and P by $P - \langle P \rangle$, a canonical transformation which does not change the canonical commutation relations. Put otherwise, one does not talk of location and momentum but their fluctuations. One may then write

$$\sigma(X)\,\sigma(P) \geq \frac{1}{2}. \tag{7.201}$$

Heisenberg's uncertainty relation (7.201) is valid for an arbitrary state, pure or mixed. It is optimal since the equal sign occurs for a pure state, a Gaussian:

$$t(x) \propto \mathrm{e}^{-cx^2}. \tag{7.202}$$

7.12.4 Quasi-Eigenfunctions

An eigenstate is characterized by the property that the corresponding observable can be measured with vanishing uncertainty.

The eigenvalue equation for the location operator reads

$$(Xf)(x) = xf(x) = af(x). \tag{7.203}$$

Its formal solution—generalized functions allowed—is

$$\xi_a(x) = \delta(x - a). \tag{7.204}$$

ξ_a is a generalized functions, it does not belong to the Hilbert space, and in particular not to $\mathcal{S}$, the domain of definition. However, in a certain sense, these quasi-eigenfunctions form a complete set of mutually orthogonal functions:

$$\int \mathrm{d}x \, \xi_b{}^*(x) \, \xi_a(x) = \delta(b - a). \tag{7.205}$$

$\delta(b - a)$ and the integral replace the Kronecker delta symbol and the sum over discrete values.

The momentum operator behaves similarly. The eigenvalue equation reads

$$(Pf)(x) = -\mathrm{i}f'(x) = pf(x) \tag{7.206}$$

the solution of which is

$$\pi_p(x) = \frac{1}{\sqrt{2\pi}} \mathrm{e}^{\mathrm{i}px}. \tag{7.207}$$

Now π_p is at least a function. However, it does not belong to $\mathcal{H}$ and also not to the domain of definition $\mathcal{S}$. Nevertheless, the momentum quasi-eigenfunctions form a complete ortho-normal set:

$$\int \mathrm{d}x \, \pi_q{}^*(x) \, \pi_p(x) = \delta(q - p). \tag{7.208}$$

7.13 Ladder Operators

In this section we refer to two operators X and P obeying the canonical commutation rules. These operators might be the location and the linear momentum of a particle (moving along a line). But they may also stand for something completely different. Since we employ the canonical commutation rules only, the result does not depend on the meaning of X and P.

7.13.1 Raising and Lowering Operators

Our starting point is the commutation relation

$$[X, P] = \mathrm{i}I \tag{7.209}$$

for two self-adjoint operators X and P. Their interpretation is irrelevant. The raising operator is declared as

$$A_+ = \frac{X - \mathrm{i}P}{\sqrt{2}}, \tag{7.210}$$

the lowering operator by

$$A_- = \frac{X + \mathrm{i}P}{\sqrt{2}}. \tag{7.211}$$

Note that the lowering and the raising operators are <u>not</u> self-adjoint; they are related by

$$A_-{}^\star = A_+ \text{ and } A_+{}^\star = A_-. \tag{7.212}$$

The raising and lowering operators do not commute,

$$[A_-, A_+] = I, \tag{7.213}$$

which is simple to check.

The operators A_+ and A_- are not normal and we therefore do not ask for their eigenvalues. However, the operator

$$N = A_+ A_- \tag{7.214}$$

is selfadjoint. More, it is of the form $B^\star B$ and therefore non-negative, $N \geq 0$. Its commutation rules with the lowering and raising operators are

$$[N, A_-] = -A_- \text{ and } [N, A_+] = A_+. \tag{7.215}$$

For checking this the Jacobi identity is helpful:

$$[AB, C] = A[B, C] + [A, C]B. \tag{7.216}$$

7.13.2 Ground State and Excited States

If ϕ is an eigenstate of N with eigenvalue λ, we may show easily

$$NA_+\phi = (A_+N + A_+)\phi = (\lambda + 1)A_+\phi, \tag{7.217}$$

i.e. $A_+\phi$ is an eigenstate with eigenvalue $\lambda + 1$. Likewise, $A_-\phi$ is an eigenstate with eigenvalue $\lambda - 1$. However, stepping down the eigenvalue ladder must terminate since N cannot have negative eigenvalues. There must be a normalizable state Ω such that $A_-\Omega = 0$.

We call Ω with $(\Omega, \Omega) = 1$ the ground state or vacuum state. Clearly,

$$N\Omega = 0. \tag{7.218}$$

The ground state is an eigenstate of N with eigenvalue $n = 0$. Consider

$$\phi_n = \frac{1}{\sqrt{n!}}(A_+)^n\Omega. \tag{7.219}$$

They are eigenstates of N with eigenvalue $n \in \mathbb{N}$, i.e. $n = 0, 1, 2, \ldots$,

$$N\phi_n = n\,\phi_n. \tag{7.220}$$

This can be shown by complete induction with $\phi_0 = \Omega$ as the base, or anchor case. The eigenstates (7.219) are properly normalized,

$$||\,\phi_n\,|| = 1. \tag{7.221}$$

Equation (7.221) is true for $n = 0$. Assume that it is also true for $n \in \mathbb{N}$. Then

$$\begin{aligned} ||\,\phi_{n+1}\,||^2 &= \frac{||\,A_+\phi_n\,||}{n+1} = \frac{(\phi_n, A_-A_+\phi_n)}{n+1} \\ &= \frac{(\phi_n, (N+I)\phi_n)}{n+1} = ||\,\phi_n\,||^2 \end{aligned} \tag{7.222}$$

shows that it is true in general.

The eigenvalues of N are the natural numbers. Therefore we call it a number operator.

7.13.3 Harmonic Oscillator

In many cases the energy of the system is $H = (P^2 + X^2)/2$. The first term is kinetic energy, the second the potential energy close to a stable configuration in lowest order approximation. Because of

$$N = A_+A_- = \frac{1}{2}(X + \mathrm{i}P)(X - \mathrm{i}P) = \frac{1}{2}(X^2 + P^2 - I) \tag{7.223}$$

we obtain

$$H = N + \frac{1}{2}\,I. \tag{7.224}$$

The eigenvalues of the harmonic oscillator therefore are $n + 1/2$ for $n \in \mathbb{N}$.

7.13.4 Quantum Fields

In this subsection we use the accepted parlance of many body quantum physics. Instead of raising the quantum number n we speak of creating a particle, and lowering n is associated with particle annihilation, or destruction. The mathematics behind it is the same.

We mimic the unlimited three-dimensional space by the Cartesian product of three circles of circumference L. In order to keep translation invariance we impose periodic boundary conditions. Functions have to obey

$$f(x_1, x_2, x_3) = f(x_1 + L, x_2, x_3) \tag{7.225}$$

and the same for the second and third dimension. Plane waves,

$$e_{\boldsymbol{k}}(\boldsymbol{x}) = \frac{1}{\sqrt{L^3}}\,\mathrm{e}^{\,\mathrm{i}\boldsymbol{k}\cdot\boldsymbol{x}} \tag{7.226}$$

form a complete orthonormal set where $\boldsymbol{k} = 2\pi\boldsymbol{j}/L$ with $\boldsymbol{j} \in \mathbb{Z}^3$. Orthonormal here means

$$\int \mathrm{d}^3x\; e_{\boldsymbol{q}}{}^*(\boldsymbol{x})\, e_{\boldsymbol{k}}(\boldsymbol{x}) = \delta^3_{\boldsymbol{q},\boldsymbol{k}}. \tag{7.227}$$

The Kronecker delta vanishes unless $\boldsymbol{q} = \boldsymbol{k}$; it then evaluates to 1. Note that the integral in the above expression is over a cube of edge length L with volume $V = L^3$. The sum over all allowed wave vectors gives

$$\sum_{\boldsymbol{k}} e_{\boldsymbol{k}}(\boldsymbol{y})^*\, e_{\boldsymbol{k}}(\boldsymbol{x}) = \delta^3(\boldsymbol{y} - \boldsymbol{x}). \tag{7.228}$$

We now associate with each plane wave, or mode, a ladder operator $A_{\boldsymbol{k}}$. $A_{\boldsymbol{k}}{}^\star$ creates a particle with wave vector $\boldsymbol{k}$ and $A_{\boldsymbol{k}}$ annihilates it. This is so because of the commutation relations

$$[A_{\boldsymbol{q}}, A_{\boldsymbol{k}}] = [A_{\boldsymbol{q}}{}^\star, A_{\boldsymbol{k}}{}^\star] = 0 \tag{7.229}$$

and

$$[A_{\boldsymbol{q}}, A_{\boldsymbol{k}}{}^\star] = \delta_{\boldsymbol{q},\boldsymbol{k}}\, I. \tag{7.230}$$

The number of particles with the spatial properties of e_k is

$$N_k = A_k{}^\star A_k. \tag{7.231}$$

The state where there are n_k particles with wave vector $\boldsymbol{k}$ is described by

$$\phi = \prod_k \frac{\left(A_k{}^\star\right)^{n_k}}{\sqrt{n_k!}}\,\Omega. \tag{7.232}$$

It obeys

$$N_k\phi = n_k\phi. \tag{7.233}$$

The vacuum state Ω is among them. There are no particles whatsoever which characterizes it.

States of type (7.233) are normalized and mutually orthogonal. They span a linear space, the Fock space. Its closure is the Hilbert space under discussion.

We finally can define the quantum field

$$\Phi(\boldsymbol{x}) = \sum_k e_k(\boldsymbol{x}) A_k. \tag{7.234}$$

It is a field because it depends on location $\boldsymbol{x}$. It is a quantum field since its value is an operator, not a real or complex number as is the case for classical fields.

We find the following commutation relations:

$$[\Phi(\boldsymbol{x}), \Phi(\boldsymbol{y})] = [\Phi^\star(\boldsymbol{x}), \Phi^\star(\boldsymbol{y})] = 0 \tag{7.235}$$

and

$$[\Phi(\boldsymbol{x}), \Phi^\star(\boldsymbol{y})] = \delta^3(\boldsymbol{y}-\boldsymbol{x}). \tag{7.236}$$

The quantum field is local. Values at different locations are uncorrelated.

The number of particles N turns out to be

$$N = \sum_k N_k = \int \mathrm{d}^3x\, \Phi^\star(\boldsymbol{x})\, \Phi(\boldsymbol{x}). \tag{7.237}$$

This is an operator! The particle number is not fixed, but variable.

We here have to stop the introduction into quantum field theory. More has been said in this book in the sections on quantum gases. We should continue with the Fock space, anti-commutators for fermions instead of commutator for bosons, with defining observables in terms of fields and with time evolution. All this can be formulated in such a way that quantum field theory is compatible with relativity.

7.14 Transformation Groups

The mathematical concept of a group plays an important rule in quantum physics. Concepts such as 'left is as good as right', or 'space does not have a preferred direction' express symmetries which can be combined to form groups.

7.14.1 Group

Mathematically speaking, a group is a set G of elements equipped with a mapping $G \times G \rightarrow G$. For each pair of elements $e_1, e_2 \in G$ there is an element $e \in G$ which we denote by $e = e_2 \cdot e_1$. The mapping shall have the following properties:

$$e_3 \cdot (e_2 \cdot e_1) = (e_3 \cdot e_2) \cdot e_1. \tag{7.238}$$

The combination of group elements is associative. Moreover, there is a unit element $I \in G$ such that

$$e \cdot I = e \tag{7.239}$$

holds true for all $e \in G$. Note $I \cdot I = I$. It is also required that for each $e \in G$ there is an inverse e^{-1} such that

$$e \cdot e^{-1} = I. \tag{7.240}$$

It follows, by multiplying (7.240) from the right by e, that

$$I \cdot e = e. \tag{7.241}$$

We conclude that the right unity I, as postulated by (7.239), is as well a left unity, as stated by (7.241). The unit element is unique since $Ie = I'e = e$, by multiplying from the right with e^{-1}, implies $I = I'$.

Although the group axioms are so simple, a surprisingly large field of applications is covered. In the following we present a few examples.

The most simple group consists of one element only which one may denote as I. The composition law says $I \cdot I = I$. Quite formally this defines a group, although a boring one.

7.14.2 Finite Groups

The number of group elements may be finite or infinite. In the latter case, the group elements are usually indexed by real parameters.

7.14.2.1 A Group with Only Two Elements

Let us define space inversions P by $(Pf)(\boldsymbol{x}) = f(-\boldsymbol{x})$ where f is a function of three real arguments. Obviously, $P \cdot P = I$. I here stands for 'do nothing with the function'. The set $\{I, P\}$ with this multiplication law forms a group. The composition law is that of transformations—first apply A, then B, this makes up $C = B \cdot A$—which automatically fulfills the requirement of associativity. Each group element has an inverse, i.e. $P^{-1} = P$. Quantum physics refers to this simple group again and again, let it be space inversion P, time reversal T, charge conjugation C, and many more operations.

7.14.2.2 Permutations

Consider an arrangement of N different objects. We label them by an index $i = 1, 2, \dots, N$. For $N = 4$, $(1, 2, 4, 3)$ is an example, $(2, 1, 4, 3)$ another one. A rearrangement of the objects is called a permutation P, such as $P(1, 2, 4, 3) = (2, 1, 4, 3)$. Permuting first by P' and then by P'' defines $P = P'' \cdot P'$. Again, the composition law is that of transformations, hence associativity is guaranteed. 'Do not permute the objects' defines the unit element I, and every permutations can be undone. It follows that the set of all permutations of N objects is a group, the symmetric groups S_N.

7.14.2.3 More Complex Finite Group

There are 32 crystal symmetry groups, all of them finite. We just present one example. The 3 m symmetry group 3 m has a three-fold rotation axis and a reflexion plane through this axis, but no inversion center. See Fig. 7.3 for a sketch.

It consists of a rotation R by 120°, a reflexion Π, and products thereof. See Table 7.2 for the group structure.

Not all elements of the group commute, it is not Abelson. For instance, $\Pi R \hat{\boldsymbol{u}} = -\hat{\boldsymbol{w}}$ while $R \Pi \hat{\boldsymbol{u}} = -\hat{\boldsymbol{v}}$.

7.14.3 Topological Groups

There are groups with non-countably many members. The group elements are labeled by real or complex numbers with the standard topology of neighborhood. Such groups therefore are called topological. The index set can be compact[4] or not, giving rise to compact or non-compact topological groups. We shall discuss three examples which are intimately related to time and space.

---

[4]In the sense of topology.

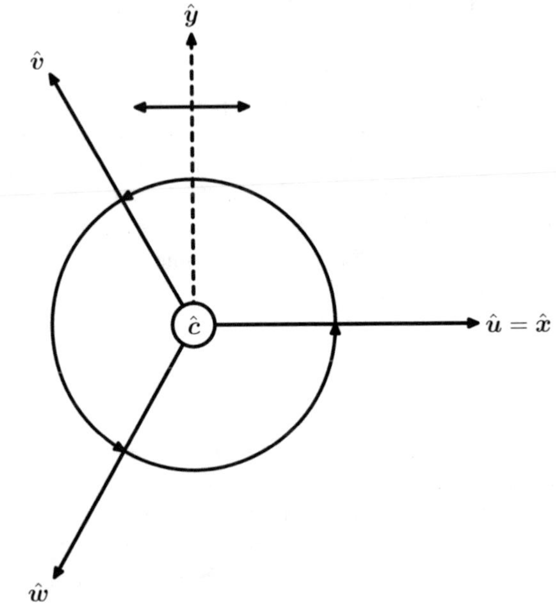

Fig. 7.3 3-m Symmetry. The plane orthogonal to the $\hat{z} = \hat{c}$ axis. $\hat{x}$ and $\hat{y}$ as well as $\hat{u}$, $\hat{v}$, $\hat{w}$ are unit vectors. Mirroring with respect to the $\hat{y}$, $\hat{c}$ plane (or $\hat{x} \to -\hat{x}$) as well as $\hat{u} \to \hat{v} \to \hat{w} \to \hat{u}$ are symmetry operations. These transformations define the 3 m space point group.

Table 7.2 Multiplication table of the 3 m symmetry group. The group is made up of the mirror transformation Π and a rotation R by 120°, and products thereof. Row B, column A contains the product $B \cdot A$

$B \cdot A$	I	Π	R	R^{-1}	ΠR	$R\Pi$
I	I	Π	R	R^{-1}	ΠR	$R\Pi$
Π	Π	I	ΠR	$R\Pi$	R	R^{-1}
R	R	$R\Pi$	R^{-1}	I	Π	ΠR
R^{-1}	R^{-1}	ΠR	I	R	$R\Pi$	Π
ΠR	ΠR	R^{-1}	$R\Pi$	Π	I	R
$R\Pi$	$R\Pi$	R	Π	ΠR	R^{-1}	I

7.14.3.1 Translations

Consider a smooth function $f = f(t)$ where $f \in \mathbb{C}$ and $t \in \mathbb{R}$. This function can be translated according to

$$(T_\tau f)(t) = f(t + \tau), \tag{7.242}$$

for real τ. Obviously,

$$(T_{\tau''} T_{\tau'}) f = T_{\tau'' + \tau'} f, \tag{7.243}$$

the T_τ form an Abelian group. It is generated by the Lie algebra consisting of I and H:

$$f(t+\tau) = f(t) + \frac{\tau}{1!} f'(t) + \frac{\tau^2}{2!} f''(t) + \cdots = \mathrm{e}^{-\mathrm{i}\tau H} f(t). \tag{7.244}$$

H is to be identified with

$$H = \mathrm{i}\frac{\mathrm{d}}{\mathrm{d}t}. \tag{7.245}$$

This one-parameter topological group is non-compacts since $\mathbb{R}$ is not. The minus sign is a convention. The generator H is identified with the energy observable.

Spatial translations in the 1, 2, and 3-direction form a non-compact three-parameter topological Abelian group. The generators

$$P_j = -\mathrm{i}\nabla_j \tag{7.246}$$

are called linear momenta. Again, the sign is a convention. We have covered translations in a section of its own, in particular how to define the linear momentum so that it becomes self-adjoint.

7.14.3.2 Rotations

Consider smooth functions $f : \mathbb{R}^3 \to \mathbb{C}$. Let R with $RR^\dagger = I$ be an orthogonal Matrix. Since $|\, R\boldsymbol{x} \,| = |\, \boldsymbol{x} \,|$, the matrix R induces a rotation

$$f_R(\boldsymbol{x}) = f(R\boldsymbol{x}). \tag{7.247}$$

Such rotations $f \to f_R$ form a group.

Consider

$$R = \begin{pmatrix} \cos\alpha & -\sin\alpha & 0 \\ \sin\alpha & \cos\alpha & 0 \\ 0 & 0 & 1 \end{pmatrix}. \tag{7.248}$$

It describes a rotation around the 3-axis by an angle α. For a small angle $\mathrm{d}\alpha$ we may write

$$f_R(\boldsymbol{x}) = f(x_1 - \mathrm{d}\alpha\, x_2, \mathrm{d}\alpha\, x_1 + x_2, x_3) = f(\boldsymbol{x}) - \mathrm{i}\mathrm{d}\alpha\, J_3 f(\boldsymbol{x}) \tag{7.249}$$

where

$$J_3 = -\mathrm{i}\{x_1\nabla_2 - x_2\nabla_1\}. \tag{7.250}$$

Therefore, an arbitrary rotation around the 3-axis is given by

$$f(R\boldsymbol{x}) = f_R(\boldsymbol{x}) = \mathrm{e}^{-\mathrm{i}\alpha J_3} f(\boldsymbol{x}). \tag{7.251}$$

Rotations around the 1- and 2-axes are defined likewise. The most general rotation R results in

$$f_R(\boldsymbol{x}) = \mathrm{e}^{-\mathrm{i}\boldsymbol{\alpha}\cdot\boldsymbol{J}} f(\boldsymbol{x}), \tag{7.252}$$

where $\boldsymbol{\alpha} = \alpha\hat{\boldsymbol{n}}$. This is a rotation around an axis $\hat{\boldsymbol{n}}$ by an angle α. The rotations $R : \mathbb{R}^3 \to \mathbb{R}^3$, in the vicinity of I, are $R = I - \mathrm{i}\boldsymbol{\alpha}\cdot\boldsymbol{J} + \cdots$ Their commutation rules read $[J_1, J_2] = \mathrm{i}J_3$ and cyclic permutations thereof, or

$$[J_i, J_j] = \mathrm{i}\epsilon_{ijk} J_k. \tag{7.253}$$

Rotations of three-dimensional space $\mathbb{R}^3$ form a non-Abelian compact topological group. Its three generators J_i are angular momenta.

7.14.3.3 Special Group E_2

The energy-momentum four-vector of a particle can be transformed to rest, $(mc, 0, 0, 0)$, where $m > 0$ is the mass. Transformations which do not change this standard vector form the little group, rotations in this case.

Massless particles behave differently. There is no frame of reference where they are at rest. A possible standard vector describes that they move with momentum=1, say, in 3-direction, $(1, 0, 0, 1)$. The corresponding little group consists of translations in the 2, 2-plane and rotations around the 3-axis. This is the symmetry group of the Euclidean two-dimensional plane which is denoted as E_2. In contrast to the rotation group its parameter space is non-compact. Consequently, the unitary irreducible representations are either one- or infinite-dimensional.

7.14.4 Angular Momentum

We focus on the rotation group. As mentioned above it is generated by three self-adjoint operators J_1, J_2 and J_3 commuting as

$$[J_1, J_2] = \mathrm{i}J_3 \tag{7.254}$$

and cyclic permutations thereof. Not all three components can be diagonalized simultaneously, i.e. by one and the same decomposition of unity.

However, there is another operator, namely $\boldsymbol{J}^2$, which commutes with all three angular momentum components,

$$[\boldsymbol{J}^2, J_k] = 0. \tag{7.255}$$

Hence, $\boldsymbol{J}^2$ and J_3, say, may be diagonalized simultaneously.

7.14.4.1 Eigenspaces

Let λ be an eigenvalue of the self-adjoint operator $\boldsymbol{J}^2$ and $\mathcal{D}$ the corresponding eigenspace. Any $f \in \mathcal{D}$ satisfies

$$\boldsymbol{J}^2 f = \lambda f. \tag{7.256}$$

We shall see later which values λ are allowed. The angular momentum operators map $\mathcal{D}$ into itself, i.e. $L_k\mathcal{D} \subset \mathcal{D}$, because of $\boldsymbol{J}^2 J_k f = J_k \boldsymbol{J}^2 f = \lambda J_k f$.

Our goal is to decompose $\mathcal{D}$ into eigenspaces of J_3. For this we define ladder operators

$$J_+ = J_1 + \mathrm{i}J_2 \text{ and } J_- = J_1 - \mathrm{i}J_2. \tag{7.257}$$

Note that one is the adjoint of the other. The following commutation rules hold true:

$$\begin{aligned} [J_3, J_+] &= J_+ \\ [J_3, J_-] &= -J_- \\ [J_+, J_-] &= 2J_3. \end{aligned} \tag{7.258}$$

The operators J_+ and J_- map $\mathcal{D}$ into itself as well.

J_3, restricted to $\mathcal{D}$, is a self-adjoint operator and therefore has eigenvectors in $\mathcal{D}$. Let $f \in \mathcal{D}$ be such a normalized eigenvector with eigenvalue μ. Because of

$$J_3 J_+ f = (J_+ J_3 + J_+) f = (\mu + 1) J_+ f \tag{7.259}$$

and

$$J_3 J_- f = (J_- J_3 - J_-) f = (\mu - 1) J_- f \tag{7.260}$$

one has found two new eigenvectors. The eigenvalues have increase or decreased by 1, respectively. With J_+ one may step up the angular momentum ladder, with J_- down. Because of

$$\lambda = (f, \boldsymbol{J}^2 f) \geq (f, J_3^2 f) = \mu^2, \tag{7.261}$$

stepping up or down will not work ad infinitum.

There is a maximal eigenvalue j of J_3 in $\mathcal{D}$ with eigenvector f_j. It obeys

$$J_3 f_j = j f_3 \text{ and } J_+ f_j = 0. \tag{7.262}$$

We calculate

$$\boldsymbol{J}^2 = J_- J_+ + J_3(J_3 + I) = J_+ J_- + J_3(J_3 - I) \tag{7.263}$$

which, if applied to f_j, gives

$$\lambda = j(j+1) \tag{7.264}$$

We step down from f_j by J_- repeatedly and will arrive at a vector f_k with the smalles J_3 eigenvalue. By applying the second expression of (7.263) to f_k one obtains

$$\lambda = k(k-1). \tag{7.265}$$

Because of $j \geq k$ we conclude $j \geq 0$ and $k = -j$. The difference $j - k$ is the number of down-steps, a natural number. We have found out that $j = 0, 1/2, 1, 3/2 \ldots$ Let us summarize:

- The eigenspaces of $\boldsymbol{J}^2$ have the dimension $d = (2j+1) \in \mathbb{N}$, i.e. $d = 1, 2, 3, \ldots$ The angular momentum quantum number j assumes integer or half-integer values.
- In a $(2j+1)$-dimensional eigenspace the component J_3 assumes eigenvalues $m = -j, -j+1, \ldots, j-1, j$. The eigenvalue m of J_3 is sometimes called the magnetic quantum number.
- The common eigenvectors of $\boldsymbol{J}^2$ and J_3 are characterized by

$$\boldsymbol{J}^2 f_{j,m} = j(j+1) f_{j,m} \text{ and } J_3 f_{j,m} = m f_{j,m}. \tag{7.266}$$

- In addition we know

$$J_+ f_{j,j} = 0 \text{ and } J_- f_{j,-j} = 0. \tag{7.267}$$

One should continue by calculating the spherical harmonics $Y_{l,m}$ and work out the Laplacian as an expansion into spherical harmonics. The rules for combining angular momenta is likewise of interest. We have studied these topics in the text if need be.

Glossary

The entries in this list of key words and persons are ordered alphabetically. ▹ refers to another glossary entry. Persons are briefly described by their origin, field of activity, and life span.

A

Abel Niels Henrik: Norwegian mathematician, 1802–1829

Abelian group A ▹ *Group* the elements of which commute.

Adjoint linear operator Every ▹ *Linear operator* A is associated with another linear operator $A^\star$ such that $(g, Af) = (A^\star g, f)$ holds true for all vectors f, g. This is strictly true for bounded operators only. Otherwise the respective domains of definition must be taken into account.

Alignment of dipoles The dipole moments of atoms or molecules are partially aligned by external fields. The degree of alignment depends on the temperature: the colder, the better aligned. The relationship differs for electric and magnetic dipole moments because the latter are quantized, the former not.

Ammonia molecule The ammonia molecule consists of a unilateral triangle of hydrogen ions and a nitrogen ion localized on the central axis, either above or below. Both configurations have the same energy, and there is a transition amplitude. As a consequence, the ground state splits into a symmetric and an antisymmetric combination of the two equivalent configurations. The transition energy $2\pi\hbar f$ between the symmetric and the antisymmetric combination defines the microwave standard of $\lambda = 1.255932$ cm. Both energy levels shift in an external electric field which allows to select excited molecules. A weak signal of the right frequency triggers a phase-synchronized output signal, ▹ *Maser*.

P. Hertel, *Quantum Theory and Statistical Thermodynamics*,
Graduate Texts in Physics, DOI 10.1007/978-3-319-58595-6

Angular momentum Rotations of a system are generated by the three self-adjoint operators $\boldsymbol{J}$ of the angular momentum. Since rotations do not commute in general, the angular momentum components also do not commute. Instead, they obey $[J_1, J_2] = \mathrm{i}\hbar J_3$ (and cyclic permutations thereof). The square $\boldsymbol{J}^2$ commutes with the three operators J_k, hence $\boldsymbol{J}^2$ and J_3, say, may be diagonalized simultaneously. The eigenvalues of $\boldsymbol{J}^2$ are $j(j+1)\hbar^2$, where $j = 0, 1/2, 1, \ldots$ For given j the eigenvalues of J_3 are $m\hbar$ where $m = -j, -j+1, \ldots, j-1, j$. These properties are consequences of the commutation relations only.

Annihilation operator The counterpart of a ▷ *Creation operator*.

Atomic units A system of units tailored to the needs of atomic and molecular physics by setting $\hbar = e = m_\mathrm{e} = 4\pi\epsilon_0 = 1$. Energies then are in units of $E_* = 27.2$ eV (Hartree), lengths in units of Bohr's radius $a_* = 52.9$ pm.

Avogadro Amadeo: Italian physicist, 1776–1856.

Avogadro's number One mole of ^{12}C contains N_A particles. More generally, one mole is the amount of substance, measured in gram, as indicated by its molecular weight (here 12). The value of Avogadro's number is $N_\mathrm{A} = 6.0221409 \times 10^{23}$.

B

Balance equation An additive, transportable quantity Y is characterized by its ▷ *Density* $\varrho(Y)$, its ▷ *Current density* $\boldsymbol{j}(Y)$ and by the volumetric ▷ *Production rate* $\pi(Y)$; all three are classical fields. The balance equation says that the amount of Y in a unit volume increases because there is a net inflow of Y or because Y is produced: $\partial_t \varrho(Y) + \nabla \cdot \boldsymbol{j}(Y) = \pi(Y)$. There are balance equations for particles of a certain species, for mass and electric charge, for linear and angular momentum and for internal energy and entropy. The latter two are precise statements of the ▷ *First main law* and the ▷ *Second main law*.

Basov Nicolay G.: Soviet physicist, 1922–2001, Nobel prize 1964 together with Prokhorov and Townes for 'fundamental work in the field of quantum electronics, which has led to the construction of oscillators and amplifiers based on the maser–laser principle'.

Black body radiation ▷ *photon gas*

Black hole A singularity in space-time. The gravitational field bends geodesic lines to such an extent that not even light may escape. Black holes suck in surrounding matter which emits X rays when atoms are smashed. A black hole is a center of strong gravitational attraction which may force a star on a Kepler orbit. This effect allows to determine the mass of a black hole, for instance 4.2 million sun masses for the suspected black hole near the center of the Milky Way. If a large star has exhausted its nuclear fuel, it will collapse either to become a ▷ *White dwarf*, or a ▷ *Neutron star*, or a black hole.

Boltzmann Ludwig: Austrian physicist, 1844–1906.

Boltzmann gas The limit of a quantum gas for low chemical potential. In this limit there is no difference between fermions and bosons, the particles should just be identical. The Boltzmann gas is well described by ▹ *Classical statistical mechanics* even if its particles interact.

Born Max: German physicist, 1882–1954

Born approximation An approximation method to solve the scattering problem. The Schrödinger equation when tailored to a scattering experiment (▹ *Scattering Schrödinger equation*) says that the solution ψ is a sum of an incoming plane wave and a scattered wave. The latter depends on the not yet know solution ψ. Born suggested the following approximation: in lowest order, replace ψ by an incoming plane wave. The depletion of the incoming wave due to scattering is neglected. This procedure might be repeated, and one obtains a power series in the interaction potential. Inspecting the solution in the far zone provides the ▹ *Scattering amplitude*. Essentially, it is the Fourier transform of the interaction potential.

Born-Oppenheimer approximation An approximation method for calculating the properties of molecules and crystals. The lightweight electrons will attain their best configuration practically instantaneously, the corresponding minimal energy defines a potential for the ions. When resorting to the ▹ *Raleigh-Ritz principle*, the infimum over all wave functions is approximated by products where one factor depends on the ionic, the other on the electronic configuration. Thereby one has to take into account the proper symmetry behavior when interchanging identical particles.

Bose Satyendra Nath Indian physicist, 1894–1974

Bose-Einstein condensation A Bose gas may accommodate particles up to a certain temperature-dependent density only. Particles in excess condensate coherently in the ground state. This phenomenon is responsible for the supra-fluidity of Helium at very low temperatures and the super-conductivity of certain materials below a critical temperature. The bosons in question are strongly correlated pairs of quasi-electrons.

Boson Particle with integer spin including photons and phonons. Their creation-annihilation operators obey commutation relations, therefore the eigenvalues of the particle number observable are $n = 0, 1, 2, \dots$

Brioullin Léon N.: French physicist 1889–1969

Brioullin zone In a periodic and infinite crystal the allowed wave vectors are restricted to and homogeneously distributed in a finite region, the Brioullin zone. This is a polyhedron in wave vector space respecting the crystal's symmetry.

de Broglie Raymond French physicist 1892–1987, Nobel prize 1929 for 'his discovery of the wave nature of electrons'.

Brownian motion ▷ *Fluctuating* forces on a small, but still visible particle cause it to walk at random. Its motion is passive, driven by the suspending liquid, and not a manifestation of life. The square of the displacement grows linearly with time, the constant of proportionality is the diffusion constant of the Brownian particle. ▷ *Diffusion* is nothing else but Brownian motion en masse.

Bunyakovsky Viktor Yakovlevich Russian mathematician, 1804–1889

C

C* **algebra** An algebra with a star operation. Its elements may be multiplied by scalars, usually complex numbers. They may be added and multiplied. In general, the multiplication is non-commutative. With respect to multiplication, there is a neutral element I. To each element A of the C* algebra there is a starred element $A^\star$ obeying the rules of adjoining. There is a norm such that $\| A \|^2$ is the largest value λ for which $AA^\star - \lambda I$ is not invertible. The algebra is closed with respect to this norm. The observables of both the classical and the quantum framework form a C* algebra, the former commutative, the latter not.

Callen-Welton theorem The imaginary part $\chi''(\omega)$ of a susceptibility is an integral over all frequencies u where the integrand contains the factor $S(u)$, the spectral intensity. Since it is never negative, the dissipative contribution χ'' is never negative as well. The Callen-Welton theorem is a consequence of the ▷ *Wiener–Khinchin theorem* (relating ▷ *Correlations* with ▷ *Fluctuations*) and the ▷ *Kubo-Martin-Schwinger formula* characterizing the Gibbs state.

Canonical commutation rules The three components X_j of location and of linear momentum P_k obey the canonical commutation rules $[X_j, P_k] = \mathrm{i}\hbar\delta_{jk}I$. I is the unity operator. The canonical commutation relations reflect the structure of three-dimensional Euclidean space.

Canonical relation Generalized coordinates and momenta parametrize the ▷ *Phase space* of classical mechanics. There is a set of observables equipped with a structure as described by ▷ *Poisson* brackets. The relations between generalized coordinates and momenta are 'canonical' in the sense of 'setting the rules'.

Cauchy Augustin Louis French mathematician, 1789–1857

Cauchy–Schwarz-Bunyakovsky inequality ▷ *Cauchy–Schwarz inequality* and ▷ *Bunyakovsky*

Causal function A function $t \to f(t)$ is causal if it vanishes for negative times t. There is an intimate relation between the real part and the imaginary part of its Fourier transform as explained in ▷ *Dispersion relations*.

Center of mass frame A scattering experiment must be described with respect to a frame of reference. This can be chosen such that the momentum vectors of the incoming particles add up to zero. In this case, the center of mass is at rest. The center of mass frame is to be preferred for theoretical considerations. The ▷ *Laboratory frame*—the scatterer is initially at rest—is better suited for describing the actual experiment. A simple Galilei or Lorentz transformation allows to relate laboratory with center of mass data.

Chemical affinity A chemical reaction of type r produces ν^{ra} particles of species a; ▷ *Stoichiometric coefficients*. With μ_a as the chemical potential for a-particles, the affinity for one reaction of type r is $A_r = \sum_a \nu^{ra} \mu_a$. The chemical affinity may be thought of as the driving force for the reaction. $A_r = 0$ indicates equilibrium.

Chemical potential The chemical potential μ_a of identical particles of species a is a ▷ *Lagrange* parameter, just like temperature. It takes care of the auxiliary condition that the number N_a of a-particles (an observable) has a prescribed expectation value $\bar{N}_a$, when looking for the entropy maximum. If particles can be exchanged between the system and its environment, the chemical potentials of the system and the environment are the equal when in equilibrium. ▷ *Chemical affinity*.

Chemical reaction In a chemical reaction, some particles vanish (because they react) and others appear (reaction products). A reaction of type r is characterized by integer numbers ν^{ra} without common divisor, for a fixed index r. The ▷ *Chemical affinity* A_r determines whether there will be a forward ($A_r > 0$), a backward ($A_r < 0$) or no reaction at all ($A_r = 0$).

Classical hydrogen atom The classical hydrogen atom is a model where a massive positive charge attracts a light-weight negative charge by the Coulomb force. Moreover, the system emits radiation because the small mass is accelerated. The power loss gets so strong that the system would crash after a finite time. According to classical mechanics there are no stable atoms.

Classical limit for power loss An atom or a molecule in an excited state spontaneously decays into a less excited state by emitting a photon of fitting energy $\hbar\omega$. One may calculate the lifetime of the excited state and the energy loss per unit time. The latter expression coincides with its counterpart from classical electrodynamics if the expectation value $\omega^4(f, \boldsymbol{d}^2 f)$ is replaced by $q^2 \ddot{\boldsymbol{x}}^2$. Here $\boldsymbol{d} = q\boldsymbol{x}$ is the dipole moment, f the wave function of the system.

Cluster expansion The expression for the free energy of interacting particles may be evaluated by grouping terms into k-clusters with k particles in interaction range. The contributions of two particles in interaction range give rise to the second virial coefficient $b_2(T)$. ▷ *Virial coefficients*.

Cold solid The heat capacity of a solid has two contributions. One comes from the phonon gas which describes crystal lattice vibrations (▷ *Debye model*). For temperatures well below the ▷ *Debye temperature* Θ_D, only acoustic phonons get excited and store heat proportional to T^3. The second contributions is provided by the Fermi gas of quasi-free electrons. Its low temperature heat capacity is proportional to the density of states at the Fermi level; it grows linearly with the temperature. The heat capacity of a cold solid material therefore is $C = aT + bT^3$. Only at temperatures below one Kelvin or so the electronic contribution will dominate. Note that electric insulators have no linear term because the density of states vanishes at the Fermi level.

Conduction In continuum physics the current density $\boldsymbol{j}(Y)$ of a certain quantity Y may be split into a convection and a conduction contribution. The former takes into account that a streaming material carries the said property with it, $\varrho(Y)\,\boldsymbol{v}$, where $\boldsymbol{v}$ is the streaming velocity. The rest is transported by conduction, i.e. by interactions without anything streaming. We therefore write $\boldsymbol{j}(Y) = \varrho(Y)\,\boldsymbol{v} + \boldsymbol{J}(Y)$. Only the conduction current density $\boldsymbol{J}(Y)$ transforms correctly as a vector field.

Continuum physics A particular view on physics centered around the ▹ *Material point*. A material point is a tiny region with nanometer dimensions, say. It can be characterized by its location $\boldsymbol{x}$ and a few other properties, but it has no internal structure. Hence a point. This is the view on matter of an engineer. For the physicist the material point is a large system. It contains so many particles that fluctuations may safely be neglected. More, it is a good approximation that the material point be always in equilibrium with its environment. The properties of a system are described by classical fields $f = f(t, \boldsymbol{x})$ where the field value $f(t, \boldsymbol{x})$ pertains to the material point at $\boldsymbol{x}$ at time t. The equations of continuum physics are of two types. Balance equations are valid for all systems irrespective of the kind of matter we are dealing with. These equations must be augmented by material equations which describe special classes of matter or particular materials like gases, liquids, solids, electric conductors, dielectrics, and so on. Applications reach from aerodynamics to optics, from acoustic to stellar matter, from process engineering to sustainable energy production. We have included in this book some topics because material equations, if not purely phenomenological, must be calculated in the framework of quantum physics.

Correlation function Let M be an observable and $\Delta M = M - \langle\, M \,\rangle$ its fluctuation. The expectation value is calculated with respect to a stationary state, i.e. a state which commutes with the Hamiltonian. In the ▹ *Heisenberg picture* observables change with time according to the ▹ *Heisenberg equation*. The correlation function $K = K(s, t)$ for the observable M is the expectation value of the symmetrized product $(\Delta M_s \Delta M_t + \Delta M_t \Delta M_s)/2$. Since the state under discussion is stationary, the correlation function depends on the time difference $\tau = s - t$ only, i.e. $K = K(\tau)$. The ▹ *Wiener–Khinchin theorem* says that the correlation function is the Fourier transform of a positive spectral intensity.

Coulomb scattering Scattering of two point-like charged particles interacting via the Coulomb potential $\propto 1/r$. The differential cross section can be worked out analytically and in Born approximation; both expressions coincide. The integrated differential cross section is infinite because the Coulomb potential has an infinite range.

Courant Richard German mathematician, 1888–1972

Creation operator Counterpart of an annihilation operator; ▹ *Ladder operators*.

Cross section A beam of incoming particles hits a target of randomly positioned scattering centers and suffers scattering. The flux Φ of the incident beam decreases exponentially with depth x as $\Phi(x) = \Phi(0)\exp(-n\,\sigma\,x)$ where n is the density of scattering centers and σ the cross section, an area.

Curie–Weiss law The electric susceptibility of a dielectric substance falls off with temperature like $1/T$, so the Curie law. The same applies to a paramagnetic material. A ferromagnet is characterized by its Curie temperature Θ_C. Above this temperature the material is paramagnetic and falls off like $1/(T - \Theta_C)$, the Curie–Weiss law. Below Θ_C the material behaves as a ferromagnet. If the external magnetic induction is slowly switched off there remains a finite spontaneous magnetization.

Current density The density of an additive, transportable quantity Y, denoted by $\varrho(Y)$, may change for two reason. On the one hand, Y can be redistributed, it flows. On the other hand, Y may be produced or it may vanish. The second mechanism is described by the volumetric (per unit volume) production rate $\pi(Y)$. Redistribution, or flow, is characterized by the direction of flow and by the amount of Y which passes through an area element $\mathrm{d}\boldsymbol{A}$. This amount can be written as $\mathrm{d}\boldsymbol{A} \cdot \boldsymbol{j}(Y)$ and defines the current density. ▷ *Conduction* for splitting the current density into the conduction term $\boldsymbol{J}(Y)$ and a convective part $\varrho(Y)\boldsymbol{v}$. Here $\boldsymbol{v}$ is the streaming velocity.

D

Debye Peter: Dutch physicist, 1884–1966, Nobel prize in chemistry 1936 for '…investigations on dipole moments and the diffraction of X-rays and electrons in gases'

Debye model Approximation for the heat capacity of crystal lattice vibrations. The density of states for the phonon gas is replaced by its low frequency limit (acoustic waves). It is cut off at $\hbar\omega = k_B\Theta_D$ such that the number of degrees of freedom of lattice vibrations is met. Hence, the low frequency behavior and the integral of the density of states are correct. This explains the behavior of a ▷ *Cold solid* and the ▷ *Dulong–Petit law*. The Debye temperature Θ_D of ice has a value of 192 K. For a refinement ▷ *Debye–Einstein model.*

Debye–Einstein model Improvement of the ▷ *Debye model.* Only the acoustic branches of the dispersion relation are taken into account for defining the ▷ *Debye temperature*. Optical branches are approximated by constant average frequencies. Still the low frequency behavior and the integral of the density of states are correct which implies the T^3 law (▷ *Cold solid*) and the ▷ *Dulong–Petit law.*

Debye temperature ▷ *Debye model.*

Defect Imperfections in the otherwise completely regular crystal structure are unavoidable. They may be point-like or zero-dimensional, one or two dimensional. Point defects comprise isolated vacancies, interstitials and substitutions which must be classified further. Dislocations are line defects which may affect the mechanical properties substantially. The existence of a surface is a planar defect. A defect, because the environment of an ion is not symmetric. Grain boundaries are another form of defect. Avoiding defects for instance in silicon and inserting defects (doping) are of utmost technical importance.

Deflection angle In a scattering experiment an incoming particle is removed from the beam for various reasons. At low energies, when particle production is impossible, we encounter elastic scattering. The scatterer deflects the incoming particle by the deflection, or scattering angle. Its definition depends on the frame of reference, either the laboratory or the ▷ *Center of mass frame*.

Degeneracy pressure A fermi gas of non-interacting particles exerts a pressure even at zero temperature. In a solid this pressure is balanced by the Coulomb attraction of the ions, in a ▷ *White dwarf* or ▷ *Neuron star* by gravity. The dependency of degeneracy pressure p_0 on electron density n is $p_0 \propto n^{5/3}$ for non-relativistic electrons and $p_0 \propto n^{4/3}$ for ultra-relativistic fermions.

Delta function A generalized function, or distribution. Defined for continuous functions f by $\int \mathrm{d}x f(x)\delta(x-y) = f(y)$. Invented by ▷ *Dirac*.

Density fluctuation The particle number variance of an arbitrary open system is $\sigma^2(N) = \bar{N} k_\mathrm{B} T \kappa_T$, with $\bar{N} = \langle N \rangle$ and the isothermal compressibility κ_T. From this expression follows the standard deviation for density fluctuations.

Density of a quantity The content of an additive, transportable quantity Y in a region $\mathcal{V}$ at time t is a volume integral $\int_{x \in \mathcal{V}} \mathrm{d}V \, \varrho(Y; t, \boldsymbol{x})$ over the Y-density $\varrho(Y)$.

Density of states The number of one-particle states per unit volume and per unit energy interval.

Detailed balance An oscillating electric field $\mathcal{E}(t)$ causes ▷ *Forced transitions* between to states with energies E_1 and E_2. For the same field, the transition probabilities $W_{1\to 2}$ and $W_{2\to 1}$ are the same, according to the principle of detailed balance.

Dielectric A material with negligible electric conductivity. The absorptive part $\chi''(\omega)$ of the ▷ *Susceptibility* is very small as compared with the refractive part $\chi'(\omega)$. Because of the ▷ *Kramers–Kronig relation*, dielectric behavior is restricted to certain transparency windows.

Dielectric displacement In electrodynamics, the dielectric displacement field is defined by $\boldsymbol{D} = \epsilon_0 \boldsymbol{E} + \boldsymbol{P}$ where $\boldsymbol{P}$ is the polarization. The dielectric displacement is generated by the density of freely mobile charges.

Differential cross section ▷ *Deflection angle*. The distribution of elastically scattered particles over the solid angle $\mathrm{d}\Omega$. $\sigma_\mathrm{diff}(\theta) = \mathrm{d}\sigma/\mathrm{d}\Omega$ is the differential cross section, it depends on the deflection angle θ only. The expressions for the center of mass and the laboratory frame differ in general, but can be converted by standard formulas.

Diffusion Copious ▷ *Brownian motion*. Technically, the conduction contribution to the transport of a-particles as described by $\boldsymbol{J}_a = \boldsymbol{J}(N_a)$, the diffusion currents. They are related to non-constant particle densities by $\boldsymbol{J}_a = -\sum_b D_{ab} \nabla n_a$. The matrix of diffusion constants D_{ab} is almost always diagonal.

Diffusion constant ▷ *Diffusion*. If a particle suffers an average displacement s and if collisions occur after a time span τ on the average, the diffusion constant can be estimated by $D \approx s^2/2\tau$. Einstein has derived an expression for the diffusion constant in terms of particle radius R, viscosity η of the suspending liquid and the temperature, $D = k_B T / 6\pi\eta R$.

Dirac Paul English physicist 1902–1984, Nobel prize 1933 together with ▷ *Schrödinger*

Dirac equation A relativistically covariant Schrödinger-like equation for a four-component wave function. It describes a particle-antiparticle pair of spin = 1/2 fermions. It couples minimally to an external electromagnetic field, meaning that the four-momentum p_μ is replaced by $p_\mu - qA_\mu$ where q is the particle charge and A_μ the four-potential.

Dispersion relation Linear or linearized wave equations describe plane waves with wave vector $\boldsymbol{k}$ and angular frequency ω. There is a functional relationship $\omega = \omega(\boldsymbol{k})$. Whenever this relationship is not linear—other than for sound waves or light in vacuum—the phenomenon of dispersion is encountered. In an inhomogeneous medium, waves of different frequency (color) move along different trajectories. This is reflected by a frequency-dependent refractive part $\chi'(\omega)$ of the susceptibility. It is related to the absorptive part as described by the ▷ *Kramers-Kronig relation* a consequence of which is that χ' cannot be constant. The Kramers-Kronig relation reflects that the response of a system to perturbations is causal.

Dissipation When a system in thermodynamic equilibrium is perturbed for a finite time by a rapidly external field, one has to spend work. The field energy is dissipated, it is transformed into internal energy. This is a consequence of the dissipation-fluctuation theorem.

Dissipation-fluctuation theorem The absorptive part $\chi''(\omega)$ of the susceptibility is proportional to the spectral intensity $S(\omega)$ with a frequency dependent positive factor. The proof requires the ▷ *Wiener–Khinchin theorem* and the ▷ *Kubo-Martin-Schwinger formula*.

Dulong–Petit law The contribution of lattice vibrations to the heat capacity of a solid is $3k_B$ per lattice ion, for temperatures much larger the the ▷ *Debye temperature*. In other words, the molar heat capacity is $3R$ at high temperatures.

E

Effective mass The effective mass of a quasi-particle can be read off its dispersion relation $\omega = \omega(k)$ or $\epsilon = \epsilon(p)$, where ω, $\epsilon = \hbar\omega$, k and $p = \hbar k$ are angular frequency, energy, wave number and momentum, respectively. Acceleration and force are portional with a constant of proportionality $1/m^*(k)$, the inverse effective mass, which is the second derivative of the dispersion relation. Only for a quadratic relationship the effective mass is a constant, the mass. The effective mass of a photon vanishes.

Eigenspace For an ▷ *Eigenvalue* ν of a certain normal operator N the solutions f of $Nf = \nu f$ form a linear subspace. This subspace is characterized by its ▷ *Projector*.

Eigenvalue A ▹ *Normal operator* N has at least one eigenvalue ν which obeys $Nf = \nu f$, where the eigenvector f does not vanish. The ▹ *Eigenvalue* is real if the operator is self-adjoint, hence normal.

Einstein Albert German physicist, 1879–1955, Nobel prize 1921 'for his services to theoretical physics, and especially for his discovery of the law of the photoelectric effect'.

Einstein's summation convention If in an expression the same index $i = 1, 2, 3$ appears twice, summation over it is silently understood. Thus, $\boldsymbol{a} \cdot \boldsymbol{b}$ may be written as $a_i b_i$.

Elasticity module Appears in ▹ *Hooke's law*, a linear relation between stress and strain.

Electric charge A conserved quantity; its volumetric production rate vanishes. The corresponding balance equation, or continuity equation, can be extracted from Maxwell's equations. Electric charge comes in units of the elementary charge e. The electric charge serves as a coupling constant for the interaction between a charged particle and the electromagnetic field.

Electric dipole moment The electric potential ϕ of a charge distribution ϱ may be expanded as $\phi(r\boldsymbol{n}) = (1/4\pi\epsilon_0)\left\{Q/r + \boldsymbol{d} \cdot \boldsymbol{n}/r^2 + \cdots\right\}$. This expansion makes sense only if the total charge Q vanishes. The electric dipole moment is given by $\boldsymbol{d} = \int \mathrm{d}^3x \, \varrho^{\mathrm{e}}(\boldsymbol{x})\boldsymbol{x}$.

Electron diffraction Scattering of electrons on crystals is called diffraction. The beam of incoming particles produces a diffraction beam if ▹ *Laue's conditions* are met, because of maximal constructive interference. There is practically no intensity in random directions because the secondary waves interfere destructively. Electron or neutron diffraction results in the same interference pattern as X-ray diffraction.

Electronic band structure Assume that the primitive identical cells of a crystal were isolated. There would be sharp energy levels. Each one is N-fold degenerate if there are N such isolated cells. The crystal is built by letting the cells approach each other closer and closer. The formerly N-fold degenerate energy levels will split and form bands. The states within one energy band are labeled by wave vectors from the ▹ *Brillouin zone*. Each one-particle state is occupied at most twice (fermions, spin 1/2). At zero temperature, the electrons occupy the lowest levels up to the certain ▹ *Fermi energy*. In an insulator the conduction band is completely filled and the valence band above it is separated by a large band gap. A semiconductor is characterized by a small band gap. Some electrons may be thermally excited into the valence band thereby producing a small, temperature-dependent electric conductivity. In metals the valence band and the conduction band overlap and the Fermi energy lies in the overlapping region. Even at zero temperature there are mobile electrons.

Energy The generator of time translations. After having prepared a state, waiting for a time span t and thereafter measuring M amounts to measuring the observable M_t. The transformation from M to M_t is unitary because the possible values of M and M_t are the same. Since waiting for t_1 and then for t_2 is the same as waiting for $t_2 + t_1$, the waiting operators form a group and can be written as $U_t = \exp(-\mathrm{i}tH/\hbar)$. The minus sign and the $\hbar$ is a convention. H is a ▷ *Self-adjoint operator*, the energy observable. Its eigenvalues or expectation values are likewise called energy.

Enthalpy A ▷ *A thermodynamic potential*. With $H = F + ST$ one may write $\mathrm{d}H = -S\mathrm{d}T + V\mathrm{d}p$ with the usual meaning of the symbols. The enthalpy $H = H(T, p)$ of a certain amount of streaming gas remains constant.

Entropy States, as represented by ▷ *Probability operators* ϱ, are either pure or mixed. The degree of mixing is measured by the entropy $S(\varrho) = -k_{\mathrm{B}}\mathrm{Tr}\,\varrho \ln \rho$. The entropy vanishes for a pure state and is positive otherwise. Even if a system is well isolated—its internal energy $\langle H \rangle$ being kept constant—the unavoidable interactions with the environment will mix the system's state more and more so that its entropy increases, according to the ▷ *Second main law*. The entropy increases until it has reached its maximum. The corresponding Gibbs state G describes the thermal equilibrium between system and environment. Not only the functional $S = S(\varrho)$, but also the equilibrium value $S(G)$ goes by the name of entropy.

Expectation value The outcome m of a single measurement of the property M cannot be predicted with certainty. The measurement must be repeated many times under identical conditions, i.e. in the same state. We call such a procedure an experiment. The results of the first N repetitions allow to estimate the mean by $(m_1 + m_2 + \cdots + m_N)/N$ which converges with $N \to \infty$ towards $\bar{M}$. Thus an experiment is to be described as follows. The system is prepared in a state ϱ and one and the same observable M is measured repeatedly. The average or mean value then is $\bar{M} = \langle M \rangle = \mathrm{Tr}\,\varrho M$. The expectation value $\langle M \rangle$ can be calculated, $\bar{M}$ can be measured in an experiment. They should coincide. Recall that the outcome of a single measurement cannot be predicted in general.

Experiment A sequence of independent repetitions of a measurement. For each repetition, the system must be prepared in one and the same state ϱ, and one and the same observable M has to be measured. ▷ *Expectation value*. With M, any function $f(M)$ can be measured in the same experiment. The state ϱ is an abstraction of a certain preparation procedure, and M stands for the machinery to measure the property M.

External parameter An external parameter acts on the system, but the system's reaction on the parameter is either negligibly small or can be compensated by a smart controlling mechanism. For example, the external field of a plate capacitor may be kept constant by maintaining a certain voltage even if the material between the plates undergoes changes such as heating.

F

Fermi Enrico: Italian physicist, 1901–1954, Nobel prize 1938 for 'his demonstrations of the existence of new radioactive elements produced by neutron irradiation, and for his related discovery of nuclear reactions brought about by slow neutrons'

Fermi gas A gas of non-interacting ▷ *Fermions*, particularly electrons (▷ *White dwarf*), quasi-electrons (crystals) or neutrons (▷ *Neutron star*). Particles try to avoid each other, therefore there is a pressure even at zero temperature. This ▷ *Degeneracy pressure* counterbalances gravitation (white dwarf or neutron star) or electrostatic forces in a solid.

Fermi energy The ▷ *Chemical potential* of quasi-electrons at zero temperature. Only single particle states below ϵ_F are occupied at zero temperature.

Fermion Particle with half-integer spin including electrons, proton and neutrons. Their creation-annihilation operators obey anti-commutation relations, therefore the eigenvalues of the particle number observable are $n = 0$ or $n = 1$ only.

Feynman Richard US-American physicist, 1918–1988, Nobel price 1965 together with Tomonaga and Schwinger 'for their fundamental work in quantum electrodynamics'.

Feynman rules A few simple rules for combining probability amplitudes. The probability of a process is proportional to the absolute square of an amplitude. If the process runs via an intermediate stage, the respective amplitudes must be multiplied. If the process may run via alternative intermediate stages without leaving a trace, the amplitudes should be added. These rules, if cleverly applied, may explain many quantum phenomena. However, what is the precise meaning of 'alternative' or 'intermediate stage'? Feynman's rules cannot replace the mathematical framework of quantum physics.

First main law The internal energy $U = \text{Tr}\,\varrho H$ can change because the state or the Hamiltonian changes, $\text{d}U = \text{Tr}\,\text{d}\varrho\, H + \text{Tr}\,\varrho\,\text{d}H$. The first contribution is heat $\text{d}Q$, the second work $\text{d}W$. Note that heat in this context is not the expectation value of a 'heat' observable. There is no such thing.

First main law for polarizable matter We describe matter which is exposed to a dielectric displacement field $\boldsymbol{D}$ or a magnetic induction field $\boldsymbol{B}$. It requires the amount $\text{d}W = \int \text{d}V\,\{\boldsymbol{E}\cdot\boldsymbol{D} + \boldsymbol{H}\cdot\text{d}\boldsymbol{B}\}$ of work to change the displacement of the induction. One should split off the energy of the electromagnetic field the expression for which is not unique in the presence of polarizable matter. One of them is such that the the differential of the free energy of matter is $\text{d}F = -S\text{d}T - \int \text{d}V\,\boldsymbol{P}\cdot\text{d}\boldsymbol{E} - \mu_0 \int \text{d}V\,\boldsymbol{M}\cdot\text{d}\boldsymbol{H}$ in standard notation. The advantage of this form of the first main law is that the electric and magnetic field strengths as external parameters are easy to control.

Fluctuation Fluctuations usually refer to the Gibbs state G. The fluctuation $\Delta M = M - \langle M \rangle$ is the deviation of an observable M from its expectation value $\langle M \rangle = \text{Tr}\, GM$. Clearly, the expectation value of the fluctuation vanishes while its variance $\sigma^2(M) = \langle (\Delta M)^2 \rangle$ is never negative. Fluctuations at different times define the correlation function for the process $t \to M_t$, ▹ *Wiener–Khinchin theorem*. The variance of some fluctuations may be derived directly from a ▹ *Thermodynamic potential*. A particle suspended in a liquid is driven by fluctuating forces, ▹ *Brownian motion*. Density fluctuations in the atmosphere cause ▹ *Rayleigh scattering* and are responsible for the blue day sky.

Free particle A free particle moves without forces acting on it. Non-relativistic free particles moving in a vacuum are characterized by the dispersion relation $\hbar\omega = (\hbar \boldsymbol{k})^2/2m$. If electrons move in a perfect crystal they are free as well, but the branches of their dispersion relation is never a quadratic function. They are free quasi-particles and react on electric forces with an ▹ *Effective mass*.

Forced transitions A rapidly oscillating external electric field causes transitions between two energy levels if the energy difference fits and if symmetry allows it. A transition from the lower level to the higher has the same probability than the reverse process, ▹ *Detailed balance*.

Form factor Scattering of point-like charged particles, such as electrons, on a scatterer with extended charge distribution is described by a ▹ *Scattering amplitude* for Coulomb scattering multiplied by a correction, or form factor. The form factor depends on the wave vector transfer Δ only; it is the Fourier transform of the ground state charge distribution (wave function squared). High momentum transfer experiments on electron-proton or electron-neutron scattering provide form factors for protons and neutrons. They suggest that these ‘elementary’ particles are composed of three ▹ *Quarks* with electric charges $e/3$ (up-quark) and $-2e/3$ (down-quark).

Free charge In polarizable matter, some charges are elastically bound and others are mobile or free. The density of the latter ϱ^{f} generates the dielectric displacement field $\boldsymbol{D}$. The mobile electrons may stream which is described by a current $\boldsymbol{j}^{\text{f}}$. Maxwell's equations for the electromagnetic field in polarizable matter allow to derive the continuity equation for the free charge alone, not just for the total electric charge.

Free energy The free energy F is introduced as a Lagrange parameter such that the Gibbs state becomes normalized. It depends on the temperature T and on the external parameters; it may serve as a ▹ *Thermodynamic potential*. Its differential is $\text{d}F = -S\text{d}T + \text{d}W$ with $S = \mathcal{S}(G)$ as ▹ *Entropy* and $\text{d}W$ as ▹ *Work*. There is also a functional $\mathcal{F}(\varrho) = \text{Tr}\, \rho \{H + k_{\text{B}}T \ln \varrho\}$ which attains its minimum at the Gibbs state. For $T = 0$ this boils down to the ▹ *Raleigh-Ritz principle*. Note that $\mathcal{F}(G)$ gives the free energy as a thermodynamic potential and that the functional is compatible with $F = U - TS$. The trace converges for $T > 0$ only.

G

Gibbs Josiah Willard US-American chemist and physicist, 1839–1903

Gibbs potential The thermodynamic potential $G = F + pV$ has the differential $dG = -SdT + Vdp$. Therefore, the Gibbs potential $G = G(T, p)$ depends on temperature and pressure. The Gibbs potential, or Gibbs free energy, is useful for describing chemical reactions at constant temperature and constant pressure, for instance in biology.

Gibbs state The Gibbs state $G = \exp(\{F - H\}/k_B T)$ maximizes the ▷ *Entropy* functional. F and T are Lagrange parameters assuring that the Gibbs state is a probability operator and that the internal energy remains constant. T has all properties of a ▷ *Temperature*. F, the ▷ *Free energy*, serves as a ▷ *Thermodynamic potential*. The Gibbs state describes the state of a system which is in thermal equilibrium with its environment.

Gluon ▷ *Quarks* are bound to baryons (three quarks) or to mesons (quark-antiquark) by strong interactions. These are mediated by the exchange of massive vector bosons. They are also called gluons because they serve as glue between quarks. Gluons play a similar role as the photon which mediates the electromagnetic forces between charged particles.

Green's function A linear relationship between two functions f and g is described by an integral $g(x) = \int dy\, G(x, y)\, f(y)$. This expression is similar to $g_i = \sum_j G_{ij} f_j$ for n-tuples of numbers. The Green's function is usually the solution of a liner differential equation with auxiliary conditions incorporated.

Group A set of elements which may be multiplied with each other. Multiplication is associative, there is a neutral element e such that $fe = ef = f$ holds true, and for every f there is a $g = f^{-1}$ with $fg = e$. Groups in physics usually consist of transformations where multiplication means executing one after the other (composition). This assures associativity and the existence of the neutral element, the identical transformations. It is necessary that the transformation be invertible. gf need not be the same as fg.

Group velocity A non-linear ▷ *Dispersion relation* $\omega = \omega(\boldsymbol{k})$ makes ▷ *Wave packets* spread in space. The center of the wave packet travels with group velocity $\boldsymbol{v} = \nabla\omega(\boldsymbol{k})$, averaged over the waves which form the packet.

H

Hamilton William Rowan Irish physicist and mathematician, 1805–1865

Hamiltonian In classical mechanics, the system's energy as a function depending on generalized coordinates and momenta. In quantum physics, the energy is an observable. In both cases, it governs the time evolution either of observables (▷ *Heisenberg picture*) or states (▷ *Schrödinger picture*).

Heat ▹ *First main law*. If the state ϱ of a system changes there is a corresponding change $\mathrm{d}Q = \mathrm{Tr}\,\mathrm{d}\varrho\, H$ of internal energy. Heat is a form of energy transfer, not the expectation value of an observable. In particular, the concept of heat is not restricted to equilibrium processes.

Heat capacity The heat capacity C is the amount of energy $\mathrm{d}U$ required to change the temperature by $\mathrm{d}T$. Either the volume or the pressure are kept constant while adding energy, therefore there is a heat capacity C_p at constant pressure and a heat capacity C_V at constant volume.

Heisenberg Werner: German physicist 1901–1976, Nobel prize 1932 for 'the creation of quantum mechanics, ...'

Heisenberg equation In the ▹ *Heisenberg picture* the states are fixed while the observables change with time according to $\mathrm{i}\hbar\dot{M}_t = [M_t, H]$.

Heisenberg ferromagnet A model for ferromagnetic behavior introduced by Heisenberg. Freely rotatable spins may interact in such a way that neighbors tend to align parallel or anti-parallel. In the former case the material behaves as a ferromagnet. Although the problem has not yet been solved analytically in three dimensions, the following effective field approximation allows for a solution. One spin factor in a term $\boldsymbol{s}_a \cdot \boldsymbol{s}_b$ is replaced by its expectation value. The approximate Hamiltonian H_{ef} now contains expectation values $\langle\, \boldsymbol{s}_a \,\rangle$ which are to be calculated with the aid of H_{ef}. This leads to a self-consistency equation which has two solutions for $T < \Theta_{\mathrm{C}}$ and one above. For temperatures above the Curie temperature Θ_{C} the material is paramagnetic, its magnetic susceptibility varies with temperature as $\chi_{\mathrm{m}} \propto 1/(T - \Theta_{\mathrm{C}})$. ▹ *Curie–Weiss law*. Below the Curie temperature there is a spontaneous magnetization $M^*(T)$ even if there is no external magnetic field. The spontaneous magnetization is largest at $T = 0$ and decreases to zero at $T = T_{\mathrm{C}}$. The Heisenberg ferromagnet model and the effective field approximation describe the phenomenon of ferromagnetism reasonably well, as compared with computed solutions.

Heisenberg picture Waiting for a time span t before measuring M defines a new observable M_t. The time evolution $t \to M_t$ is governed by the ▹ *Heisenberg equation*; it is driven by the ▹ *Hamiltonian*. ▹ *Schrödinger picture* for an alternative view. The time development of expectation values is the same, irrespective of the viewpoint, or picture.

Heisenberg uncertainty relation Momentum P and location X cannot be measured simultaneously with absolute precision. For the standard deviations, the inequality $\sigma(X)\sigma(P) \ge \hbar/2$ holds true for any state. This is a consequence of the ▹ *Canonical commutation rules*.

Helium atom The helium atom consists of the nucleus and two electrons. The total spin of the electrons is either $S = 0$ (singlet) of $S = 1$ (triplet), the former being antisymmeric, the latter symmetric under particle exchange. The spatial part of the wave function therefore must be symmetric or antisymmetric, respectively. The spin singlet ground state has a total energy of $E_s = -2.904\ E_*$. This has to be compared with $-2.000\ E_*$ for ionized helium; $0.904\ E_*$ is the binding energy of the second electron. For the spin triplet state the binding is weaker, $E_t = -2.175\ E_*$. The helium triplet state cannot decay into the singlet state by an electric dipole transition. The magnetic dipole transition is allowed, but rare. The lifetime of a helium triplet ground state is about 2 h.

Hermite Charles: French mathematicien, 1822–1901

Hermite polynomials The Hermite polynomials $H_0, H_1, \dots$ form a complete set of orthogonal polynomials. 'Orthogonal' here refers to the scalar product $(g, f) = \int \mathrm{d}x\, \mu(x)\, g^*(x)\, f(x)$ with a weight $\mu(x) = \exp(-x^2)$. The eigenfunctions of the harmonic oscillator are proportional to Hermite polynomials.

Hidden variables Only probabilities can be predicted, not the outcome of a single measurement. Certain theories propose that the usual description of a physical system is yet incomplete; there are hidden, unobserved variables which re-establish determinism. If more than speculation, hidden variable theories should allow verifiable consequences, such as Bell's inequalities. Up to now, all hidden variable theories could be refuted while standard quantum theory always passed the experimental test.

Hilbert David: German mathematician, 1862–1943

Hilbert space A complex linear space with scalar product which is complete in its natural topology. Linear mappings of the Hilbert space $\mathcal{H}$ into itself—we call them operators—form the backbone of quantum theory. Observable properties M and states ϱ are represented by operators, and a linear functional, the ▹ *Trace*, defines expectation values $\langle\, M \,\rangle = \mathrm{Tr}\, \varrho M$. In general, operators do not commute, unlike the observables of classical mechanics.

Hopping model A one-dimensional ring of equally spaced ions and a single electron. The electron is loosely bound by ions with an energy ϵ and may hop to the next neighbor with amplitude V. Both ϵ and V do not depend on the site index (translational symmetry). The energy levels of the electron in the entire crystal (quasi-electron) are labeled by k, and one finds $\epsilon(k) = \epsilon - V \cos ka$ where a is the distance between two neighbors. This is a ▹ *Dispersion relation* allowing to calculate the group velocity and effective mass. Wave packets represent normalized states. The effect of an external electric field can be handled: the quasi-electron reacts with an acceleration times the effective mass. An impurity causes scattering or trapping. The model can be modified to allow for direct hopping to the next but one neighbor, and so on. In this way any dispersion relation may be simulated.

Hooke's law The deformation of an elastic solid medium is described by the ▹ *Strain tensor*. This strain tensor is linearly related to the ▹ *Stress tensor*. Hooke's law for an isotropic medium contains two material constants, the ▹ *Elasticity module* and ▹ *Poisson's ratio*.

Hydrogen atom Electron bound to a proton. The ▹ *Classical hydrogen atom* is not stable. Schrödinger could show that his wave theory provided the correct energy levels and thus predicted the system of spectral lines. The wave function for a bound state is a product of an angular part (spherical harmonic) and a radial eigenvalue equation which could be solved analytically. There are the angular momentum quantum numbers l, m as well as a radial quantum number ν_{r} counting the nodes of the radial wave function. The energy levels are $E_{nlm} = -1/2n^2 E_*$ where the principal quantum number is $n = \nu_{\mathrm{r}} + l + 1$. Later the hydrogen spectrum could be explained by ▷ *Dirac* without non-relativistic approximations. Some energy levels with the same principal quantum number differ slightly.

Hydrogen molecule Two hydrogen atoms may bind to an H_2 molecule. One discusses a cloud of two light electrons in the force field of two protons. The electrons may be in a singlet or a triplet state, but only the former allows a bound molecule. ▷ *Helium atom*. In the spirit of the ▷ *Born-Oppenheimer approximation*, there are two protons interacting by a molecular potential with a minimum below the ionized states. The problem can be reformulated as a freely moving center of mass and relative motion. One may now study the vibrational degrees of freedom and the rotational spectrum.

I

Incoherent radiation A beam of photons with random phases. At a fixed point, the ▷ *Correlation function* of the electric field strength is proportional to a delta function; the ▷ *Spectral intensity* is constant (white noise).

Interaction picture Assume a Hamiltonian $H_t = H + V_t$, a sum of a manageable part H and a perturbation V_t. Transforming to the interaction picture facilitates an approximate solution as a power series of perturbations. Beginning in the Schrödinger picture, one translates in time with H only. Without perturbation, the transformed state were constant. With perturbation, the change rate of the transformed state is small, namely proportional to V_t. The inital condition and the Schrödinger equation are rewritten as an integral equation which easily lends itself to a power series expansion. At the end, the transformation to the interaction picture should be undone.

Internal energy The energy density of a ▷ *Material point* consists of kinetic energy $\varrho \boldsymbol{v}^2/2$, potential energy—electric and gravitational—and a rest which is aptly called internal. Traditionally, internal energy is denoted by U. It is the expectation value $\mathrm{Tr}\, \varrho H$ of the energy observable H in a frame where the system is at rest. ϱ may be an arbitrary state or the equilibrium, i.e. ▷ *Gibbs state*. The ▷ *First main law* $\mathrm{d}U = \mathrm{d}Q + \mathrm{d}W$ is valid in the general cases. The conduction current density $\boldsymbol{J}^{\mathrm{u}}$ of internal energy goes under the name of heat current density.

Internal friction Momentum transport in continuously distributed matter is described by T_{jk}, the symmetric stress tensor. It can be split into a normal part T'_{jk} which transforms under time reversal like the convection contribution $\varrho\, v_j v_k$, and the rest T''_{jk} which acquires a minus sign. The latter causes internal friction. For an incompressible Newtonian liquid (like water), internal friction is described by $T''_{jk} = \eta(\nabla_j v_k + \nabla_k v_j)$, with η as viscosity. $\boldsymbol{v}$ is the streaming velocity. Internal friction is caused by velocity gradients.

Irreversible process A thermodynamic process is reversible if its state ϱ_t is always close to an equilibrium state $G(T_t, H_t)$. If, however, the temperature or the external parameters of H_t are changed rapidly, the process is irreversible and entropy is produced. Continuum physics teaches that there are six reasons for irreversible behavior, namely heat conduction, diffusion, internal friction, electric conduction and chemical reactions.

J

Joule James Prescott: British physicist, 1818–1889

Joule's heat The internal energy U may change with time because of five causes: inflow of internal energy or heat conduction, compression of the medium, polarization, internal friction and the irreversible motion of charges counter to an electric field. The latter effect is called Joule's heat, it is the contribution $\boldsymbol{J}^{e''} \cdot \boldsymbol{E}$ to the volumetric production rate $\pi(U)$ for internal energy. $\boldsymbol{J}^{e''}$ is the irreversible part of the electric conduction current. ▹ *Ohm's law* and ▹ *Internal friction*.

Joule-Thomson effect If a fluid is made to pass a throttle, the temperature will change by $\Delta T = \mu_{\mathrm{JT}}\,\Delta p$. The Joule-Thomson coefficient μ_{JT} vanishes for the ideal gas; it may be positive or negative for real gases. The Joule-Thomson coefficient can be calculated if the inter-molecular potential is known. Above the inversion temperature of 51 K, diminishing pressure decreases the temperature of helium. Below the inversion temperature the effect is opposite. Air at room temperature always cools down when expanding, its inversion temperature is much higher.

K

Khinchin Aleksandr Soviet mathematician, 1894–1959

Khinchin's theorem ▹ *Wiener–Khinchin theorem*

Kramers-Kronig relations ▹ *Dispersion relation* for the dielectric susceptibility. Its refractive part is an integral over the absorptive, or dissipative part which is always positive according to the ▹ *Callen-Welton theorem*. Absorption within a certain frequency band causes dispersion and refraction at another frequency range.

Kubo Ryogo Japanese physicist, 1920–1995

Kubo-Martin-Schwinger (KMS) formula An alternative characterization of the equilibrium state. Observables can be translated by complex numbers z instead of time via analytical continuation. Kubo, Martin and Schwinger could prove the relation $\mathrm{Tr}\,\varrho A_z B = \mathrm{Tr}\,\varrho B A_{z-\mathrm{i}\hbar\beta}$ with $\beta = 1/k_{\mathrm{B}}T$ the inverse temperature. The so-called KMS-formala is valid if ϱ is the Gibbs-state.

L

Laboratory frame A scattering arrangement where the target is at rest. The scatterer suffers a recoil. The energy of the scattered particle and the ▹ *Deflection angle* refer to the laboratory frame; they must be transformed into ▹ *Center of mass frame* values.

Ladder operator A step-down, or annihilation operator A and its adjoint, the step-up or creation operator. These properties follow from the commutation relations $[A, A^\star] = I$. The eigenvalues of the corresponding number operator $N = A^\star A$ are the natural numbers. With anti-commutators instead of commutators, N has the eigenvalues zero or one only.

Lagrange Joseph-Louis: French mathematician, 1736–1813

Lagrange multiplier Certain optimization problems with constraints may be solved by introducing Lagrange multipliers. As an example we consider how to find a stationary value of the entropy functional $S(\varrho) = -k_B T \mathrm{Tr}\, \varrho \ln \rho$. Not all variations $\mathrm{d}\varrho$ are allowed, they are restricted by $\mathrm{Tr}\, \varrho = 1$ and by $\mathrm{Tr}\, \varrho H = U$, in standard notation. Multiplying the first restriction $\mathrm{Tr}\, \mathrm{d}\varrho = 0$ by the Lagrange multiplier a, the second restriction $\mathrm{Tr}\, \mathrm{d}\varrho H = 0$ by b, and adding to $-k_B T \mathrm{Tr}\, \mathrm{d}\varrho \ln \rho$ results in $\mathrm{Tr}\, \mathrm{d}\varrho\{-k_B T \ln \varrho + aI + bH\} = 0$ where now the variations $\mathrm{d}\varrho$ are no more restricted. The solution to our problem is therefore $-k_B T \ln \varrho + aI + bH = 0$ which describes the Gibbs state.

Lagrangian In classical mechanics, the difference $L = T - V$ between kinetic and potential energies, expressed in generalized coordinates and velocities. The action, an integral over an arbitrary trajectory, is stationary at the physical solution. This solution is characterized by Lagrange's equations, one for each degree of freedom. The entire procedure guarantees that a transformation to new generalized coordinates result in the transformed solution. The concept may be transferred to classical and quantum field theory where the action is an integral over space and time of a Lagrangian density.

Langevin Paul: French physicist, 1872–1946

Langevin equation A differential equation for random variables instead of functions. ▹ *Brownian motion* for example is described by a Lagrange equation for the location of the Brownian particle which is driven by a fluctuating force described by its ▹ *Correlation function*.

Laplace Pierre-Simon: French astronomer, physicist and mathematicien, 1749–1827

Laplacian $\Delta = \nabla_x^2 + \nabla_y^2 + \nabla_z^2$, may be expressed in ▹ *Spherical coordinates* such that the derivatives with respect to angles is described by the square of the orbital angular momentum.

Lattice vibrations The ions of a crystal lattice vibrate about their equilibrium configuration. Since this is characterized by minimal potential energy, the Hamiltonian for lattice vibrations is a quadratic form in the deviations from the equilibrium positions. This quadratic form can be diagonalized into independent normal modes. Such modes may be exited either not at all, by one quantum of oscillation (phonon) and so on. The system therefore similar to a phonon gas where the details are hidden in the branches of the dispersion relation. In thermal equilibrium, the system of lattice vibrations is described by a density of states for which there are plausible approximations. ▷ *Debye model* and ▷ *Debye-Einstein model*.

von der Laue Max: German physicist, 1879–1960, Nobel prize 1914 for 'his discovery of diffraction of X-rays by crystals, …'

Laue condition A crystal is constructed by mounting copies of a unit cell at $\boldsymbol{\xi}_j = j_1\boldsymbol{a}^{(1)} + j_2\boldsymbol{a}^{(2)} + j_3\boldsymbol{a}^{(3)}$. The j are three integer numbers and the $\boldsymbol{a}$ are three vectors spanning the unit cell. There is an incoming beam of particles with wave vector $k\,\boldsymbol{n}^{\text{in}}$. One will detect an intensive beam of diffracted particles in direction $\boldsymbol{n}^{\text{out}}$ if the Laue condition $\boldsymbol{\Delta}\cdot\boldsymbol{a}^{(r)} = 2\pi\nu_r$ is fulfilled with three integer numbers ν. Here $\boldsymbol{\Delta} = k\,\boldsymbol{n}^{\text{out}} - k\,\boldsymbol{n}^{\text{in}}$ is the wave vector transfer from the incoming to the outgoing particle beam

Legendre Adrien-Marie: French mathematician, 1752–1833

Legendre transformation Let $f = f(x)$ be a convex function. We define its Legendre transform by $f_{\text{L}}(y) = \inf_x\{f(x) - xy\}$. This function is concave. If f is sufficiently smooth and strictly convex, the infimum is a minimum at $\bar{x}$ such that $f'(\bar{x}) = y$ holds true. This equation defines the inverse $X = X(y)$ of the derivative by $f'(X(y)) = y$. The Legendre transformed function therefore is $f_{\text{L}}(y) = f(X(y) - yX(y)$. All this amounts to $\mathrm{d}f(x) = y\mathrm{d}x$ and $\mathrm{d}f_{\text{L}}(y) = -x\mathrm{d}y$. The role of argument and derivative have been interchanged. The Legendre transformation of a concave function should be defined with the supremum instead of the infimum.

Lennard-Jones potential A potential $\phi(r) = \phi_0\{(a/r)^{12} - 2(a/r)^6\}$ simulating strong repulsion at short distance and attraction at large distance between the molecules. If the second virial coefficient of argon is fitted to a Lennard-Jones potential, it predicts the ▷ *Jule-Thomson effect* rather well. Of course it cannot bind an argon molecule.

Lifetime of an excited state An excited state i of an atom or molecule can spontaneously decay into levels j of lower energy, if symmetry allows. Summing up the decay rates $\Gamma_{i\to f}$ leads to a total decay rate Γ_i for the excited state i. The inverse of it is the lifetime τ_i of the level. After a time t, the fraction of particles not yet decayed is given by $\exp(-t/\tau_i)$.

Light as a perturbation The external electromagnetic field and matter interact mainly through electric dipoles. The electric field is considered an external parameter, and the Hamiltonian is $H_t = H - \int \mathrm{d}V\, \boldsymbol{P}(\boldsymbol{x})\cdot\boldsymbol{E}(t,\boldsymbol{x})$ depends explicitely on time. Note that $\boldsymbol{P} = \boldsymbol{P}(\boldsymbol{x})$, the polarization, is a field of observables which in the ▷ *Schrödinger picture* do not depend on time.

Linear chain A one-dimensional sequence of equally spaced ions may serve as a model crystal. There is either a finite number the ends of which are joined by periodic boundary conditions, or the linear chain is infinite. Springs between the ions simulate the potential energy, and loosely bound electrons may jump from one place to neighboring sites. In this way, the electronic band structure of real crystals can be simulated as well as lattice vibrations. By studying a chain of type $\dots ABAB \dots$ acoustical and optical phonon branches show up, and a single ion being different mimics scattering and trapping of electrons at impurities.

Linear momentum The three components P_j of linear momentum, or simply momentum, are the generators of spatial translations. Since translations form a commutative topological ▹ *Group*, the momentum operators commute. They are represented by $\hbar/\mathrm{i}$ times the partial derivative in j-direction. These operators are not bounded and therefore not defined on the entire Hilbert space of square-integrable functions. However, if restricted to absolutely continuous functions, the P_j are self-adjoint.

Linear operator A linear mapping of the Hilbert space into itself. Linear operators, or operators for short, can be multiplied by complex numbers, they can be added and multiplied, i.e. one applied after the other. Operators do not commute in general. Operators have a norm based on the natural norm of the Hilbert space. Bounded operators cause a continuous mapping. Each operator A is associated with an adjoint operator $A^\star$ characterized by $(g, Af) = (A^\star g, f)$ for all vectors f and g. Operators which commute with their adjoints are normal. This important class of operators comprises ▹ *Unitary*, ▹ *Self-adjoint*, ▹ *Positive* and ▹ *Probability operators*. They can be diagonalized.

Linear response Assume that a system is perturbed according to $H_t = H - \lambda(t)V$. The initial condition shall be $\varrho_t \to G$ for $t \to -\infty$. The solution can be expanded in a power series in V. The first order, or linear response is $\mathrm{Tr}\, \varrho_t M = \mathrm{Tr}\, GM + \int \mathrm{d}\tau\, \theta(\tau)\lambda(t-\tau)\Gamma(\tau)$ with an explicit expression for the response function Γ in terms of M and V. The jump function guarantees that only preceding perturbations contribute.

Liouville Joseph: French mathematician, 1809–1882

Liouville theorem An important theorem in classical mechanics. The dynamics of a system with f degrees of freedom may be thought of as streaming in ▹ *Phase space*. A certain region moves in the course of time, but its volume does not change. Put otherwise, Hamilton's equations guarantee that the points of phase space stream like an incompressible fluid. This is true if the volume is measured as $\mathrm{d}\Gamma \propto \mathrm{d}q_1 \mathrm{d}p_1 \dots \mathrm{d}q_f \mathrm{d}p_f$.

Local quantum field ▹ *Quantum field*

Location Three operators $\boldsymbol{X}$ which transform under a translation by $\boldsymbol{a}$ as $\boldsymbol{X} \to \boldsymbol{X} - \boldsymbol{a}\, I$. Obviously, they commute with each other. They do not commute with the generators of translations, the ▷ *Linear momentum operators*. Instead, location and momentum obey the ▷ *Canonical commutation rules*.

Lorentz Hendrik: Dutch physicist, 1853–1928, Nobel prize 1902 with Pieter Zeeman for '...their researches into the influence of magnetism upon radiation phenomena'.

M

Magnetic dipole moment The induction field $\boldsymbol{B}$ is the curl of a vector potential $\boldsymbol{A}$. It is generated by a current distribution $\boldsymbol{j}$. If the latter is centered around the origin, the vector potential in the far zone is given by $\boldsymbol{A}(r\boldsymbol{n}) = (\mu_0/4\pi)\boldsymbol{m} \times \boldsymbol{n}/r^2$ where $\boldsymbol{m} = (1/2)\int d^3y\, \boldsymbol{y} \times \boldsymbol{j}(\boldsymbol{y})$ is the magnetic dipole moment of the charge-current distribution. The induction field falls off as $1/r^3$.

Magnetization The density $\boldsymbol{M}$ of magnetic dipole moment, a field. The magnetic field is defined as $\boldsymbol{H} = (1/\mu_0)\boldsymbol{B} - \boldsymbol{M}$.

Mandelstam variables For relativistic particles, energy and momentum are combined into a four momentum vector p_μ with $p_0 = \sqrt{m^2c^2 + \boldsymbol{p}^2}$. The particle's energy is $E = cp_0$. The 'square' p^2 is defined as $p_0^2 - \boldsymbol{p}^2$. In a scattering reaction $1 + 2 \rightarrow 3 + 4$, energy and momentum must be conserved. The corresponding four momentum vectors obey $p^{(1)} + p^{(2)} = p^{(3)} + p^{(4)}$. The Mandelstam variable $s = (p^{(1)} + p^{(2)})^2$ is an invariant, it describes the energy squared of the incoming particles. The variable $t = (p^{(3)} - p^{(1)})^2 = (p^{(4)} - p^{(2)})^2$, another Mandelstam variable, characterizes the square of the momentum transfer from particle 1 to particle 3. It is likewise an invariant. 'Invariant' in this context means that the value does not depend on the frame of reference. Differential cross sections should be expressions in s and t. Then a distinction between ▷ *Center of mass frame* and ▷ *Laboratory frame* is unnecessary.

Maser A maser (Microwave Amplifiction by Stimulated Emission of Radiation) is a device to generate and amplify microwave radiation. Molecules should have an excited state which emits microwaves in the transition to the ground state, such as the ▷ *Ammonia molecule*. If a beam of such molecules, a mixture of ground and exited states, passes through an inhomogeneous electric field, the molecules in the ground state and in the excited states will follow different paths. It is therefore possible to collect the excited molecules in a resonant cavity. If triggered by a weak signal of matching wave length, all excited molecules undergo a ▷ *Forced transition* to the ground state with the emitted photons in phase. Since the transition energy depends on an external electric field (▷ *Stark effect*), the maser may be tuned. The field also exerts a force on the molecule which is different for the two states. This effect allows pumping the system by separating and selecting the excited states.

Mass The mass of an elementary particle is an immutable relativistic invariant quantity. In continuum physics, the mass is regarded as an additive, transportable quantity M with density $\varrho(M) = \rho$ and current density $\boldsymbol{j}(M) = \varrho\, \boldsymbol{v}$. In solid state physics the quasi-free electrons are described by their dispersion relation. The factor

m^* between acceleration and electrostatic force is the effective mass; its inverse is given by the second derivative of the dispersion curve $\omega = \omega(\boldsymbol{k})$ (aside from a factor $\hbar$).

Material point The central concept of ▷ *Continuum physics*. A region so small that it has no further structure, but properties like electric charge, chemical composition and internal energy as well as entropy. A region so large that its N particles form an open thermodynamic system close to the thermodynamic limit $N \to \infty$. Both idealizations are realistic if relative ▷ *Fluctuations* remain undetected at eight digits precision or better. The material point, because it is so small, is always in an equilibrium state and therefore has a temperature and a chemical potential. The continuum, although locally in equilibrium, may not be in global equilibrium.

Measurement An experiments is a sequence of independent repetitions of a measurement. Idealized, the system is prepared by some measurement and selecting the desired outcomes. Think of splitting a beam of particles in a strong induction field into beams of particles with well defined energy and momentum. The preparation procedure is summarized as a state described by a probability operator ϱ. Once the state has been prepared, an observable property M can be measured. Observables are represented by self-adjoint operators. Their eigenvalues correspond to the possible results of a measurement. When the measurement is performed, one of the eigenvalues is registered. If repeated, a different result will be registered. However, if repeated very often, the mean of the measured values converges towards the expectation value $\langle M \rangle = \mathrm{Tr}\, \varrho M$.

Metal A crystal with mobile quasi-electrons (▷ *Electronic band structure*) close to the ▷ *Fermi energy*. Metals are good electric and heat conductors because the mobile electrons transport charge as well as energy.

Microwave standard The frequency $f = 23.87012$ GHz of transitions between the antisymmetric and symmetric configuration of an ▷ *Ammonium molecule*.

Minimal coupling The interaction with matter and an external electromagnetic field is described by minimal coupling. The four-momentum p_μ of a particle is replaced by $p_\mu - qA_\mu$ where q is the particle's charge and A_μ the four-potential of the electromagnetic field. ▷ *Dirac equation*. The electromagnet field is obtained from the four-potential by $\boldsymbol{E} = -\boldsymbol{\nabla}\phi - \nabla_t \boldsymbol{A}$ and $\boldsymbol{B} = \boldsymbol{\nabla} \times \boldsymbol{A}$, with $\phi = cA_0$.

Minimax theorem Assume that an operator like the energy H has—below a continuous spectrum—a number of discrete eigenvalues $E_1 < E_2 < \cdots$ in increasing order. They may be calculated by the following procedure. Select an n-dimensional linear space $\mathcal{D}_n$. Calculate the maximum (d, Hf) for all normalized f in $\mathcal{D}_n$. The minimal such value coincides with the eigenvalue E_n. The eigenvalue is the minimum of a maximum which explains the name. For E_1 the minimax theorem is nothing else but the ▷ *Raleigh-Ritz principle*. The minimax theorem was derived by ▷ *Courant*.

Mixed state A state is a continuous linear functional $M \to \varrho(M)$ with $\varrho(I) = 1$ on the set of observables M. In standard quantum physics it is realized by a

▹ *Probability operator* ϱ such that $\varrho(M) = \text{Tr}\, \varrho M$. ▹ *Trace of an operator*. States can be mixed: $\varrho_\alpha = (1-\alpha)\varrho_1 + \alpha\varrho_2$ for $0 \le \alpha \le 1$. $\varrho = \Pi$ is a ▹ *Pure state* for any one-dimensional projector Π. The degree of mixedness is given by the ▹ *Entropy* $S(\varrho) = -k_B \text{Tr}\, \varrho \ln \rho$. The entropy is never negative; it vanishes if and only if the state is pure.

Molecular potential According to the ▹ *Born-Oppenheimer approximation* the electron cloud settles practically instantaneously in a fixed ionic configuration. Its energy is the molecule's potential energy, or molecular potential. The heavy ions move slowly in this potential giving rise to ▹ *Molecular vibrations* and ▹ *Molecular rotations*.

Molecular rotations A molecule in its minimal energy configuration allows for rotations since the ▹ *Molecular potential* is rotationally invariant. It is a good approximation that even in highly excited rotational states the deviations from this minimizing configuration remain rather small; the molecule behaves as a rigid rotator. A diatomic molecule is characterized by it rotational energy levels $l(l+1)\hbar^2/2\mu R_0^2$ with μ as reduces mass and R_0 as equilibrium distance. $l = 0, 1, \dots$ is the angular momentum quantum number, each level is $2l+1$-fold degenerate. The transition between the lowest and the first excited level is in the far infrared.

Molecular vibrations The ▹ *Molecular potential* has a stable minimum. In lowest order the minimum energy is a constant plus a quadratic form in the deviations from the equilibrium locations of the molecule's ions. This quadratic form may be diagonalized such that the Hamiltonian becomes a sum of independent normal harmonic oscillations. ▹ *Lattice vibrations*. The vibrational excitations of each normal mode are equidistant. For a hydrogen molecule the transition energy is 0.54 eV (near infrared).

Momentum transfer The transfer $\Delta \boldsymbol{p} = \hbar \Delta \boldsymbol{k}$ of momentum (wave vector) from the incoming particle onto the scattered particle. ▹ *Mandelstam variables* for a relativistic description.

Multipoles A charge-current distribution located at the origin produces potentials ϕ and $\boldsymbol{A}$ which can be expanded in powers of $1/r$ in the far zone. The total electric charge Q causes an $1/r$ term. Dipole moments, electric or magnetic, lead to $1/r^2$ contributions, quadrupole fields vanish as $1/r^3$, and so on. ▹ *Electric dipole moment*, ▹ *Magnetic dipole moment*.

N

Negative hydrogen ion A proton can bind two electrons. A helium like trial function for the spin singlet configuration with nuclear charge 1 instead of 2 yields a ground state energy of −0.513 a.u. (atomic units), the experimental value is −0.528 a.u. This is well below the ionization threshold of 0.5 atomic energy units. There is no bound triplet state. The negative hydrogen ion, or hydrogen anion, was first detected in the sun's photosphere.

Neutron diffraction If a beam of mono-energetic slow neutrons hits a crystal (not a target of randomly positioned nuclei), a strong secondary beam of diffracted neutrons can be observed. The ▷ *Laue conditions* must be fulfilled. The diffraction pattern is the same for X ray or electron diffraction provided the wave vector $\boldsymbol{k}$ is replaced by the momentum $\boldsymbol{p} = \hbar\boldsymbol{k}$. ▷ *de Broglie*.

Neutron-molecule scattering If neutrons are scattered on a target of randomly oriented molecules, the differential cross sections is not isotropic as expected for single nuclei. The structure of the extended molecule leads to interferences, ranging from constructive to partially destructive up to no interference at all for high momentum transfer. ▷ *Form factor*.

Neutron star In the course of a star's life the normal burning of hydrogen to helium will come to an end and other nuclear cycles set in. They finally end up in iron and nickel because their binding energies are the largest. At the end of nuclear burning the star will get colder and colder until the thermal pressure can no more withstand gravitation. A sudden collapse leads to a dense inner core region while outer layers are blown off. Such a supernova transition either leads to a white star or a neutron star. In the first case atoms are ionized, there is a plasma of nuclei and electrons. The electrons exert a ▷ *Degeneracy pressure* which may be able to stabilize the star (▷ *White dwarf*) even at zero temperature. However, the electronic degeneracy pressure may be too weak, and electrons then will creep into the nuclei and finally form a degenerate neutron gas. Its degeneracy pressure then stabilizes the neutron star. In case that the original star was too massive and its core cannot be stabilized by electronic or neutron degeneracy pressure, the gravitational collapse cannot be stopped, and a ▷ *Black hole* will form. Black holes may grow beyond all proportions by swallowing dust, stars and maybe entire galaxies.

Normal mode It is a common situation that a the potential energy may be approximated by a positive quadratic form. The deviations X_r from the stable equilibrium configuration may be linearly transformed such that the ▷ *Canonical commutation rules* remain untouched. One speaks of normal modes if the transformed Hamiltonian becomes a sum of independent (commuting) oscillators. ▷ *Molecular vibrations*, ▷ *Linear chain*.

Normal operator A normal operator N commutes with its adjoint operator, $[N, N^\star] = 0$. Normal operators can be diagonalized: there is a decomposition $I = \Pi_1 + \Pi_2 + \cdots$ into mutually orthogonal ▷ *Projection operators* and a sequence of eigenvalues $\nu_1, \nu_2, \dots$ such that $N = \sum \nu_j \Pi_j$. In other words, the Hilbert space $\mathcal{H}$ can be decomposed into mutually orthogonal ▷ *Eigenspaces* $\mathcal{L}_j = \Pi_j\mathcal{H}$ such that the normal operator N, if restricted to $\mathcal{L}_j$, just causes the multiplication of each vector in $\mathcal{L}_j$ by the eigenvalue ν_j. The eigenvalues of a normal operator are complex numbers.

Number operator A self-adjoint operator N with $n = 0, 1, 2, \dots$ as eigenvalues. For each pair of ▷ *Ladder operators* A and $A^\star$ there is a number operator $N = A^\star A$.

Nyquist formula An ohmic resistance R at temperature T is the source of a fluctuating voltage U. The spectral intensity of this source of noise is $S_U(f) = 4Rk_B T$, as derived by Nyquist. The power $\mathrm{d}P = S_U(f)\,\mathrm{d}f$ is proportional to the frequency window, the resistance and to the resistor temperature. Since it does not vary with frequency one speaks of 'white noise', in analogy to white light. ▷ *Wiener–Khinchin theorem* and ▷ *Spectral intensity*.

O

Observable Observable properties of a system are described by ▷ *Self-adjoint* operators. Such an observable M is characterized by a sequence $m_1, m_2, \dots$ of possible outcomes of a ▷ *Measurement*, the eigenvalues of M, which are real numbers. If m_j has been measured, the system will be in the corresponding eigenspace $\mathcal{L}_j = \Pi_j \mathcal{H}$. In other words, $M = \sum_j m_j \Pi_j$ where the Π_j are a set of mutually orthogonal ▷ *Projection operators* which add up to $\sum_j \Pi_j = I$. Observables may be multiplied by real numbers and can be added. The product $M_2 M_1$ of two observables is not necessarily another observable since $(M_2 M_1)^\star = M_1 M_2$. Observables need not commute. Commuting observables are in a certain sense independent, their product is an observable, they can be measured simultaneously. A function $f(M)$ of an observable has the same eigenspaces as M, but the eigenvalues are given by $f(M_j)$. Here $f = f(x)$ is an arbitrary real valued function of a real argument.

Ohm's law Recall that the electric current density is split into the convection part $\varrho^e \boldsymbol{v}$ and the conduction contribution $\boldsymbol{J}^e$. The latter has an elastic part $\boldsymbol{J}^{e\prime} = \dot{\boldsymbol{P}}$ and an inelastic part $\boldsymbol{J}^{e\prime\prime}$ which is proportional to the electric field strength: $\boldsymbol{J}^{e\prime\prime} = \sigma \boldsymbol{E}$. This material equation goes under the name of Ohm's law. Conductivities and field strengths may vary widely.

Oppenheimer J. Robert: US-American physicist, 1904–1967

Orbital Suitably scaled hydrogen atom wave functions. They are labeled by the letters $s, p, d, f, \dots$ for angular momentum $l = 0, 1, 2, 3, \dots$ and by $n = 1, 2, \dots$, the main quantum number. l ranges from 0 to $n - 1$. The hydrogen atom orbitals are $1s(2), 2s(2), 2p(6), 3s(2), 3p(6), 3d(10)$ and so on. The number of orthogonal one-particle states (spin included) is given in parentheses.

Orbital angular momentum The total angular momentum $\boldsymbol{J}$ is split into the orbital angular momentum $\boldsymbol{L} = \boldsymbol{X} \times \boldsymbol{P}$ and the remainder, the spin angular momentum $\boldsymbol{S}$. The spin is the angular momentum if the system is at rest. All three: total, orbital and spin angular momentum obey the same angular momentum commutation relations. While the total angular momentum j is integer or half integer, the orbital angular momentum quantum number l is always integer. This is so because a rotation by 2π around any axis must be the identity transformation. The eigenfunctions of orbital angular momentum are the well-known spherical harmonics Y_{lm}, they depend on angles only. In the classical framework, angular momentum is the same as orbital angular momentum.

P

Parity Parity Π is a self-adjoint and unitary operator which obeys $\Pi^2 = I$. Any state can be decomposed into eigenstates of Π with eigenvalues $+1$ or -1, i.e. even or odd parity.

Pauli Wolfgang Austrian/Swiss physicist, 1900–1958, Nobel prize 1945 for 'the discovery of the exclusion principle'.

Pauli principle The wave function of identical particles acquires a minus or plus sign if the position and spin arguments of two particles are interchanged. The plus sign refers to ▹ *Bosons*, the minus sign to ▹ *Fermions*.

Perrin Jean Baptiste: French physiciste, 1870–1942, Nobel prize 1926 for 'his work on the discontinuous structure of matter, and especially for his discovery of sedimentation equilibrium'.

Perturbation Think about a system with Hamiltonian H which is considered to be well understood. A perturbation is a small addition V_t which may vary with time. Examples are the interaction with an external field or the electron's repulsion in atomic and molecular physics. A common situation is that the system, for example a crystal, is acted upon by the electric field of light where the initial state ϱ is prescribed. The perturbation $V_t = -\int \mathrm{d}V\, \boldsymbol{E}(t, \boldsymbol{x}) \cdot \boldsymbol{P}(\boldsymbol{x})$ is best dealt with by transforming to the ▹ *Interaction picture*.

Phase space Consider a classical system with f degrees of freedom. Its configuration is described by generalized coordinates $q = (q_1, q_2, \dots, q_f)$. There is a ▹ *Lagrangian* $L = L(q, \dot{q})$ which depends on generalized coordinates and generalized velocities. Define the generalized momenta by $p_j = \partial L / \partial \dot{q}_j$. There is a Hamiltonian $H(q, p) = \sum_j p_j \dot{q}_j - L(q, \dot{q})$, such that the equations of motion are $\dot{q}_j = \partial H / \partial p_j$ and $\dot{p}_j = -\partial H / \partial q_j$. The $2f$-dimensional manifold Ω of points $\omega = (q, p)$ is the system's phase space. ▹ *Liouville theorem*. Observables are real-valued functions $M = M(\omega)$ and states are probability distributions $\varrho = \rho(\omega)$ living on the phase space. The expectation value is defined by $\langle M \rangle = \int \mathrm{d}\Gamma\, \varrho(\omega)\, M(\omega)$.

Phenomenological thermodynamics Description of the equilibrium behavior of matter without recourse to quantum or classical statistical physics.

Phonon dispersion ▹ *Lattice vibrations*

Photoelectric effect A metal, if irradiated with light of sufficiently large frequency, will emit electrons. Their energy distribution of them depends on the wave length λ of the light, but not on its intensity. The linear relationship between maximal electron energy and frequency $f = c/\lambda$ has an offset which was interpreted by Einstein as the binding energy of an electron in the metal. The photoelectric effect could easily be explained easily if light were made up of photons with energy hf.

Photon gas The radiation within a cavity consists of photons. They form a ▹ *Bose* gas with zero chemical potential because the number of particles cannot be controlled. The gas has a temperature equal to the temperature of the confining walls. All experimental findings on the black body radiation could easily be explained, in particular ▷ *Planck's* expression for the spectral intensity.

Planck Max: German physicist, 1845–1947, Nobel prize 1918 for '...his discovery of energy quanta'.

Plane wave A wave with planes as surfaces of constant phase. Changes harmonically (like a sine function) in space and time. A plane wave is characterized by an angular frequency ω and a wave vector $\boldsymbol{k}$ being perpendicular to the constant phase planes. The dependency $\omega = \omega(\boldsymbol{k})$, the ▷ *Dispersion relation*, characterizes ▷ *Wave packets*.

Polarizability If an atom or molecule has no permanent electric dipole moment, it may be induced by applying an electric field. Normally, the induced dipole moment is proportional to the field strength, $\boldsymbol{d} = \alpha \boldsymbol{E}$, where α is the polarizability. The polarizability of a hydrogen atom can be worked out exactly, it is 4.5 a.u.

Polarization The density $\boldsymbol{P}$ of electric dipole moments. It serves to define the ▷ *Dielectric displacement.*

Poisson Siméon Denis: French physicist and mathematician, 1781–1840

Poisson bracket Consider two classical observables A and B defined as real-valued functions living on the system's ▷ *Phase space*. They are assigned a Poisson bracket $C = \{A, B\}$ by

$$C(\omega) = \sum_j \frac{\partial A(\omega)}{\partial q_j} \frac{\partial B(\omega)}{\partial p_j} - \sum_j \frac{\partial B(\omega)}{\partial q_j} \frac{\partial A(\omega)}{\partial p_j}$$

Poisson brackets define the structure of a classical system and are the counterpart of commutators in quantum physics. ▷ *Cannonical relation.*

Poisson distribution The probability of finding n particles in an interval if the density is given. Useful for calculating density fluctuations in an ideal gas.

Poisson's ratio A dimensionless number appearing in ▷ *Hooke's law.* The ratio of lateral to longitudinal deformation.

Positive operator A positive (non-negative) operator P can be characterized by $(f, Pf) \geq 0$ for all $f \in \mathcal{H}$ or by $P = A^{\star}A$. Both definitions are equivalent. The latter says that a positive operator is normal. Its eigenvalues are never negative.

Poynting vector The current density $\boldsymbol{S}$ of electromagnetic field energy, $\boldsymbol{S} = \boldsymbol{E} \times \boldsymbol{H}$ in standard notation.

Pressure The ▷ *Stress tensor* T_{ij} for a fluid medium at rest cannot support shear forces. It must therefore be proportional to the unit tensor, $T_{ij} = -p\delta_{ij}$. $p = p(t, \boldsymbol{x})$ is the pressure, a field. In equilibrium, the pressure field is constant within the medium, and in this sense pressure is a thermodynamic variable defined by $\mathrm{d}F = -S\mathrm{d}T - p\mathrm{d}V$ with V as the volume of the confining vessel.

Principal quantum number The ▷ *Orbitals* of the hydrogen atom are characterized by the angular momentum $l = 0, 1, \dots$ and a radial quantum number ν_{r} counting the nodes (zeros) of the radial wave function. It turns out that the energy $-1/2n^2$ a.u. depends on the principal quantum number $n = l + \nu_{\mathrm{r}} + 1$ only.

Probability operator A probability operator ϱ is a ▷ *Positive operator* with $\mathrm{Tr}\,\varrho = 1$. It describes a ▷ *Mixed state*. The expectation value of an ▷ *Observable* M in state ϱ is $\langle M \rangle = \mathrm{Tr}\,\varrho M$.

Production rate The ▷ *Balance equation* of an additive, transportable quantity Y says that the amount $Q(Y; \mathcal{V})$ in a certain volume $\mathcal{V}$ may change because there is an inflow $I(Y; \partial\mathcal{V})$ through the surface and production $\Pi(Y; \mathcal{V})$ within the volume. The latter is the production rate of Y, it is a volume integral over $\pi(Y) = \pi(Y; t, \boldsymbol{x})$, the volumetric production rate. It plays a role in describing chemical reactions.

Projector A projector Π is an self-adjoint operator with $\Pi^2 = \Pi$. The set of vectors $\mathcal{L} = \Pi\mathcal{H}$ is a linear subspace. Two projectors Π_1 and Π_2 are orthogonal if $\Pi_2\Pi_1 = 0$. A decomposition of the Hilbert space $\mathcal{H}$ into mutually orthogonal subspaces $\mathcal{L}_j = \Pi_j\mathcal{H}$ is described by the sum $I = \sum_j \Pi_j$ where the projectors are mutually orthogonal.

Prokhorov Aleksandr M. Soviet physicist, 1916–2002, Nobel prize 1964 with ▷ *Basov* and Townes.

Pure state One-dimensional projectors Π_f project on a one-dimensional subspace which is spanned by a normalized vector f. For an arbitrary vector $g \in \mathcal{H}$ we find $\Pi_f g = (f, g) f$. One-dimensional projectors are probability operators, they represent pure states. The ▷ *Entropy* of a one-dimensional projector vanishes.

Q

q-Number Old name for a linear operator as contrasted with a c-number, a classical real or complex number. c-numbers commute, q-numbers do not.

Quantum field A quantum field is constructed as follow. Space is simulated by a cube with periodic boundary conditions. There is a complete set $f_{\boldsymbol{k}} = f_{\boldsymbol{k}}(\boldsymbol{x})$ of normalized and mutually orthogonal plane waves, or modes. For each mode there is a pair $a_{\boldsymbol{k}}, a^\star{}_{\boldsymbol{k}}$ of ▷ *Ladder operators* with $[a_{\boldsymbol{k}''}, a^\star{}_{\boldsymbol{k}'}] = \delta_{\boldsymbol{k}'',\boldsymbol{k}'}$. Creation or annihilation operators for different modes are uncorrelated, they commute. We define the quantum field $A(\boldsymbol{x}) = \sum_{\boldsymbol{k}} f_{\boldsymbol{k}}(\boldsymbol{x})\, a_{\boldsymbol{k}}$. Its commutation relations are $[A(\boldsymbol{x}''), A(\boldsymbol{x}')] = 0$ and $[A(\boldsymbol{x}''), A^\star(\boldsymbol{x}')] = \delta^3(\boldsymbol{x}'' - \boldsymbol{x}')$. We speak of a field because $A = A(\boldsymbol{x})$ depends on location. It is a quantum field because its values are operators, not numbers. The quantum field is local since the field value and its adjoint commute at different locations. $N_A = \int \mathrm{d}^3x\, A^\star(\boldsymbol{x}) A(\boldsymbol{x})$ is the total number of A-particles. There is a normalized vacuum state Ω with $A(\boldsymbol{x})\Omega = 0$. $N_A\Omega = 0$ says that the vacuum is void of any particles. Similar things can be said of fermion fields. One must replace commutators by anti-commutators. Observables like energy, momentum and so on may be expressed in quantum fields.

Quarks There are three generations of structure particles. The first consists of elementary particles of which normal matter is composed. The particles of the second and third generation are massive and are created artificially. The first generation consist of the electron e^- (charge -1), the anti-up quark $\bar{u}$ ($-2/3$), the down quark d ($-1/3$), the neutrino ν_e and the anti-neutrino $\bar{\nu}_e$ (charge 0), the anti-down quark $\bar{d}$ (1/3), the up quark u (2/3) and the positron e^+ with charge $+1$. The proton, for example, is a bound (uud) state while the neutron is (udd).

R

Radial Schrödinger equation The two body problem with a spherical symmetric potential is reformulated for relative motion. The wave function may be factorized as $f(\boldsymbol{x}) = u(r)Y(\theta, \phi)$ where the r, θ, ϕ are spherical coordinates. The angular part is a ▷ *Spherical harmonic* function $Y_{lm}(\theta, \phi)$. The differential equation for the radial part is an ordinary differential equation, an eigenvalue problem. It inherits a repulsive centrifugal barrier from the angular part which is proportional to $l(l+1)/r^2$.

Random walk A trajectory $\boldsymbol{x}(t)$ where the increments Δt and $\Delta \boldsymbol{x}$ are random variables. Δt is equally distributed with mean τ while Δx is normally distributed with mean 0 and variance a^2. A concept used by Einstein to explain ▷ *Brownian motion*.

Lord Rayleigh formerly John William Strutt British physicist, 1842–1919, Nobel prize 1904 for 'his investigations of the densities of the most important gases and for his discovery …of argon'.

Rayleigh scattering Scattering of light on particles or irregularities much smaller than the wave length $\lambda = 2\pi c/\omega$. The primary light beam induces—via ▷ *Polarizability*—dipole moments $\boldsymbol{d}$ which in turn generate secondary light with an intensity proportional to $\| \ddot{\boldsymbol{d}} \|^2$. This explains why Rayleigh scattering grows as ω^4. Density ▷ *Fluctuations* cause Rayleigh scattering.

Rayleigh-Ritz principle The ground state energy and its wave function may be determined approximately by choosing a suitable normalized ▷ *Trial wave function* f and looking for the minimal expectation value. The trial wave function might be a linear combination of manageable functions or it may depend non-linearly on parameters. The Rayleigh-Ritz principle is based on the inequality $E_1 \le (f, Hf)$ meaning that the ground state energy is the smallest possible expectation value.

Relativistic hydrogen atom The electron as described by the ▷ *Dirac equation* interacts via ▷ *Minimal coupling* with the four-potential A_μ of a point charge at rest. The problem can be solved analytically by the same method as for the non-relativistic case: expansion into spherical harmonics and interpolation by a power series the recursion relation for the coefficients of which must stop. The orbitals are characterized by a principal quantum number n and the total angular momentum

j. The so-called fine structure is the lowest order correction in the fine structure constant $\alpha = e^2/4\pi\epsilon_0\hbar c$:

$$E_{nj} - mc^2 = -\frac{1}{2n^2}\left(1 + \frac{\alpha^2}{4n^2}\frac{4n-2}{j+1/2} + \cdots\right)E_*$$

Reversible process The Gibbs state $G(T, \lambda)$ depends on the temperature T and other external parameters λ, like volume. If these parameters change slowly enough with time t, the current state is always a Gibbs state, i.e. $\varrho_t \approx G(T_t, \lambda_t)$. As the name says, such a process can be reverted.

Ritz Walther: Swiss physicist, 1878–1976

Röntgen Wilhelm: German physicist, 1845–1923, Nobel prize 1901 for '…the discovery of the remarkable rays subsequently named after him'.

Rotation group A rotation in three-dimensional space is described by a real 3×3-matrix R obeying $R^\dagger R = I$. As a consequence, $R\boldsymbol{x} \cdot R\boldsymbol{y} = \boldsymbol{x} \cdot \boldsymbol{y}$ holds true for any pair of vectors. Lengths $\| \boldsymbol{x} \| = \sqrt{\boldsymbol{x} \cdot \boldsymbol{x}}$ are not changed, and angles (defined by $\boldsymbol{x} \cdot \boldsymbol{y} = \| x \| \| y \| \cos\alpha$) also keep their values. Rotation matrices form a topological ▷ *Group* of transformations called O_3. The determinant of $R \in O_3$ is either $+1$ or -1. Rotations with $\det R = 1$ form the special rotation group SO_3. Any rotation in SO_3 is continuously connected with $I \in SO_3$. Rotations are generated by three complex matrices M_1, M_2, M_3 which obey $[M_1, M_2] = \mathrm{i}M_3$, and cyclic permutations thereof.

Rutherford Ernest: British physicist, 1871–1937, Nobel prize in chemistry 1908 for 'his investigations into the disintegration of the elements …'

Rutherford scattering The scattering of two charged point-like particles is described by the Rutherford scattering formula. The scattering amplitude is $f = -2m(q_1 q_2/4\pi\epsilon_0)\, t^{-1}$ with t the ▷ *Mandelstam variable* for the momentum transfer squared. The long range of the Coulomb potential causes the cross section to be infinite.

S

Scalar product A ▷ *Hilbert space* $\mathcal{H}$ is a linear space with scalar product. To each pair of vectors $g, f \in \mathcal{H}$ a complex number (g, f) is assigned such that $f \to (g, f)$ is linear. Moreover, $(g, f) = (f, g)^*$ must hold true. Therefore the mapping $g \to (g, f)$ is anti-linear meaning $(g_1 + g_2, f) = (g_1, f) + (g_2, f)$ and $(\lambda g, f) = \lambda^*(g, f)$. One also requires $(f, f) \geq 0$ and that $(f, f) = 0$ implies $f = 0$. With $\| f \| = \sqrt{(f, f)}$, a norm is defined which defines the natural topology of the Hilbert space. For square-integrable function $f : \mathbb{R} \to \mathbb{C}$ the standard scalar product is $(g, f) = \int \mathrm{d}x\, g^*(x)\, f(x)$. Any scalar product guarantees the ▷ *Cauchy–Schwarz inequality* $|\,(g, f)\,| \leq \| g \| \| f \|$.

Scattering amplitude The solution of the ▷ *Scattering Schrödinger equation* in the far zone is of the form $\exp(ikx) + f(\theta)\exp(ikr)/r$. The incoming particles have a wave number k and run in x-direction. The outgoing particles propagate as a spherical wave and are observed at the ▷ *Deflection angle* θ. Their amplitude $f = f(\theta)$ is the scattering amplitude. Its square is the ▷ *Differential cross section*.

Scattering Schrödinger equation The ▷ *Schrödinger equation* must be solved with the boundary condition that there is a plane wave of incoming particles. The differential equation plus boundary condition can be rewritten into an integral equation

$$\psi(\boldsymbol{x}) = \mathrm{e}^{ik\boldsymbol{n}_{\text{in}} \cdot \boldsymbol{x}} - \frac{2m}{\hbar^2} \int \mathrm{d}^3 y \, \frac{\mathrm{e}^{ik|\,\boldsymbol{y} - \boldsymbol{y}\,|}}{4\pi|\,\boldsymbol{y} - \boldsymbol{y}\,|} V(\boldsymbol{y})\psi(\boldsymbol{y})$$

where $V = V(\boldsymbol{y})$ is the scattering potential. Note that the unknown solution ψ appears on both sides of the equation. Evaluating it in the far zone yields the scattering amplitude which depends on the entire solution.

Schrödinger Erwin: Austrian physicist 1887–1961, Nobel prize 1933 with Dirac for 'the discovery of new productive forms of atomic theory'.

Schrödinger equation The Schrödinger equation describes the time evolution of states while the observables remain unaffected. There is a version for mixed states ϱ, namely $i\hbar\dot{\varrho} = [H, \rho]$, and a specialization $i\hbar\dot{f} = Hf$ pertaining to pure states, the latter being represented by wave functions $f = f_t$. H is the energy observable, or ▷ *Hamiltonian*. The time-independent Schrödinger equation describes stationary states. No expectation value (f_t, Mf_t) depends on time. Stationary pure states are characterized by $Hf = Ef$ where E is an energy eigenvalue. Note that we refer to the ▷ *Schrödinger picture* of time.

Schrödinger picture Waiting for a time span t after preparing the state ϱ defines a new state ϱ_t. The time evolution $t \to \varrho_t$ is governed by the ▷ *Schrödinger equation*; it is driven by the ▷ *Hamiltonian*. ▷ *Heisenberg picture* for an alternative view. The time development of expectation values is the same, irrespective of the viewpoint, or picture.

Schwarz Hermann German mathematician, 1843–1921

Schwarz inequality ▷ *Cauchy–Schwarz-Bunyakovsky inequality*

Schwinger Julian: US-American physicist, 1918–1994, Nobel prize 1965 together with Tomonaga for 'fundamental work in quantum electrodynamics, with deep-ploughing consequences for the physics of elementary particles'.

Second main law The ▷ *Entropy* of a system the energy content of which is kept constant will grow until it cannot grow no longer since it has reached its maximum. The reason for the increase of mixedness are unavoidable interactions with the system's environment. Thus the equilibrium, or Gibbs state is a solution of $\max_\varrho S(\rho) = S(G)$. In ▷ *Phenomenological thermodynamics* there is an additive and transportable quantity S, the entropy, which is described by its ▹ *Balance equation*. The volumetric production rate $\pi(S)$, the amount of entropy produced per unit time and unit volume, is never negative. This form of the second main law of thermodynamics is rather concrete because $\pi(S)$ consists of only five terms (friction, heat conduction, diffusion, chemical reactions and Joule's heat) all of which cause ▹ *Irreversible processes*.

Second quantization The theoretical foundation ▹ *Quantum field* theory and of many-body theory. The spatial wave functions $f_{\sigma\boldsymbol{k}}$, the modes, may be populated by none, one, and so on ▹ *Bosons* or by none or one ▹ *Fermion*. σ is a spin index, $\hbar\boldsymbol{k}$ the momentum. The ▹ *Ladder operators* $a_{\sigma\boldsymbol{k}}$ of different modes commute (bosons) or anti-commute(fermions). Observables are expressions in terms of ladder operators. The Fourier transform of the ladder operators defines a local quantum field. The observables of the theory may also be expressed in terms of quantum fields. The particle number, as an observable, has eigenvalues $n = 0, 1, \ldots$ In first quantization one restricts the Hilbert space to an eigenspace for a particular number of particles.

Self-adjoint operator A linear operator A which coincides with its adjoint, $A = A^\star$. Self-adjoint operators are normal and can be diagonalized. The eigenvalues are real numbers. They represent the ▹ *Observables* of quantum theory. Unfortunately, many important observables are represented by unbounded operators which cannot be defined for all vectors in $\mathcal{H}$.

Semiconductor A crystal with gap between the conduction and the valence band, ▹ *Electronic band structure*. If the band gap is small, at finite temperatures some electrons will be found in the valence band where they are mobile. The conductivity of semiconductors is small and depends strongly on temperature. The band gap can be manipulated by doping or by applying external fields. Semiconductor physics has become a branch of its own.

Shannon Claude: US-American mathematician, 1916–2001

Shannon entropy Originally, the number of yes/no question to be asked if a message containing N letters is to be decoded. The relative frequency of letters in the respective language is known. If the letter labeled by j occurs N_j times, the Shanon entropy of the message is defined as $S = -N\sum_j p_j \mathrm{ld} p_j$ where $p_j \approx N_j/N$. Translated to physics: the message is the protocol of an ▹ *Experiment* and the letters are the possible outcomes of a ▹ *Measurement*. For historical reasons, the Shannon entropy is multiplied by the ▹ *Boltzmann* constant, and the base 2-logarithm is replaced by the natural logarithm. The information gain per measurement is $\mathcal{S} = -k_\mathrm{B}\sum_j p_j \ln p_j$. The ▹ *Entropy* $\mathcal{S} = \mathcal{S}(\varrho)$ indicates to which degree a state ϱ is mixed.

Sound velocity ▹ *Hooke's law* in the balance equation for momentum allow for wave solution. There are one longitunially and two transversally polarized solution for the displacement of ▹ *Material points* with two different sound velocities $c_\parallel$ and $c_\perp$. The ▹ *Dispersion relation* for these acoustical waves are linear. In the ▹ *Debye model* for lattice vibrations it is assumed that the corresponding ▹ *Density of states* holds true up to an energy $k_\mathrm{B}\Theta_\mathrm{D}$.

Space translation A translation of the measuring equipment by a vector $\boldsymbol{a}$. This is described by a ▷ *Unitary operator* $U_{\boldsymbol{a}} = \exp(\mathrm{i}\boldsymbol{a} \cdot \boldsymbol{K})$ where $\hbar \boldsymbol{K} = \boldsymbol{P}$ are the three commuting components of momentum.

Spectral intensity The spectral intensity $S = S(\omega) \geq 0$ is the Fourier transform of the ▷ *Correlation function* $K = K(\tau)$. ▷ *Wiener–Khinchin theorem*. It describes the power distribution over frequencies in a fluctuating process $t \to \Delta M_t$. The spectral density of voltage fluctuations produced by an ohmic resistance is given by the ▷ *Nyquist formula*. If the spectral intensity does not depend on frequency one speaks of ▷ *White noise*.

Spherial harmonics A function $Y = Y(\theta, \phi)$ living on the unit sphere $r = 1$. r, θ, ϕ are ▷ *Spherical coordinates*. In particular the eigenfunctions of ▷ *Orbital angular momentum*: $L^2 Y_{lm} = \hbar^2 l(l+1) Y_{lm}$ and $L_3 Y_{lm} = \hbar m Y_{lm}$. The angular momentum l assumes values $0, 1, 2, \dots$ and $m = -l, -l+1, \dots, l$.

Spherical wave A wave of type $f(\boldsymbol{x}) \propto \exp(\mathrm{i}kr)/r$ describing the spherically symmetric motion of a particle with wave number k. Note that f is not square-integrable. Integration of $|f|^2$ over a sphere of radius R gives 4π: the value does not depend on R.

Spin angular momentum The total angular momentum $\boldsymbol{J}$ is the generator of rotations of the measuring equipment. $\boldsymbol{J}$ is split into the ▷ *Orbital angular momentum* $\boldsymbol{L} = \boldsymbol{X} \times \boldsymbol{P}$ and a remainder, the spin angular momentum $\boldsymbol{S}$. It obeys angular momentum commutation rules $[S_1, S_2] = \mathrm{i}\hbar S_3$ (and cyclic permutations thereof) and commutes with $\boldsymbol{X}, \boldsymbol{P}$ and therefore with $\boldsymbol{L}$. The spin squared $\boldsymbol{S}^2$ has eigenvalues $\hbar^2 s(s+1)$, with $s = 0, 1/2, 1$ and so on. The spin is an immutable property of a particle. Particles with integer spin are ▷ *Bosons* while ▷ *Fermions* are characterized by half-integer spin.

Spin singlet Two spin 1/2 states may be linearly combined to total spin $s = 0$ and $s = 1$. The former state has one component which is an anti-symmetric combination of the two spins. It is a spin singlet. The $s = 1$ combination is symmetric and consists of three components, therefore a triplet state. The singlet level does not split in an external magnetic induction field while the triplet splits into three levels.

Spin triplet ▷ *Spin singlet*

Spontaneous magnetization A ferromagnet at a temperature below the ▷ *Curie temperature* is partially magnetized without the presence of an external induction field, i.e. spontaneously. The spontaneous magnetization $M^*(T)$ is largest at zero temperature and falls off to zero at $T = \Theta_C$. The ▷ *Heisenberg ferromagnet* model solved by the effective field method describes the effect reasonably well.

Spontaneous transition ▷ *Forced transitions* between two energy levels occur if an external electric field oscillates with frequency $\hbar\omega_0 = E_2 - E_1$. The transition probability is the same for $E_2 \to E_1$ and $E_1 \to E_2$, i.e. upwards and downwards. The energy balance is $\hbar\omega_0 + E_1 = E_2$ (upwards) and $-\hbar\omega_0 + E_2 = E_1$ (downwards).

There is also a spontaneous downward transition $E_2 = E_1 + \hbar\omega_0$ the probability rate for which can only be calculated in the framework of quantum electrodynamics. Einstein reasoned that the black body radiation law is compatible with the transition laws only if the rate for spontaneous transition is

$$\Gamma^{\text{sp}}_{2\to 1} = \frac{4}{3}\frac{\omega_0^3}{4\pi\epsilon_0\hbar c^3}\sum_{j=1,2,3} |\,(f_2, d_j f_1)\,|^2$$

with f_1 and f_2 being normalized wave functions and $\boldsymbol{d}$ the dipole moment operator.

Spreading of free particles In one dimension, the Hamiltonian of a free particle is $H = P^2/2m$, with P the linear momentum operator. In the ▷ *Schrödinger picture*, the particle's wave function obeys $\mathrm{d}f_t/\mathrm{d}t = -(\mathrm{i}/\hbar)Hf_t$. The average location is $\langle X \rangle_t = \int \mathrm{d}x\, x |\, f_t\,|^2$. It follows that $\mathrm{d}\langle X \rangle_t/\mathrm{d}t = (1/m)\langle P \rangle_t$. Likewise, $\mathrm{d}\langle P \rangle_t/\mathrm{d}t = 0$ and $\mathrm{d}^2\langle X \rangle_t/\mathrm{d}t^2 = 0$. This seems to be classical mechanics: the free particle moves with constant velocity and unaccelerated, the momentum is conserved. In the following, we choose a frame of reference where the particle is at rest. One calculates $\mathrm{d}^2\langle X^2 \rangle_t/\mathrm{d}t^2 = (2/m^2)\langle P^2 \rangle$ or $\sigma^2(X)_t = \sigma^2(X)_0 + (\sigma(P)\,t/m)^2$ for the variance. The spread $\sigma(X)_t$ grows at first quadratically, later linearly with time. The minimal spread cannot be zero because of $\sigma(X)\sigma(P) \geq \hbar/2$. ▷ *Heisenberg uncertainty principle*.

Square-integrable A function $f = f(x)$ is square-integrable if the integral $\int \mathrm{d}x\,|\,f(x)\,|^2$ has a finite value. According to the ▷ *Cauchy–Schwarz inequality* the scalar product of two square-integrable functions, namely $(g, f) = \int \mathrm{d}x\, g^*(x)\, f(x)$, is well defined. The limit of a converging sequence of square-integrable functions exists and is a square-integrable function. Hence, square-integrable functions form a Hilbert space. To be pedantic, the integral is a Lebesgue integral. A 'function' is really an equivalence class of functions which are equal except on a set of measure 0.

Stationary perturbation A ▷ *Perturbation* which does not depend on time. In atomic or molecular physics, the interaction with a quasi-static electric or magnetic field is a stationary perturbation. Another example is electron repulsion.

Statistical thermodynamics In contrast with ▷ *Phenomenological thermodynamics*, the field is fully embedded into the quantum theory of matter. The concepts of phenomenological thermodynamics like entropy, equilibrium state, thermodynamic potentials and so on can be given a precise meaning and many material equations—like the ideal gas law, the contributions of molecule vibration and rotation to the heat capacity, the Joule-Thomson coefficient and many more—can be derived.

Stationary state In the ▷ *Schrödinger picture*, states ϱ_t change with time, observables do not. $\varrho_t = U_t \varrho U^{\star}{}_t$ is transformed from time $t = 0$ to t by the waiting operator U_t which is a function of the Hamiltonian. It follows that states ϱ which commute with the Hamiltonian do not change with time, they are stationary. Note that the Gibbs state, also a function of the Hamiltonian, is stationary. But the converse is not true: there are stationary states which do not describe thermodynamic equilibrium.

The eigenfunctions of the Hamiltonian describe stationary pure states. This explains why the eigenfunctions of the Hamiltonian are so much in the focus of quantum physics.

Stark Johannes: German physicist 1874–1975, Nobel prize 1919 for 'his discovery of the Doppler effect in canal rays and the splitting of spectral lines in electric fields'.

Stark effect The energy levels of an atom will split if an external electrostatic field is applied. The dependency on field strength is linear if two levels with different ▹ *Parity* are degenerate, and quadratic otherwise.

Star operation Observables are embedded in a C* algebra. They can be multiplied with complex numbers, added, and multiplied just as numbers, but they do not necessarily commute. A unity with $IA = AI$ belongs to the algebrah. In addition, there is a star operation $A \to A^\star$ with $A^{\star\star} = A$. It behaves like conjugation for complex numbers, however with $(AB)^\star = B^\star A^\star$. There is a norm defined as the largest number λ (in modulus) for which $(A^\star A - \lambda I)$ is not invertible. With respect to this norm, the C* algebra shall be complete. Observables M obey $M = M^\star$.

State Abstractly, a state is a continuous positive linear functional ϱ on a C* algebra with $\varrho(I) = 1$ and $\varrho(A^\star A) \geq 0$. More concretely, a state is described by a ▹ *Probability operator* ϱ such that $\varrho(A) = \text{Tr}\,\varrho A$. The probability operator obeys $\varrho \geq 0$ and $\text{Tr}\,\varrho = 1$. In general, the state is mixed. A pure state is given by a one-dimensional ▹ *Projection operator* on a normalized vector f. The ▹ *Entropy* $S(\varrho)$ of a pure state vanishes, it is positive otherwise.

Stoichiometric coefficients ▹ *Chemical reactions*.

Strain tensor An elastic solid medium upon deformations is described by the displacement field $\boldsymbol{u}$. The material point originally at $\boldsymbol{x}$ is displaced by $\boldsymbol{u}(\boldsymbol{x})$. For small displacement gradients the strain tensor is defined by $S_{jk} = (\nabla_j u_k + \nabla_k u_i)/2$. It vanishes for a rigid translation or for a rotation of the entire medium and in fact describes deformations.

Stress tensor The current density $j_i(P_k)$ of momentum $\boldsymbol{P}$ may be split into the convection part $\varrho v_i v_k$ and the conduction part $-T_{ik} = -T_{ik}(\boldsymbol{x})$, the stress tensor. Angular momentum conservation requires that the stress tensor be symmetric. For small deformations, stress and strain are related linearly, as stated by ▹ *Hooke's law*.

Substantial time derivative A change in a streaming medium as seen by a co-moving observer. It is defined as $(\mathrm{D}_t f)(\boldsymbol{x}) = \dot{f}(\boldsymbol{x}) + \boldsymbol{v}(\boldsymbol{x}) \cdot \nabla f(\boldsymbol{x})$.

Susceptibility A dielectric material in an external electric field $\boldsymbol{E}$ acquires a polarization $\boldsymbol{P} = \chi_\mathrm{e}\boldsymbol{E}$. Likewise, a paramagnetic or diamagnetic substance in an external magnetic field has a magnetization $\boldsymbol{M} = \chi_\mathrm{m}\boldsymbol{H}$. Both susceptibilities χ depend on temperature, ▹ *Curie–Weiss law*. If the external fields are not quasi-static but oscillating rapidly, the above relations apply to the Fourier components. Consequently the susceptibilities $\chi_{\mathrm{e,m}}(\omega)$ depend on the angular frequency. The theory of ▹ *Linear response* allows to calculate explicit expressions for susceptibilities and to derive some of their fundamental properties. In crystalline media, $\boldsymbol{E}$ and $\boldsymbol{P}$ need not be parallel, the susceptibility therefore is a tensor. The same applies to the magnetic case.

T

Temperature A parameter T characterizing the thermodynamic equilibrium, i.e. the ▹ *Gibbs state* of a system. Formally it is introduced as a ▹ *Lagrange multiplier* when maximizing the ▹ *Entropy* with the auxiliary condition that the internal energy is kept fixed, $\langle H \rangle = U$. It can be shown that two systems in thermal contact have the same temperature and that $U(T)$ grows with increasing T. Moreover, an ideal gas has a pressure which is proportional to T. This law defines the temperature; the scale is fixed by the triple point of vapor, water, and ice; its temperature is 273.16 K exactly.

Thermal noise An ohmic resistance R produces a fluctuating voltage the spectral intensity of which is described by the ▹ *Nyquist formula*. The power of this source of noise is proportional to R. It is called thermal noise because the power is also proportional to temperature T.

Thermodynamic potential In ▹ *Phenomenological thermodynamics* one has variables like energy, temperature, volume, pressure, entropy and so on which characterize thermodynamic equilibrium. The variables are related by being partial derivatives of certain potentials. The free energy may serve as an example: $\mathrm{d}F = -S\mathrm{d}T - p\mathrm{d}V$. There is a function $F = F(T, V)$ such that $S = -\partial F/\partial T$ and $p = -\partial F/\partial V$, with the usual meaning of the symbols. By a ▹ *Legendre transformation* the roles of variable and derivative may be interchanged. Enthalpy H, internal energy U or the Gibbs potential G are more thermodynamic potentials. The second partial derivatives commute which leads to unexpected results, such as $\partial p/\partial T = \partial S/\partial V$. That there are thermodynamic potentials is a triviality in ▹ *Statistical thermodynamics*.

Thomson Joseph John British physicist, Nobel prize 1906 for 'his theoretical and experimental investigations on the conduction of electricity by gases'.

Time translation The laws of physics are the same here and today and elsewhere and tomorrow. No law of nature should refer to a time, only time differences count. Putting forward all clocks by an amount τ amounts to a time translation. Since the possible outcomes of an observable are not changed, a time translation is represented by a unitary transformation U_τ, by a waiting operator. Since first waiting by τ' and then by τ'' is the same as waiting by $\tau' + \tau''$, the waiting operators form a one-parameter topological group. The continuous solution of the composition law $U_{\tau'}U_{\tau''} = U_{\tau'+\tau''}$ results in an exponential function $U_\tau = \exp(-\mathrm{i}\tau\Omega)$. The minus sign is a convention. The angular frequency operator Ω must be self-adjoint to make U_τ a unitary operator. $H = \hbar\Omega$ is the energy; it has been known long before quantum physics.

Townes Charles H.: US-American physicist, 1915–2015, Nobel prize 1964 with ▷ *Basov* and ▷ *Prokhorov*.

Trace of an operator For an operator A and a complete orthonormal set $e_1, e_2, \dots$ the trace is defined as $\text{Tr } A = \sum_j (e_j, Ae_j)$. Another base $\bar{e}_j = Ue_j$ (U is a ▷ *Unitary operator*) yields $\sum_j (\bar{e}_j, A\bar{e}_j)$ which expression coincides with $\sum_j (e_j, Ae_j) = \text{Tr } A$. In other words, the trace of an operator is calculated with the aid of a complete orthonormal set, but any other such set results in the same trace. $A \to \text{Tr } A$ is a linear functional with $\text{Tr } I = 1$. The trace is commutative in the sense of $\text{Tr } AB = \text{Tr } BA$.

Trapping An impurity in a crystal may bind, or trap a mobile electron, as demonstrated by the ▷ *Hopping model*. The localized energy level must be lower than normal. Loosely bound electrons spread out wider than tightly bound, i.e. with large binding energy.

Trial wave function A trial wave function incorporates the essential features of the true ground state wave function; it contains a few adjustable parameters κ. It should be constructed in such a way that the expectation value $E(\kappa)$ of the energy H can be calculated easily. The minimum over the adjustable parameters κ is expected to be a good approximation to the ground state wave function and energy. Anyhow, the minimal energy is an upper bound to the true ground state energy $E_1 \le \min_\kappa E(\kappa)$. ▷ *Rayleigh-Ritz principle*. Note that the trial wave function has to fulfill the symmetry requirements of the ▷ *Pauli principle*.

U

Unitary operator An invertible linear mapping $U : \mathcal{H} \to \mathcal{H}$ is unitary if it preserves scalar products. This means $(Ug, Uf) = (g, f)$ for all $f, g \in \mathcal{H}$. $U^\star U = I$, $U^\star = U^{-1}$ and $UU^\star = I$ are consequences. Therefore, U is a ▷ *Normal operator*. There is a decomposition $I = \Pi_1 + \Pi_2 + \cdots$ of unity into mutually orthogonal ▷ *Projection operators* such that $U = \sum_j u_j \Pi_j$ holds true with eigenvalues u_j on the unit circle of the complex plane: $|u_j| = 1$. Unitary operators describe symmetries.

V

Vacuum In ▷ *Second quantization* or ▷ *Quantum field theory* the state Ω which is void of any particle. It is characterized by $\phi(t, \boldsymbol{x})\Omega = 0$ where ϕ is any quantum field, boson or fermionic.

Vector Norm In a ▷ *Hilbert space* the expression $\| f \| = \sqrt{(f, f)}$ defines a norm. Indeed, $\| f \| \ge 0$, $\| f \| = 0$ implies $f = 0$, $\| \lambda f \| = |\lambda| \| f \|$ and $\| f + g \| \le \| f \| + \| g \|$, the triangle inequality, are fulfilled. Therefore, the Hilbert space has a natural topology. A mapping $\mathcal{H} \to \mathcal{H}$ is continuous if norm-convergent sequences are mapped into norm-convergent sequences.

Virial coefficients The ▷ *Pressure* of a real gas (as contrasted with an ideal gas) can be expanded in a power series of the particle density n as $p(T, n) = k_B T \left\{ n + b_2(T)n^2 + b_3(T)n^3 + \cdots \right\}$. The temperature dependent coefficients $b_j(T)$

are virial coefficients. $b_1(T) = 1$ is the same for all gases. All gases behave alike at sufficiently low particle density. The second and higher virial coefficients may be calculated from the intermolecular potential. The virial power series has a radius of convergence which indicates a transition to the liquid phase.

Viscosity For a liquid, the reversible part of the stress tensor has the form $T'_{jk} = -p\delta_{jk}$ where $p = p(\boldsymbol{x})$ denotes the local pressure. The irreversible part is $T''_{jk} = \eta(\nabla_j v_k + \nabla_k v_j)$. The viscosity η is a strongly temperature dependent characteristic of the liquid under study.

Volumetric production rate ▷ *Production rate*

W

Waiting operator ▷ *Time translation*

Wave packets ▷ *Plane waves* $\exp(\mathrm{i}k\,x)$ are not normalizable. They must be superimposed as $f(x) = (1/2\pi)\int \mathrm{d}k\,\hat{f}(k)\exp(\mathrm{i}kx)$ in order to become square-integrable. This is the case if the Fourier transform $\hat{f} = \hat{f}(k)$ is square-integrable. The time dependence of a wave packet is determined by the dispersion relation $\omega = \omega(k)$. One has $f_t(x) = (1/2\pi)\int \mathrm{d}k\,\hat{f}(k)\exp(-\mathrm{i}\omega(k)t + \mathrm{i}kx)$. The wave packet travels as $\langle X \rangle_t = x_0 + vt$ where the velocity v is the averaged ▷ *Group velocity* $\mathrm{d}\omega/\mathrm{d}k$. One can also calculate the spread of the wave packet. For large times the width of the wave packet as measured by $\sigma(X) = \sqrt{\langle X^2 \rangle - \langle X \rangle^2}$ is proportional to $|\,t\,|$. This, however, is true only if the wave packet moves in homogeneous space. A potential, a deviation from homogeneity, may bind the particle such that its motion is and remains confined. The considerations presented here are easily generalized to three-dimensions.

White dwarf A white dwarf is the result of a supernova: the nuclear fuel is exhausted and there is not enough thermal pressure to balance gravitational attraction. The sudden collapse leads to an outer shell of matter which is blown away into space. The rest is a dense core, so dense, that to a large extent the electrons are no more bound to their nuclei; they form a plasma. Since the electrons are fermions, there is a strong ▷ *Degeneracy pressure* even at zero temperature. This pressure stabilizes the white dwarf against further gravitational collapse. A white dwarf is a dwarf in size, its radius is comparable to the radius of the earth, although its mass is comparable with that of the sun. If the mass is larger than 1.44 sun masses (Chandrasekhar limit), the degeneracy pressure of electrons is insufficient to balance gravitation. The core will get stabilized by the degeneracy pressure of neutron. ▷ *Neutron star*. If the white dwarf has a companion star from which it sucks in matter, the Chandrasekhar mass limit may be reached resulting in a type I supernova. All such supernovae have the same absolute brightness. They may serve as a standard candle which allows to determine its distance.

White noise According to the ▷ *Wiener–Khinchin theorem* the Fourier transform of a ▷ *Correlation function* is a non-negative ▷ *Spectral intensity* $S = S(\omega)$. One speaks of white noise if the spectral intensity does not depend on angular frequency, as is the case for white light. The noise is white if the fluctuations ΔM_t at different times are uncorrelated. An example is the fluctuating force on a Brownian particle, ▷ *Brownian motion*.

Wiener Norbert US-American mathematician, 1894–1964

Wiener–Khinchin theorem The correlation function in a stationary state for fluctuations ΔM_t and $\Delta M_{t+\tau}$ depends on the time difference τ only. As a consequence, the Fourier transformed fluctuations are uncorrelated for different frequencies. The Fourier transform of the correlation function is a never negative spectral intensity $S = S(\omega)$, so the Wiener–Khinchin theorem.

Work The expectation value is a bilinear functional. In particular, the internal energy $U = \text{Tr}\, \varrho H$ may change because either the state ϱ changes or the Hamiltonian H. $\text{d}U$ has two contributions, namely heat $\text{d}Q = \text{Tr}\, \text{d}\varrho\, H$ and work $\text{d}W = \text{Tr}\, \varrho\, \text{d}H$. See ▷ *Heat* and ▷ *First main law*.

X

X-rays also called ▷ *Röntgen* rays. Electromagnetic radiation in the keV region.

Z

Zeeman Pieter Dutch physicist, 1865–1943, Nobel prize 1902 together with ▷ *Lorentz*.

Zeeman effect The levels of an atom or molecule split proportionally to the external induction $\mathcal{B}$. The energy shift of the electron in a hydrogen bound state is $\Delta E = (e\mathcal{B}/m)\langle\, L_3 + 2S_3 \,\rangle$. The expectation value depends on the combination of orbital angular momentum and spin to total angular momentum.

Index

P. Hertel, *Quantum Theory and Statistical Thermodynamics*,
Graduate Texts in Physics, DOI 10.1007/978-3-319-58595-6

Q

R

S

Printed in the United States
By Bookmasters